Master Electrician's Review:

Based on the *National Electrical Code*® 2011

SEVENTH EDITION

Master Electrician's Review:

Based on the *National Electrical Code*® *2011*

SEVENTH EDITION

Richard E. Loyd

DELMAR
CENGAGE Learning™

Australia • Brazil • Japan • Korea • Mexico • Singapore • Spain • United Kingdom • United States

Master Electrician's Review, 7E
Richard Loyd

Vice President, Technology and
  Trade Professional Business Unit:
  Gregory L. Clayton

Director of Building Trades: Taryn Zlatin
  McKenzie

Product Development Manager: Robert Person

Development: Nobina Preston

Director of Marketing: Beth A. Lutz

Marketing Manager: Marissa Maiella

Senior Production Director: Wendy Troeger

Senior Content Project Manager:
  Elizabeth C. Hough

Senior Art Director: Benjamin Gleeksman

For product information and technology assistance, contact us at
Cengage Learning Customer & Sales Support, 1-800-354-9706

For permission to use material from this text or product,
submit all requests online at **www. cengage.com/permissions**.
Further permissions questions can be e-mailed to
permissionrequest@cengage.com

Example: Microsoft ® is a registered trademark of the Microsoft Corporation.

Library of Congress Control Number: 2011927908

ISBN-13: 978-1-4390-5960-9

ISBN-10: 1-4390-5960-8

Delmar
5 Maxwell Drive
Clifton Park, NY 12065-2919
USA

Cengage Learning is a leading provider of customized learning solutions with
office locations around the globe, including Singapore, the United Kingdom,
Australia, Mexico, Brazil, and Japan. Locate your local office at:
international.cengage.com/region

Cengage Learning products are represented in Canada by Nelson Education, Ltd.

Visit us at **www.InformationDestination.com**

For more learning solutions, please visit our corporate website at
www.cengage.com

Notice to the Reader

Publisher does not warrant or guarantee any of the products described herein or perform
any independent analysis in connection with any of the product information contained
herein. Publisher does not assume, and expressly disclaims, any obligation to obtain and
include information other than that provided to it by the manufacturer. The reader is
expressly warned to consider and adopt all safety precautions that might be indicated by
the activities described herein and to avoid all potential hazards. By following the instructions contained herein, the reader willingly assumes all risks in connection with such
instructions. The publisher makes no representations or warranties of any kind, including
but not limited to the warranties of fitness for particular purpose or merchantability nor
are any such representations implied with respect to the material set forth herein, and
the publisher takes no responsibility with respect to such material. The publisher shall
not be liable for any special, consequential, or exemplary damages resulting, in whole or
part, from the readers' use of, or reliance upon, this material.

Printed in the United States of America
1 2 3 4 5 6 7 15 14 13 12 11

I dedicate this book to my wife and business partner, Nancy. She is my inspiration and number one fan. Her assistance is invaluable. I would also like to dedicate it to all my wonderful circuit rider friends. Their expertise in the *National Electrical Code*® and the electrical industry continue to provide me with the latest information on our great industry.

Table of Contents

Foreword ix
Preface xi
About the Author xiv

CHAPTER ONE

Examinations and National Testing
    Organizations 2
Developing an Item (Question) Bank 2
Preparing for an Examination 3
Sample Examination—75 Questions 7

CHAPTER TWO

National Fire Protection
    Association Standards 25
NEC®. 25
Local Codes and Requirements 30
Testing Laboratories 31
Question Review 33

CHAPTER THREE

Basic Electrical Mathematics Review 40
Calculator Math 40
Working with Fractions 40
Working with Decimals 43
Working with Percentages 43
Powers 44
Working with Square Roots 44
Ohm's Law 45
Practice Exam—15 Questions 49
Question Review 51

CHAPTER FOUR

Introduction to the NEC® 59
Know Your Code Book 59
Question Review 80

CHAPTER FIVE

General Wiring Methods 87
Question Review 101

CHAPTER SIX

Branch Circuits 108
Feeders 121
Voltage Drop 121
Question Review 123

CHAPTER SEVEN

Services 600 Volts or Less 131
Services Over 600 Volts 137
Question Review 139

CHAPTER EIGHT

Overcurrent Protection 146
Capable of Carrying the Available
    Fault Current 147
Question Review 151

CHAPTER NINE

Grounding Electrical Systems 159
Question Review 178

CHAPTER TEN

Wiring Methods 186
Question Review 201

CHAPTER ELEVEN

General-Use Utilization Equipment 208
Question Review 215

CHAPTER TWELVE

Special Equipment and
    Occupancies 222
Question Review 233

CHAPTER THIRTEEN

Special Conditions and
    Communication Systems 240
Question Review 244

CHAPTER FOURTEEN

Electrical and *NEC*® Question Review 251
Practice Examination—50 Questions........ 253
Code Quiz 1 263
Code Quiz 2 264
Code Quiz 3 266
Code Quiz 4 268
Code Quiz 5 272
Code Quiz 6 276
Code Quiz 7 280
Code Quiz 8 284
Code Quiz 9 288
Code Quiz 10 292
Code Quiz 11 296
Practice Exam 1 300
Practice Exam 2 305
Practice Exam 3 310

Practice Exam 4 315
Practice Exam 5 320
Practice Exam 6 326
Practice Exam 7 331
Practice Exam 8 336
Practice Exam 9 341
Practice Exam 10 347
Practice Exam 11 353
Practice Exam 12 359
Final Examination 365

Appendix 1: Symbols 383
Appendix 2: Basic Electrical Formulas 386
Appendix 3: Answer Key 389

Index 413

Foreword

ELECTRIC PER SE

Electricity! In ancient times, it was believed to be an act of the gods. Not even Greco-Roman civilizations could understand this thing we know today as electricity. It was not until about 1600 C.E. that any scientific theory was recorded, when after 17 years of research William Gilbert wrote a book on the subject titled *De Magnete*. It was almost another 150 years before major gains were made in understanding electricity so that people might some day be able to control it. At that time, Benjamin Franklin, whom we often refer to as the grandfather of electricity or electrical science, and his close friend, Joseph Priestley, a great historian, began to gather the works of the many worldwide scientists who had been working independently. (One has to wonder where this wondrous industry would be today if the great minds of yesteryear had the benefit of our great communication networks.) Franklin traveled to Europe to gather this information, and with that trip, our industry began to surge forward for the first time. Franklin's famous kite experiment allegedly took place about 1752. Coupling the results of that with other scientific data gathered, Franklin decided he could sell installations of lightning protection to every building owner in the city of Philadelphia. He was very successful in doing just that!

We credit Franklin for the terms "conductor" and "nonconductor," which replaced "electric per se" and "non–electric."

With the accumulation of the knowledge gained from these many scientists and inventors of the past, many other great minds emerged and lent their names to even more discoveries. Their names are familiar as other terms related to electricity. Alessandro Volta, James Watt, André Marie Ampère, George Ohm, and Heinrich Hertz are just a few of the great minds that enabled our industry to come together.

The next major breakthrough came only a little more than 100 years ago. Thomas Edison, the holder of hundreds of patents promoting direct current (DC) voltage, and Nikola Tesla, the inventor of the 3-phase motor and the promoter of alternating current (AC) voltage, began a bitter competition in the late 1800s.

Electricity is here! *(Reprinted with permission from Underwriters Laboratories, Inc.®)*

The 1893 Columbian Exposition at the Chicago World's Fair. The Palace of Electricity astonished crowds; it also astonished electricians by repeatedly setting fire to itself. *(Reprinted with permission from Underwriters Laboratories, Inc.®)*

Edison was methodically making inroads with direct current all along the Eastern Seaboard, and Tesla had exhausted his finances for his experimental project with alternating current in Colorado. Then, the city of Chicago requested bids to electrically light the Columbian Exposition of 1893. Lighting the world's fair in Chicago, which was celebrating 400 years since Columbus discovered America, required building generators and nearly a quarter-million lights. Edison teamed up with General Electric and submitted a bid based on the Edison lightbulb powered by direct current. George Westinghouse, already successful in the electrified railroad, the elevator, and about 400 other inventions, joined with Tesla and submitted the lowest bid—about one-third of the Edison-General Electric bid. Westinghouse and Tesla were awarded the contract. They based their bid on the Westinghouse Stopper Light, powered by alternating current. The installation was made by Guarantee Electric of Saint Louis, Missouri. This firm is still one of the oldest and largest electrical contractors in business today. It is said that thousands came out to see this great lighting display. But almost as spectacular as the lighting were the arcing, sparking, and fires started by the display, with accidents occurring daily! It is said that Edison was embellishing the dangers of alternating current by publicly electrocuting dogs while promoting direct current as being much safer than alternating current. But the public knew better because plenty of fires and accidents were also occurring with direct current.

About this time, three very important events took place almost simultaneously. The Chicago Board of Fire Underwriters hired an electrician from Boston as an electrical inspector for the exposition, William Henry Merrill. Mr. Merrill saw the need for safety inspections and started Underwriters Laboratories in 1894. In 1895, the first nationally recommended code was published by the National Board of Fire Underwriters (now the American Insurance Association). The 1895 *Electrical Code®* was drafted through the combined efforts of the architectural, electrical, insurance, and other allied interests based on a document that resulted from actions taken in 1892 by Underwriters National Electrical Association, which met and consolidated various codes that were already in existence. The first *NEC®* was presented to the National Conference on Electrical Rules, which was composed of delegates from various national associations who voted to unanimously recommend it to their respective associations for adoption or approval. A group from the city of Buffalo, New York, visited the world's fair and, based on that visit, hired Westinghouse and Tesla to build hydro-generators to be powered by Niagara Falls. Alternating current had won out over direct current as this country's major power source. The rest is history!

Preface

The primary purpose of this book is to provide electrical students with a concise, easily understandable study guide to the 2011 edition of the *National Electrical Code®* and the application of electrical calculations. This text is written recognizing that the electrical field has many facets, and the user may have diverse interests and varying levels of experience. However, the interests of the relatively experienced electrician preparing to enter the electrical industry as a legally competent licensed electrical contractor, the master electrician, the journeyman electrician, the electrician-in-training, and especially the advanced electrical student apprentice were all carefully considered in the text's preparation.

This book provides a ready source of the basic information on the *NEC®*. It may be used for reference or as study material for students and electricians who are preparing for an examination for licensing. It is especially designed as an aid to those studying for the nationally recognized examinations, such as the International Code Council (ICC). It will also help anyone preparing for any electrical examination, and it will provide a quick, easily understandable study guide for those needing to update themselves on the *NEC®* and basic electrical mathematical formulas and calculations. The text is brief and concise for easy application for classroom or home study.

METRICS AND THE *NEC®*

The United States is the last major country in the world not using the metric system as the primary system of measurement. We have been very comfortable using U.S. customary values. This is changing. Manufacturers are now showing both inch-pound and metric dimensions in their catalogs. Plans and specifications for governmental new construction and renovation projects begun after January 1, 1994, have been done using the metric system. You may not feel comfortable with metrics, but metrics are here to stay. You might just as well become familiar with the metric system.

The *NEC®* and other National Fire Protection Association (NFPA) standards are becoming international standards. All measurements in the 2011 *NEC®* are shown with metrics first, followed by the inch-pound value in parentheses—for example, 600 mm (24 in.). However, neither metric nor inch-pound is preferred over the other. Users are free to select the form of measurement they wish. Ease in understanding is of utmost importance in *Master Electrician's Review*. Therefore, inch-pound values are shown first, followed by metric values in parentheses—for example, 24 in. (600 mm). A *soft metric conversion* is when the dimensions of a product already designed and manufactured to the inch-pound system are converted to metric dimensions. The product does not change in size. A *hard metric measurement* is when a product has been designed to metric dimensions. No conversion from inch-pound measurement units is involved. A *hard conversion* is when an existing product is redesigned into a new size.

Each lesson is designed purposely to require the student to apply the entire *NEC*® text and not specific chapters or articles. It has been found that when studying for a timed, open-book examination, the student must gain proficiency in the Table of Contents, the Index, and the ability to move quickly from cover to cover to find the correct answer to each question in a timely fashion.

In the 2011 edition of the *NEC*®, existing inch-pound dimensions did not change. Metric conversions were made, and then rounded off. Where rounding off would create a safety hazard, the metric conversions are mathematically identical. For example, if a dimension is required to be 6 feet, it is shown in metrics in the *NEC*® as 1.8 m (6 ft). Note that the 6 ft remains the same, and the metric value of 1.83 m has been rounded off to 1.8 m. The *Master Electrician's Review* reflects these rounded off changes, except that the inch-pound measurement is shown first—that is, 6 feet (1.8 m).

TRADE SIZES

A unique situation exists. Strange as it may seem, what electricians have been referring to for years has not been correct! Raceway sizes have always been an approximation. For example, there has never been a 1 in. raceway! Measurements taken from the *NEC*® for a few types of raceways show the following:

Trade Size	Inside Diameter
½ Electrical Metallic Tubing	0.622 inch
½ Electrical Nonmetallic Tubing	0.560 inch
½ Flexible Metal Conduit	0.635 inch
½ Rigid Metal Conduit	0.632 inch
½ Intermediate Metal Conduit	0.660 inch

You can readily see that the cross-sectional areas, critical when determining conductor fill, are different. It makes sense to refer to conduit, raceway, and

tubing sizes as "**trade sizes**." *NEC*® 90.9(C)(1) states: *Where the actual measured size of a product is not the same as the nominal size, trade size designators shall be used rather than dimensions. Trade practices shall be followed in all cases.** The *Master Electrician's Review* uses the term "trade size" when referring to conduits, raceways, and tubing. For example, instead of referring to a ½ in. EMT, it is referred to as trade size ½ EMT.

The *NEC*® also uses the term **metric designator**. A trade size & times ½ in. EMT is shown as metric designator 16 (½). A trade size & times 1 in. EMT is shown as metric designator 27 (1). The numbers 16 and 27 are the metric designator values. The ½ and 1 are the trade sizes. The metric designator is a rigid steel conduit's inside diameter in rounded off millimeters (mm). Here are some of the more common sizes of conduit, raceways, and tubing. A complete table is found in the *NEC*®, Table 300.1(C). Because of possible confusion, this text uses only the term "trade size" when referring to conduit, tubing, and raceway sizes.

METRIC DESIGNATOR AND TRADE SIZE	
Metric Designator	Trade Size
12	⅜
16	½
21	¾
27	1
35	1¼
41	1½
53	2
63	2½
78	3

Conduit knockouts in boxes do not measure up to what we call them. Here are some examples.

Trade Size Knockout	Actual Measurement
½	⅞ inch
¾	1³⁄₃₂ inches
1	1⅜ inches

* Reprinted with permission from NFPA 70-2011.

Outlet boxes and device boxes use their nominal measurement as their trade size. For example, a 4 in. × 4 in. × 1½ in. box does not have an internal cubic-inch volume of 24 in.³ Table 314.16(A) shows this size box as having a volume of 21 in.³ This table shows trade sizes in two columns in both millimeters and inches.

In this text, a square outlet box is referred to as a 4 × 4 × 1½ in.², a 4 in. × 4 in. × 1½ in.², or trade size 4 × 4 × 1½. Similarly, a single-gang device box might be referred to as a 3 × 2 × 3 in. box, a 3 in. × 2 in. × 3 in. deep box, or a trade size 3 × 2 × 3 box.

Trade sizes for construction material will not change. A 2 × 4 is really a name, not an actual dimension. A 2 × 4 will still be referred to as a 2 × 4. This is its trade size.

In this text, measurements directly related to the *NEC®* are given in both inch-pound and metric units. In some instances, only the inch-pound units are shown. This is particularly true for the examples of raceway and box fill calculations, load calculations for square foot areas, and on the plans (drawings).

Because the *NEC®* rounded off most metric conversion values, a computation using metrics results in a different answer when compared with the same computation done using inch-pounds. For example, load calculations for a residence are based on 3 volt-amperes per square foot or 33 volt-amperes per square meter.

- For a 40 × 50 foot dwelling: 3 VA × 40 ft × 50 ft = 6000 volt-amperes.

- In metrics, using the rounded off values in the *NEC®*: 33 VA × 12 m × 15 m = 5940 volt-amperes.

The difference is small, but nevertheless, it is different.

To show calculations in both units throughout this text would be very difficult to understand and would take up too much space. Calculations in either metrics or inch-pounds are in compliance with *NEC®* 90.9(D). In 90.9(C)(3), we find that metric units are not required if the industry practice is to use inch-pound units.

It is interesting to note that the examples in Annex D of the *NEC®* use inch-pound units, not metrics.

ACKNOWLEDGMENTS

The author would like to give special thanks to:
Allied Tube and Conduit
American Iron and Steel Institute (AISI)
Calculated Industries, Inc.
Carlon/Lamson and Sessions Company
Cooper B-Line Inc.
Cooper Bussman
Cooper Industries
Crouse-Hinds EMC
Pyrotenax USA Inc.
Square D Company
Taymac Corporation
Underwriters Laboratories, Inc.
Unity Manufacturing

The author and Delmar Cengage Learning would like to acknowledge the time and effort of the reviewers. Our thanks go to:

Michael Forister
Charles Trout

Applicable tables and section references are reprinted with permission from the NFPA 70-2011, the *National Electrical Code®*, copyright 2010, the National Fire Protection Association, Quincy, MA 02269. This reprinted material is not the complete and official position of the NFPA on the referenced subject, which is represented only by the standard in its entirety.

About the Author

Richard E. Loyd is a nationally known author and consultant specializing in the *NEC®* and the model building codes. He is president of his own firm, R & N Associates, located in Sun Lakes, Arizona. He and his wife Nancy travel throughout the country presenting seminars and speaking at industry-related conventions. He also serves as a code expert at 35 to 40 meetings per year at International Association of Electrical Inspectors (IAEI) meetings throughout the United States. Mr. Loyd represented the National Electrical Manufacturers Association (NEMA) Section 5RN Steel Rigid Conduit and Tubing as an *NEC®* consultant from 1986 through April 1997. Mr. Loyd presently represents the Steel Tube Institute (STI) of North America as an *NEC®* and Model Codes consultant. He represents the American Iron and Steel Institute (AISI) on American National Standards Institute (ANSI)/ NFPA 70 in the *NEC®* as a member of Code Committee Eight (8), which is the panel responsible for raceways, Code Committee Five (5), which is the panel responsible for grounding and NFPA 90A and 90B air conditioning committee standards. Mr. Loyd is actively involved doing forensic inspections and investigations on a consulting basis and serves as a special expert in matters related to the application of codes as they relate to installations and safety. He is a member of the Arizona Chapter of the Electrical Inspectors Association. He is past chairman of the Electrical Section of the National Fire Protection Association (NFPA).

He is an active member in the International Association of Electrical Inspectors (IAEI); the NFPA; the Institute of Electrical and Electronics Engineers (IEEE), where he is presently a member of the Power Systems Grounding Committee (Green Book); and the International Code Conference (ICC); Mr. Loyd is currently licensed as a master contractor/electrician in Arkansas (license #1725) and Idaho (license #2077) and is an NBEE certified master electrician.

Mr. Loyd served as the chief electrical inspector and administrator for the State of Idaho and for the State of Arkansas. He has served as chairman of NFPA 79 "Electrical Standard for Industrial Machinery"; as a member of Underwriters Laboratories Advisory Electrical Council; as chairman for the Educational Testing Service (ETS); and he served as associate editor for Intertec Publications (*EC&M* magazine) for 10 years. He is a master electrician and former electrical contractor. He has served as chairman of the National Board of Electrical Examiners (NBEE). He is an accredited instructor for licensing certification courses in Idaho, Oregon, and Wyoming and has taught basic electricity and *NEC®* classes for Boise State University (BSAU) in Boise, Idaho.

CHAPTER 1

OBJECTIVES

After completing this chapter, you should know:

- The methodology of the development of standard multiple-choice examinations

- The most effective methods to successfully take this type of examination

- How to properly prepare for the examination

- Your strengths and weaknesses based on a test of your initial electrical academic skills depending on the results of the examination at the end of this chapter

EXAMINATIONS AND NATIONAL TESTING ORGANIZATIONS

Examinations are an old phenomenon. They have been administered in many disciplines to evaluate competency in the field of specialty. Until recently, they were prepared by others within the chosen field, based on the knowledge of each person preparing the examination. Today, many examinations are professionally developed by state agencies specializing in examination preparation and by national testing organizations.

In recent years, many local jurisdictions, such as city, county, and state, have dropped their electrical examinations in favor of using the national testing agencies that prepare and administer electrical examinations for those local entities. However, there are still many local jurisdictions developing and administering electrical examinations much in the same manner as has been done since the beginning of electrical examinations. These examinations are often prepared and administered by local electrical inspectors and are often developed with question material related to areas in which the inspector often finds violations or to areas in the *National Electrical Code®* (*NEC®*) that the inspector feels are most difficult to understand. Rarely do local examinations fully evaluate an entry-level craftsman regarding his or her competency in this new career field. National testing organizations, such as International Code Council (International Code Council—National Contractor/ Trades Examinations: 900 Montclair Road, Birmingham, Alabama 35213. Phone: 1-877-STD-EXAM (783-3926). Fax: 1-205-599-9884. Web site: www.iccsafe.org/contractor, have carefully taken the advantages of localized testing methodology and incorporated the latest state-of-the-art technology to develop a test that is both fair to the taker and a good evaluation of the taker's knowledge in the specific career field. These testing organizations use those experienced in the trades to develop a task analysis to determine what the examination will evaluate and then develop questions based on those tasks. These organizations strive to develop question criteria that are clear and not confusing, questions that have only one answer, and questions that an entry-level candidate entering the journeyman electrical field or the master contractor field should know as minimum standards.

These examinations are generally multiple-choice written examinations. The study arrangement in this book is designed to prepare the user to successfully pass these national examinations. However, it should also provide for those preparing for local tests on these same subjects.

DEVELOPING AN ITEM (QUESTION) BANK

The development of examinations that effectively measure the minimum competency of an entry-level candidate is a process that uses many concepts. National testing firms have developed examinations that have become standards in the industry today. The first step in developing an examination is a task or job analysis. The task analysis must evaluate the ability, knowledge, and skill needed to perform the tasks related to the job in a way that will not endanger public health, safety, and welfare. The analysis must be relevant to the actual electrical practice in the field. The first step is, then, to establish the criteria, the subject areas of the trade that are most important and need to be tested. Two acceptable methods for determining these various job-related tasks would be (1) to survey licensed persons from the construction trades who perform these tasks in the field on a day-to-day basis and (2) to assemble a group of competent experts in these fields to list the content outline and determine the degree of importance of each task to be evaluated by the examination and create a blueprint that shows which percent of the exam is designated for each subject task. This blueprint ties the test to the job performance, and the test outline shows the importance of each subject. When done properly, the task analysis will substantiate the validity of the exam.

The next task is to hold workshops or group meetings to develop an item bank for a group of questions. This task of developing the item bank is normally performed by a group of interested experts working in workshops to develop the item bank based on the *NEC®* and other electrical-related texts or reference books. As the item bank is developed, it is important that each item be clear, precise, and has only one correct answer. On completion of the item bank, there should be a pool of 500–800 questions. These questions are entered into the computer, each based on the task analysis for each job task and weighted according to the difficulty level of each question. The electrical board whose members come from contract states subscribing to these national tests then determines the passing scores, the item selection, and test form assembly. When this has been completed, the examination is ready for administration. Following the test administration to the candidates, a posttest examination analysis is then conducted in order to determine the effectiveness of the examination. In the posttest analysis, which is conducted following every exam, the effectiveness of each item is examined. Inadequate items are returned to the workshop to be rewritten, corrected, or eliminated from future exams. Each item is evaluated from developed criteria from previous exams, and the difficulty level is verified. Postexamination analysis provides critical and important information for future workshops and the validity of the exams. This analysis provides information that cannot be gained in any other manner. Information continues to be developed throughout the life of the item bank. This information is fed into the computer and continually improves the validity of future examinations. The candidates themselves provide the most important information in the continuing development and improvement of the examination.

PREPARING FOR AN EXAMINATION

The first step in preparing for an examination is to obtain the examination information from the authority having jurisdiction (AHJ) to which you are about to make application for an examination. National testing firms furnish each proctor with a bulletin containing generic material that applies nationwide, and specific, unique material related only to the jurisdiction in which you are about to take an examination. This bulletin contains information that must be read carefully about the eligibility and procedure for registering for the test. This includes your verification of experience and education requirements and usually requires an application fee. It also contains the deadlines that are important for submitting the application and fees and the examination dates. For instance, you may be required to have your application and fees submitted up to 60 days before the examination date. The fees vary from state to state. They generally require the application fee to be made to the jurisdiction and the exam administration fee to be made directly to the national testing firm. However, this procedure varies, and the instructions must be read carefully. Once your application has been approved, your name will be placed on the roster for the administration of the exam at the testing center that you have selected or has been assigned to you. This will generally be the testing center closest to your home address unless you have indicated that you would like to be tested in another location. A schedule of examination dates and locations for the year is included in the candidate bulletin or is available from the local jurisdiction. You will be mailed an admission letter. Do not lose the letter. It will be required when you arrive at the test center or test location to take the exam. If you fail to bring this letter, you will not be permitted to take the exam. Also, it is very important that you read this bulletin to determine what materials are required and what materials are allowed, usually the *NEC*® or handbook in the edition in which the test is developed.

> **Warning:** Verify which *NEC*® edition is being used for the examination (2008 or 2011). Check allowable reference books before studying for the examination. Some agencies permit only the softcover *NEC*®, others permit the *NFPA*® *Handbook* and other reference books. Verify the pencil type and the type of calculator allowed. Do not wait until test day. Be prepared.

Other electrical reference books may be permitted. You are usually required to bring your own pencils and a silent, handheld, nonprogrammable, battery-powered calculator. Some jurisdictions also require a hands-on practical test. If it is required in the area in which you are making application, the outline of that work should also appear in the bulletin. Read the outline carefully. The practical test can mean the difference between passing and failing the examination. One such jurisdiction required that the candidate bring conduit, a conduit bender, hacksaw, reaming tool, tape measure, and continuity or ohmmeter. Also contained in the candidate bulletin will be the provisions for candidates with physical disabilities. Documentation of the disability must be submitted at the time the application is made. Special considerations are generally permitted to accommodate religious or disability needs as permitted by the Americans with Disabilities Act (ADA).

Finally, the candidate bulletin contains a content outline; the number of questions in each content area; the study reference material that the test has been developed from, such as the *National Electrical Code*® book, *Alternating Current Fundamentals* by Duff and Herman (Delmar Cengage Learning), *Direct Current Fundamentals* by Loper and Tedsen (Delmar Cengage Learning), *The American Electrician's Handbook*, and a sample of questions. It is suggested that you cover the answers and take this sample test in the bulletin much in the same manner as you would take the real examination. This will give you an idea of what the test will be like and will familiarize you with the format of the test if you are not already familiar with multiple-choice examinations. This is an important part of preparing for the electrical examination. Take the candidate bulletin seriously.

The next step in preparing for the examination is to study this book. The review questions at the end of each chapter are taken from the *NEC®* and not necessarily only from the content of each chapter. The *NEC®* is a reference document. As a reference document, it is necessary that you develop a strong proficiency in using the Table of Contents and the Index to be able to quickly locate the articles and sections of the *Code* you need. For each review question that you answer, you should not only research and find the correct answer, but you should also reference the *NEC®* section in which the answer was found. As you go through all the review material, take the sample tests throughout the document. This will enable you to pass the test the first time you take it. Anything less may prove disappointing.

> Each lesson is designed purposely to require the student to apply the entire *NEC®* text and not just specific chapters or articles. It has been found that when studying for a timed, open-book examination, the student must gain proficiency in the Table of Contents, the Index, and the ability to move quickly from cover to cover in order to find the correct answer to each question.

You should be aware that if you fail the examination, some jurisdictions require a waiting period of 6 months to 1 year before you can retake it.

The next section of this book contains a sample examination similar to the examination that you will take in the jurisdiction that you are studying for and similar to the examination given by many national testing organizations. This test should be taken in a quiet room. You should time yourself for 3 hours. The examination should be taken all at one time. To do otherwise will not give you a true evaluation of your present knowledge and of the areas in which you are weakest. Learning which areas are your weakest will enable you to concentrate your efforts and study time to gain the competency needed to pass your examination.

You may have purchased this book for various reasons. Some individuals study merely to improve their understanding in their chosen field, but most do so because they are required to pass their performance evaluation on their job or to pass an evaluation in a state or local examination. The specific reason may

be unimportant. The important thing is that you are about to undertake the study of electricity that, if done properly, will improve your competency in this field. The better you prepare, the better you will do on any evaluation or examination. Preparedness has more than one effect; for example, your IQ can fluctuate 30–40 points between any given dates.

The best way to improve individual performance is to reduce the anxiety level. To improve your study habits, consistent study time is important. You should allow yourself the same amount of time each day or each week to study this course. Consistent study time improves your retention.

This book and other Delmar Cengage Learning books related to this subject are designed to improve an individual's understanding of how to use the *NEC®* as a reference document and how to use the basic electrical formulas as applicable. They duplicate as nearly as possible the test format used by many national testing organizations and local jurisdictions. Whether you are preparing for an examination or just improving your skills, good study habits will give you the maximum efficient use of your time.

Basic Minimum Preparedness Tools

By the time test day arrives, you should have achieved the following:

1. The *NEC®* is a reference book. Most examinations recognize this and are "open-book exams." However, it is still necessary that you are comfortable using it as a tool.

   • Be comfortable using the *NEC®* Table of Contents. Understand each chapter heading and where and how to find things quickly.

   • Be comfortable using the *NEC®* Index using key words.

   • Have a thorough understanding of *NEC®* Article 90—Introduction. This article gives you the Code purpose, what is covered and not covered, Code arrangement, enforcement, rules, and planning.

   • Have a thorough understanding of *NEC®* Article 100—Definitions. This article contains special definitions that are specific to this document, different from the definition you find in a standard dictionary, such as "bathroom," "accessible," and "location."

   • Have a thorough understanding of *NEC®* Article 110—Requirements for Electrical Installations.

This article covers general requirements applicable to all installations other than those covered by Chapter 8[90.3].

- Have a thorough understanding of *NEC®* Article 300—Wiring Methods. This article covers general wiring method requirements applicable to all types of wiring methods.

2. The following basic electrical mathematical formulae are a must if you are to successfully pass the test.

- Voltage drop
- Ohm's law
- Common alternating current (ac) formulae

If you have achieved this basic necessary knowledge and you have completed this book, you should have no problem passing any standard electrical examination.

Get a good night's sleep the day before the test.

ALTERNATING CURRENT		
To Find	**Single Phase**	**Three Phase**
Amperage when "HP" is known	$\dfrac{HP \times 746}{E \times \%EFF \times PF}$	$\dfrac{HP \times 746}{E \times \%EFF \times PF \times 1.73}$
Amperes when "kW" is known	$\dfrac{kW \times 1000}{E \times PF}$	$\dfrac{kW \times 1000}{E \times PF \times 1.73}$
Amperes when "KVA" is known	$\dfrac{KVA \times 1000}{E}$	$\dfrac{KVA \times 1000}{E \times 1.73}$
Kilowatts	$\dfrac{E \times I \times PF}{1000}$	$\dfrac{E \times I \times PF \times 1.73}{1000}$
Kilovolt-Amperes "KVA"	$\dfrac{E \times I}{1000}$	$\dfrac{E \times I \times 1.73}{1000}$
Horsepower	$\dfrac{E \times \%EFF \times PF}{746}$	$\dfrac{E \times I \times \%EFF \times PF \times 1.73}{746}$
Percent Efficiency $= \dfrac{Output}{Input}$		
$PF = \dfrac{Power\ (Watts)}{Apparent\ Power} = \dfrac{kW}{KVA}$		

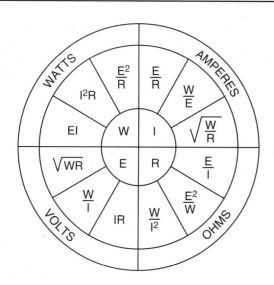

Test Day

Check your Candidate Information Bulletin to confirm allowable materials. You should have the following items to take with you to the examination.

- Admission mailer (letter).
- Photo I.D., such as a driver's license.
- *NEC®* book (applicable edition) and any other approved reference material that you feel will benefit you in taking the test. ***Note:*** Your reference material will be checked for handwritten notes and other markings not acceptable to the jurisdiction in which you are testing.
- A silent, nonprinting, nonprogrammable calculator that is not capable of alphabetic entry.
- Two No. 2 pencils.

The time has finally arrived for you to take your examination. Be sure you allow plenty of time to travel to the testing location; it is important to arrive on time. Some testing agencies will not permit admission if you arrive late. On the other hand, do not arrive too early. Discussing the test with other examinees may raise your anxiety level. Nervousness is contagious! Be sure you have your receipt for your application and your entrance ticket. Be sure you have all materials required by the jurisdiction or testing agency, including study books and a handheld calculator.

When a test is multiple choice, you generally have four alternative answers; therefore, the odds are four to one that you can get the right answer. If you can eliminate two of the alternatives, your chances are now increased to 50/50. This is extremely beneficial when referencing the *Code*, as you can quickly eliminate half of the possibilities, leaving you only two to research.

A panelboard is installed in a dwelling unit bathroom. Which of the following answers is correct?

1. It is not an *NEC®* violation.
2. It is permitted, but it must be suitable for wet locations.
3. The location of panelboards is not covered by the *NEC®*.
4. It is an *NEC®* violation.

We can quickly eliminate 2 and 3. The answer is 4. *NEC®* 240.24(E) prohibits installing overcurrent devices in dwelling unit bathrooms.

Read the directions carefully. Many mistakes are made merely because of misunderstanding. If you have any questions, discuss them with the proctor before the test time starts. Any discussions after the test starts will take away from the time you have to answer the questions. Allow yourself ample time for each question. Answer the question first before you check any alternatives. This way, you can evaluate your answer against any alternatives that you may find in the *NEC*®. Do not spend too much time on any one question. If you find the question extremely difficult and you are unsure of the answer or cannot find the answer in the *NEC*®, skip that question and go back to it only if you have time at the end of the test.

You should never spend more than 3 minutes per question. Your first choice is usually best. Read the questions very carefully. Make sure you know key words that might change the meaning of the question. Note any negatives, such as "Which of the following are not, or shall not be permitted?" It may be beneficial to underline key words as you read the question. This has a tendency to channel your thoughts in the right direction.

Sample Question

Tip: Break the question down and get to the root question.

The bathroom in a custom home includes a vanity with double sinks (basins) located on one side of the bathroom. There is an outlet located adjacent to each sink (basin), with ground-fault circuit-interrupter (GFCI) protection. On the other side of the bathroom, just opposite to and about 8 feet away, is a makeup vanity (without a sink [basin]) with a receptacle outlet located at the same height as the GFCI outlets across the room.

Would this outlet above this vanity be required to also have GFCI protection?

Answer:

The key words are: bathroom
home (dwelling)
GFCI

The question is, would an outlet not adjacent to a sink in a bathroom need GFCI protection? The answer is yes, all receptacles located in a dwelling unit bathroom are required to be protected with GFCI protection. The reference is *NEC*® 210.8(A)(1).

This question could be time-consuming because it causes you to look at the location in relation to the sink. *NEC*® 210.52(D) requires a receptacle to be located within 3 feet (900 mm) of the outside edge of each basin (sink). However, the question does not ask that! The question asks whether another receptacle located outside this dimension needs GFCI protection.

Summary

- Dress properly; wear comfortable shoes and loose-fitting clothing. The test facility should be approximately 70°F. Too many clothes or uncomfortable clothes have a tendency to make you drowsy and concentration difficult.

- Plan to arrive at the test site 20–30 minutes early. Anxiety will reduce your effectiveness. Do not be late or too early. Make sure ahead of time that you know where the test site is located and whether parking is available.

- Be sure you have your acceptance letter and proper I.D. Verify ahead of time so that you have the identification materials required by the testing agency.

- Do not forget your *NEC*® book, reference books, calculator, and other acceptable/required materials. Pack all these materials the night before in a backpack or briefcase.

- After you have answered all the questions, recheck your work.

  1. Make sure that you have not made clerical mistakes.

  2. Make sure that your answers are recorded on the correct line on the answer sheet.

  3. Go over the extremely difficult questions.

  4. Regarding questions on which you know you have guessed—you may want to mark them in some way so that if there is time at the end of the test, you can do additional research.

Good luck!

SAMPLE EXAMINATION

Examination Instructions:

For a positive evaluation of your knowledge and preparation awareness, you must:

1. Locate yourself in a quiet atmosphere (room by yourself).
2. Have with you at least two sharp No. 2 pencils, the 2011 *NEC*®, and a handheld calculator.
3. Time yourself (3 hours) with no interruptions.
4. After the test is complete, grade yourself honestly and concentrate your studies on those sections of the *NEC*® in which you missed the questions.

Caution: Do not just look up the correct answers because the questions in this examination are only an exercise and not actual test questions. Therefore, it is important that you be able to quickly find answers from throughout the *NEC*®.

Directions: Each question is followed by four suggested answers: A, B, C, and D. In each case, select the one that best answers the question.

1. Liquidtight flexible metal conduit can be used in which of the following locations?
    A. In areas that are both exposed and concealed
    B. In areas that are subject to physical damage
    C. In connection areas for gasoline-dispensing pumps
    D. In areas where ambient temperature is to be greater than 200°C

    Answer: _____   Reference: _____

2. The maximum number of quarter bends allowed in one run of polyvinyl chloride Type PVC conduit is
    A. 2.
    B. 4.
    C. 6.
    D. 8.

    Answer: _____   Reference: _____

3. The maximum size of flexible metallic tubing that can be used in any construction or installation is
    A. Trade size ⅜ (12).
    B. Trade size ½ (16).
    C. Trade size ¾ (21).
    D. Trade size 1 (27).

    Answer: _____   Reference: _____

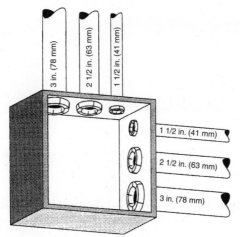

4. The size of the pull box in the diagram above should not be less than
   A. 12 in. × 14 in. (300 mm × 350 mm).
   B. 16 in. × 18 in. (400 mm × 450 mm).
   C. 22 in. × 22 in. (560 mm × 560 mm).
   D. 24 in. × 24 in. (600 mm × 600 mm).

   Answer: _____   Reference: _____

5. A metal raceway system installed in the washing area of a car wash must be spaced at least how far from the walls?
   A. ⅛ in. (3 mm)
   B. ¼ in. (6.4 mm)
   C. ¾ in. (19 mm)
   D. 1 in. (25 mm)

   Answer: _____   Reference: _____

6. Unless otherwise indicated, busways should be supported at intervals not to exceed how many feet?
   A. 3 (900 mm)
   B. 5 (1.5 m)
   C. 7 (2.1 m)
   D. 10 (3.0 m)

   Answer: _____   Reference: _____

7. Which of the following percentages of conduit fill should be used when four Type THW conductors are being installed?
   A. 55 %
   B. 53 %
   C. 40%
   D. 31%

   Answer: _____   Reference: _____

8. Flexible 18 American Wire Gauge (AWG) supply cords of listed appliances are considered adequately protected if the circuit overcurrent device is set at a maximum value of how many amperes?
   A. 15
   B. 20
   C. 30
   D. When applied within the appliance listing requirements

   Answer: _____   Reference: _____

9. All the following sizes of solid aluminum conductors shall be made of an aluminum alloy except
   A. 8 AWG.
   B. 10 AWG.
   C. 12 AWG.
   D. 14 AWG.

   Answer: _____   Reference: _____

10. An insulated bushing is required to be used on a raceway entering a cabinet if the ungrounded conductors entering the raceway are at least
    A. 10 AWG.
    B. 8 AWG.
    C. 6 AWG.
    D. 4 AWG.

    Answer: _____   Reference: _____

11. The allowable distance between supports of non-metallic-sheathed cable installed in an on-site, constructed, one-family dwelling is a maximum of how many feet?
    A. 2½ (750 mm)
    B. 3 (900 mm)
    C. 4 (1.2 m)
    D. 4½ (1.4 m)

    Answer: _____   Reference: _____

12. It is permissible to use 4 AWG type THW copper conductors for service in dwelling units to a maximum load of
    A. 100 amperes.
    B. 95 amperes.
    C. 70 amperes.
    D. 65 amperes.

    Answer: _____   Reference: _____

13. Where conductors of less than 600 volts emerge from the ground, they must be protected by enclosures or raceways that extend from below grade to a point how many feet above the finish grade?
    A. 5 (1.5 m)
    B. 6 (1.8 m)
    C. 7 (2.1 m)
    D. 8 (2.4 m)

    Answer: _____ Reference: _____

14. An installation consisting of six 3-wire cables installed in a 6-foot (1.8 m) nonmetallic wireway is being installed in a light commercial building. Which of the following statements is (are) not true?
    A. The six multiwire cables are not required to be installed together in a neat workmanlike manner.
    B. The sum of the cross-sectional areas of all contained current-carrying conductors shall not exceed 20% of the interior area of the wireway.
    C. The adjustment factors specified in 310.15(B)(3) is not applicable because there are less than 30 current-carrying conductors at 20% fill.
    D. The wireway may not be installed where subject to physical damage.

    Answer: _____ Reference: _____

15. The maximum rating for a residential service that uses 2 AWG THW copper service-entrance conductors is
    A. 95 amperes.
    B. 110 amperes.
    C. 115 amperes.
    D. 125 amperes.

    Answer: _____ Reference: _____

16. Grounding electrode conductors in metal enclosures for surge protective devices rated 1kw or less shall be
    A. bare.
    B. insulated.
    C. bonded on one end of the enclosure only.
    D. installed in compliance with 250.64(E).

    Answer: _____ Reference: _____

17. The number of overcurrent devices in a single cabinet of a panelboard shall not exceed
    A. 30.
    B. 36.
    C. 42.
    D. the number for which the panelboard is rated, designed, and listed.

    Answer: _____ Reference: _____

18. When it is impractical to locate a service head directly above the point of attachment, the maximum allowable placement distance from the point of attachment is how many feet?
    A. 1 (300 mm)
    B. 2 (600 mm)
    C. 3 (900 mm)
    D. 4 (1.2 m)

    Answer: _____   Reference: _____

19. An underground service to a small, controlled water heater is to be installed. This service requires that copper conductors be at least
    A. 12 AWG.
    B. 10 AWG.
    C. 8 AWG.
    D. 6 AWG.

    Answer: _____   Reference: _____

20. Service conductors can have less than 3 feet (900 mm) of clearance from which of the following?
    A. Doors
    B. Tops of windows
    C. Fire escapes
    D. Porches

    Answer: _____   Reference: _____

21. Service conductors that pass over rooftops with a 4 in 12 in slope or less must have a vertical clearance of not less than how many feet?
    A. 3 (900 mm)
    B. 7 (2.1 m)
    C. 8 (2.4 m)
    D. 10 (3.0 m)

    Answer: _____   Reference: _____

22. The interrupting rating shall be marked on all circuit breakers rated more than
    A. 1000 amperes.
    B. 2000 amperes.
    C. 2500 amperes.
    D. 5000 amperes.

    Answer: _____   Reference: _____

23. Illumination is mandatory for service equipment and panelboards in a dwelling unit if the service to the unit exceeds how many amperes?
    A. 100
    B. 200
    C. 400
    D. Required for all services regardless of amperage

    Answer: _____   Reference: _____

24. Ground-fault protection is required for _____-ampere, 480-volt, 4-wire, 3-phase solidly grounded wye services.
    A. 800
    B. 1000
    C. 1200
    D. above 2000

    Answer: _____   Reference: _____

25. Which of the following is the correct connection sequence for a 3-phase, delta-connected motor to be connected to 240-volt supply?

	L1	*L2*	*L3*
A.	To 1 and 7	To 2 and 8	To 3 and 9; 4, 5, and 6
B.	To 1, 7, and 6	To 2, 8, and 4	To 3, 9, and 5
C.	To 1, 7, and 4	To 2, 8, and 5	To 3, 9, and 6
D.	To 1	To 2	To 3; 7 and 4; 8 and 5; 9 and 6

    Answer: _____   Reference: _____

26. Impedance is present in which of the following circuits?
    A. ac only
    B. dc only
    C. resistance only
    D. both ac and dc

    Answer: _____   Reference: _____

27. Excluding fuses and exceptions, how many overcurrent protection devices, such as trip coils, relays, or thermal cutouts, are required on a 3-phase motor?
    A. 5
    B. 3
    C. 2
    D. 0

    Answer: _____   Reference: _____

28. All of the following are induction-type motors except
    A. wound-rotor.
    B. split-phase.
    C. capacitor.
    D. universal.

    Answer: _____   Reference: _____

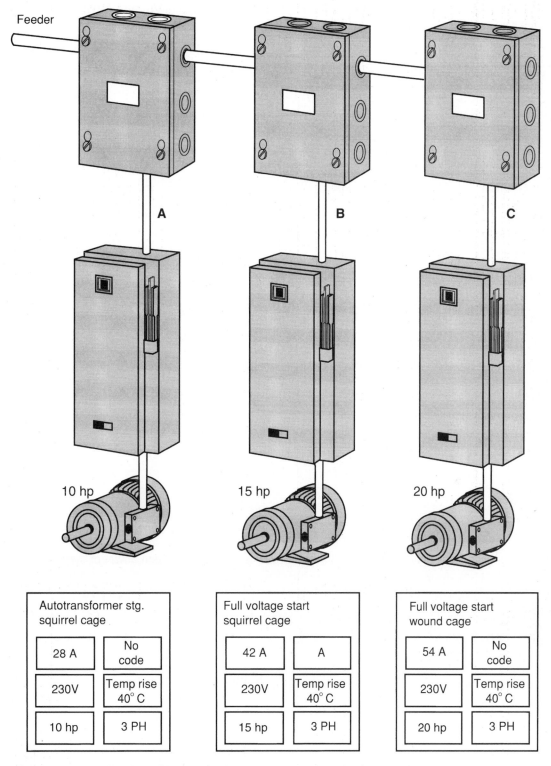

Feeder

A   B   C

10 hp   15 hp   20 hp

Autotransformer stg. squirrel cage	
28 A	No code
230V	Temp rise 40° C
10 hp	3 PH

Full voltage start squirrel cage	
42 A	A
230V	Temp rise 40° C
15 hp	3 PH

Full voltage start wound cage	
54 A	No code
230V	Temp rise 40° C
20 hp	3 PH

29. In the diagram above, the size of THW feeder conductors shall not be less than
   A. 4 AWG.
   B. 2 AWG.
   C. 1 AWG.
   D. 1/0 AWG.

Answer: _____   Reference: _____

30. In the diagram on page 13, the size of Type THW conductors for branch-circuit A shall not be less than
    A. 10 AWG.
    B. 8 AWG.
    C. 6 AWG.
    D. 4 AWG.

    Answer: _____   Reference: _____

31. In the diagram on page 13, the size of Type THW conductors for branch-circuit C shall not be less than
    A. 8 AWG.
    B. 6 AWG.
    C. 4 AWG.
    D. 3 AWG.

    Answer: _____   Reference: _____

32. The attachment plug and receptacle can be used as the controller for a portable motor that has a maximum horsepower rating of
    A. ⅓ horsepower.
    B. ½ horsepower.
    C. 1 horsepower.
    D. 2 horsepower.

    Answer: _____   Reference: _____

33. Snap switches can be grouped or ganged in outlet boxes if voltages between adjacent switches do not exceed how many volts?
    A. 200
    B. 300
    C. 400
    D. 500

    Answer: _____   Reference: _____

34. Each doorway that leads into a transformer vault from the building interior should have a tight-fitting door with a minimum fire rating of how many hours?
    A. 1
    B. 2
    C. 3
    D. 5

    Answer: _____   Reference: _____

35. The raceway run between warm and cold locations shall be
    A. sealed.
    B. sleeved.
    C. insulated.
    D. provided with drains.

    Answer: _____ Reference: _____

36. Except for points of support, non-IC Type recessed portions of luminaire enclosures shall be spaced at least how many inches from combustible material?
    A. ¼ (6 mm)
    B. ½ (13 mm)
    C. 1 (25 mm)
    D. 3 (75 mm)

    Answer: _____ Reference: _____

37. Surface-mounted fluorescent luminaires that contain a ballast and are to be installed on combustible, low-density cellulose fiberboard shall be spaced not less than how many inches from the surface of the fiberboard?
    A. ½ (13 mm)
    B. 1 (25 mm)
    C. 1½ (38 mm)
    D. 2 (50 mm)

    Answer: _____ Reference: _____

38. The minimum amount of on-site fuel for an emergency generator using an internal combustible engine as the prime mover shall be sufficient for not less than
    A. 2 hours at full demand.
    B. 4 hours at 75 % demand.
    C. 6 hours at 50 % demand.
    D. 8 hours at 25 % demand.

    Answer: _____ Reference: _____

39. Which of the following statements about dimmers installed in theaters is (are) true?
    I. Dimmers installed in ungrounded conductors must have overcurrent protection not greater than 125 % of the dimmer rating.
    II. The circuit-supplying autotransformer type dimmers must not exceed 150 volts between conductors.
    A. I only
    B. II only
    C. Both I and II
    D. Neither I nor II

    Answer: _____ Reference: _____

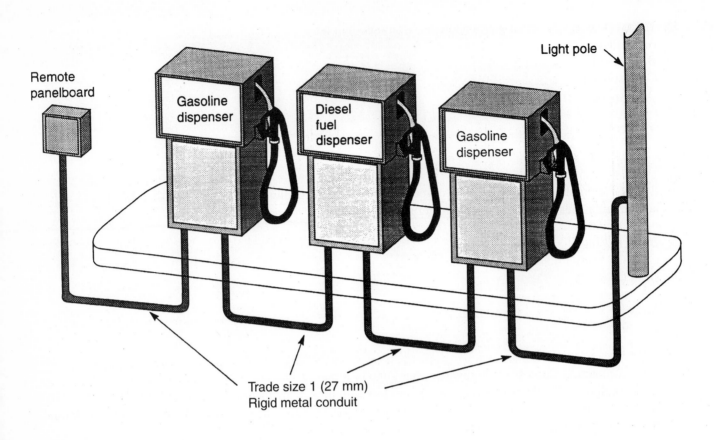

Remote panelboard

Gasoline dispenser

Diesel fuel dispenser

Gasoline dispenser

Light pole

Trade size 1 (27 mm)
Rigid metal conduit

40. In the diagram above, how many conduit seals are required?
   A. 2
   B. 4
   C. 6
   D. 8

   Answer: _____   Reference: _____

41. An area classed as a Class II hazardous location will contain which of the following?
   A. Combustible dust
   B. Flammable gases
   C. Ignitable liquid
   D. Ignitable vapors

   Answer: _____   Reference: _____

42. If the wiring is to be placed 18 in (450 mm) above a Class I location, which of the following wiring methods is prohibited for a commercial automotive repair shop?
   A. Metal-clad cable
   B. Electrical metallic tubing
   C. Rigid nonmetallic conduit
   D. non-metallic-sheathed cable

   Answer: _____   Reference: _____

43. The minimum number of receptacles in a patient bed location of a hospital's general care area shall be
    A. 1 duplex.
    B. 2 duplex.
    C. 3 duplex.
    D. 3 single.

    Answer: _____   Reference: _____

44. Which of the following statements about lightning protection is (are) true?
    I. If lightning protection is required for an irrigation machine, a grounding electrode system shall be connected to the machine at a stationary point.
    II. Lightning protection conductors should be spaced in accordance with NFPA 780 from non-current-carrying metal parts of electrical equipment.
    A. I only
    B. II only
    C. Both I and II
    D. Neither I nor II

    Answer: _____   Reference: _____

45. Which of the following is a permissible way to ground a 230-volt, single-phase, residential air-conditioning unit that replaces an older unit supplied by a 2-wire, 230-volt circuit?
    I. A new circuit containing an equipment grounding conductor
    II. An 8-foot (2.5 m) ground rod installed at the unit
    A. I only
    B. II only
    C. Both I and II
    D. Neither I nor II

    Answer: _____   Reference: _____

46. When connected to a solely rod, pipe, or plate grounding electrode, grounding electrode copper conductors need not be sized larger than
    A. 8 AWG.
    B. 6 AWG.
    C. 4 AWG.
    D. 2 AWG.

    Answer: _____   Reference: _____

47. Which of the following is a permitted grounding electrode?
    A. A ½ inch driven steel approved ground rod
    B. A well with a nonmetallic casing
    C. Building foundation
    D. An underground cold water metal piping system 10 feet in contact with earth

    Answer: _____   Reference: _____

48. Which of the following terms can be used in place of "resistance" in the phrase "resistance is measured in ohms"?
   A. Reactance
   B. Inductance
   C. Impedance
   D. Capacitance

   Answer: _____ Reference: _____

49. The *NEC*® requires a minimum of which of the following for the kitchen small-appliance load in dwelling units?
   A. Two 20-ampere circuits
   B. Two 15-ampere circuits
   C. One 20-ampere circuit
   D. One 15-ampere circuit

   Answer: _____ Reference: _____

Questions 50 and 51 refer to a 120-gallon water heater with 220-volt, 4000-watt units and an interlocking thermostat.

50. Installation of this water heater will require branch-circuit conductors of size
   A. 8.
   B. 10.
   C. 12.
   D. 14.

   Answer: _____ Reference: _____

51. The maximum branch-circuit overcurrent protection for this water heater is
   A. 20 amperes.
   B. 25 amperes.
   C. 30 amperes.
   D. 40 amperes.

   Answer: _____ Reference: _____

Questions 52 and 53 refer to a 230-volt circuit that is 500- feet long, displays a resistance of 1 ohm, and supplies a load of 10 amperes.

52. What is the wattage of the load in this circuit?
   A. 2000 W
   B. 2200 W
   C. 2300 W
   D. 2400 W

   Answer: _____ Reference: _____

53. How much power is lost in this circuit?
    A. 100 W
    B. 120 W
    C. 240 W
    D. 220 W

    Answer: _____   Reference: _____

54. Ground-fault protection of equipment is a requirement for solidly grounded wye services that are rated at more than 150 volts-to-ground but do not exceed 600 volts, phase-to-phase, for each service disconnecting means rated at 1000 amperes or more. The maximum setting for this ground-fault protection relay is
    A. 1000 amperes.
    B. 1200 amperes.
    C. 1500 amperes.
    D. 2000 amperes.

    Answer: _____   Reference: _____

Number	Size (kcmil)	Type	Function
2	500	THW	Phases A and B
1	400	THW	Neutral
1	300	THW	Phase C
1	3 AWG	THW	Ground Conductor

55. The minimum size intermediate metal conduit required for installation of the conductors above shall not be less than
    A. Trade size 2 (53).
    B. Trade size 2½ (63).
    C. Trade size 3 (78).
    D. Trade size 3½ (91).

    Answer: _____   Reference: _____

Questions 56 and 57 refer to three 3-phase induction motors, rated respectively at 28, 16, and 12 amperes.

56. The motors would require feeder conductors rated at how many amperes?
    A. 56
    B. 63
    C. 70
    D. 112

    Answer: _____   Reference: _____

57. The motors are started at line voltage and are protected with time-delay fuses. The maximum feeder time-delay fuse should be rated at how many amperes?
    A. 70
    B. 80
    C. 90
    D. 100

    Answer: _____   Reference: _____

58. The combined load of several 240-volt, fixed space heaters on a 20-ampere circuit should not exceed how many kilowatts?
    A. 2.4
    B. 2.6
    C. 3.8
    D. 4.8

    Answer: _____   Reference: _____

59. If three resistors with values of 5, 10, and 15 ohms, respectively, are connected in parallel, the combined resistance of the units will be
    A. 1.63 ohms.
    B. 2.73 ohms.
    C. 20.0 ohms.
    D. 30.0 ohms.

    Answer: _____   Reference: _____

60. Conductors that supply an 8-kilowatt, 230-volt, fixed electric space heater consisting of resistance elements should have a kilowatt rating of not less than
    A. 6.4.
    B. 8.
    C. 10.
    D. 12.

    Answer: _____   Reference: _____

61. The *NEC*® covers the installation of electrical conductors and equipment in all the following locations except
    A. public and private buildings.
    B. floating dwelling units.
    C. buildings used by a utility for warehousing.
    D. centers of transmission and distribution of electrical energy.

    Answer: _____   Reference: _____

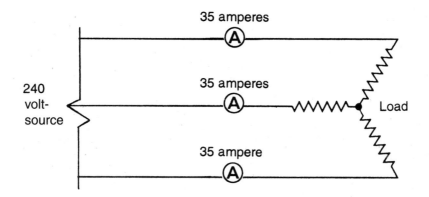

62. *Refer to the figure above.* Three 35-ampere resistive balanced resistive loads are connected to a 240-volt, 3-phase, and 3-wire circuit. What is the total power of this circuit in kilowatts?
    A. 14.6 kW
    B. 8.4 kW
    C. 7.3 kW
    D. 49.9 kW

    Answer: _____   Reference: _____

63. The most recent edition of the *NEC*® is revised every
    A. year.
    B. 2 years.
    C. 3 years.
    D. 4 years.

    Answer: _____   Reference: _____

64. Galvanized rigid conduit is used to protect conductors in a cable tray. Where the conductors enter and leave the conduit within the cable tray,
    A. junction boxes are required even if the conduit is bushed.
    B. junction boxes are not required if the conduit is bushed.
    C. no bushings are required if the conduit is smaller than 1 in.
    D. bushings are required if the conductors are smaller than #4 AWG copper.

    Answer: _____   Reference: _____

65. Surface-type cabinets for electrical equipment in damp or wet locations shall be mounted so at least a _____-inch air space between the wall or other supporting surface exists.
    A. ⅛ (3.1 mm)
    B. ¼ (6.3 mm)
    C. ⅜ (10 mm)
    D. ½ (13 mm)

    Answer: _____   Reference: _____

66. Generally, rigid metal conduit is required to be fastened in place within _____ foot/feet of each box, outlet, cabinet, or conduit termination.
    A. 1 (300 mm)
    B. 2 (600 mm)
    C. 3 (900 mm)
    D. 4 (1.2 m)

    Answer: _____   Reference: _____

67. For six size 12 AWG conductors, the minimum trade size octagonal box that can be used is
    A. 4 in. × 1¼ in. (100 mm × 32 mm).
    B. 4 in. × 1½ in. (100 mm × 38 mm).
    C. 4 in. × 2⅛ in. (100 mm × 54 mm).
    D. 4 in. × 2¼ in. (100 mm × 57 mm).

    Answer: _____   Reference: _____

68. A motor controller that is installed with the expectation of its being submerged occasionally for short periods shall be installed in a rated enclosure type No.
    A. 3
    B. 3S
    C. 4X
    D. 6

    Answer: _____   Reference: _____

69. Where recessed high-intensity discharge luminaires are installed indoors and operated by remote ballasts,
    A. thermal protection is not required.
    B. only the fixture requires thermal protection.
    C. only the ballast requires thermal protection.
    D. both the fixture and the ballast require thermal protection.

    Answer: _____   Reference: _____

70. Any pipe or duct system foreign to the electrical installation must not enter a transformer vault. The _____ is not considered foreign to the vault.
    A. vault automatic fire protection
    B. building water main
    C. air duct passing through the vault
    D. roof drain piping

    Answer: _____   Reference: _____

71. A dry-type transformer rated at 480 volts can be installed on a building column and shall not be required to be
    A. accessible.
    B. readily accessible.
    C. rated less than 112½ kVA.
    D. rated less than 25 kVA.

    Answer: _____ Reference: _____

72. A capacitor is located indoors. It must be enclosed in a vault if it contains more than _____ gallon(s) of flammable liquid.
    A. 1
    B. 2
    C. 3
    D. 5

    Answer: _____ Reference: _____

73. A single-phase hermetic refrigerant motor compressor has a rated load current of 24 amperes and a branch-circuit selection current of 30 amperes. The branch-circuit conductors are copper, Type TW. They operate at 80°F, and they are the only conductors in the conduit to this compressor. The smallest possible branch-circuit conductors must be at least size _____ AWG.
    A. 12
    B. 10
    C. 8
    D. 6

    Answer: _____ Reference: _____

74. A 3-phase, squirrel-cage motor with full-voltage reactor starting has no code letter. The calculated fuse size to provide ground-fault and short circuit protection for the branch circuit of the motor would be sufficient for the starting current of the motor. This nontime delay fuse shall be sized at _____ % of full-load current.
    A. 150
    B. 200
    C. 250
    D. 300

    Answer: _____ Reference: _____

75. A dry-type transformer is to be installed indoors. If rated more than _____ KVA, the transformer must be installed in a fire-resistant transformer room.

    A. 25.0
    B. 35.0
    C. 100.5
    D. 112.5

    Answer: _____ Reference: _____

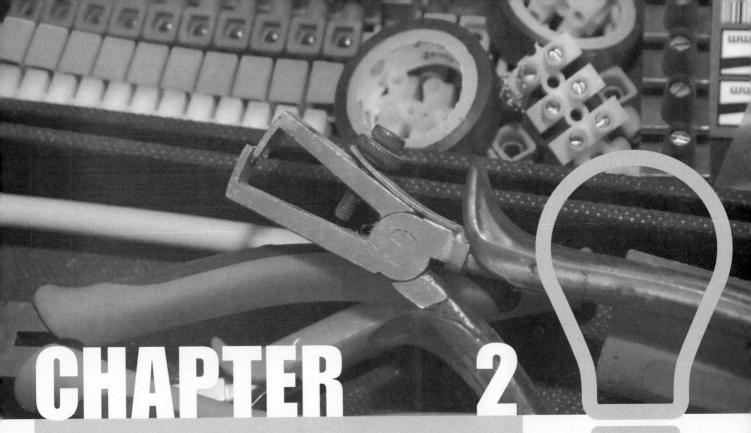

CHAPTER 2

OBJECTIVES

After completing this chapter, you should be able to:

- Identify the largest organizations that impact the electrical industry and give a brief outline of what each does and how they each impact the electrical industry (e.g., National Fire Protection Association, Nationally Recognized Testing Laboratories)

- Identify agencies responsible for enforcement and regulation of the *NEC*® and its amendments or other electrical codes

- Explain the basic authority of the *NEC*®

- Explain how the *NEC*® is organized and developed and identify segments of the industry covered by the *NEC*® and those that are not

- Explain how testing laboratories impact code enforcement

NATIONAL FIRE PROTECTION ASSOCIATION STANDARDS

The National Fire Protection Association (NFPA) has acted as the sponsor of the *NEC®* as well as many other safety standards. The most widely used electrical code in the world is the *NEC®*. The official designation for the *NEC®* is ANSI/NFPA 70.

NEC®

The *NEC®* was first developed about 100 years ago by interested industry and governmental authorities to provide a standard for the safe use of electricity. It is now revised and updated every 3 years. It was first developed to provide the regulations necessary for safe installations and to provide the practical safeguards of persons and property from the hazards arising from the use of electricity. It still provides these regulations and safeguards to the industry today. The *NEC®* is developed by 19 different Code-Making Panels (CMPs) composed entirely of volunteers. These volunteers come from industry, testing laboratories, inspection agencies, engineering, users, government, and the electrical utilities. Suggestions for the content come from various sources, including individuals like you. Anyone can submit a proposal or make a comment on a proposal submitted by another individual.

For more specifics on the *NEC®* process, you can contact the National Fire Protection Association, One Batterymarch Park, Quincy, MA 02269-9101. Request their free booklet "The NFPA Standards Making System."

The *NEC®* is advisory as far as NFPA and ANSI are concerned but is offered for use in law and regulations in the interest of life and property protection. The name *National Electrical Code®* might lead one to believe that this document is developed by the federal government. This is not so; the *NEC®* only recognizes uses of products. It approves nothing. The *NEC®* has no legal standing until it has been adopted by the *authority having jurisdiction* (AHJ) (see *NEC®* definition Article 100), usually a governmental entity. Therefore, we must first check with the local electrical inspector to see which edition of the *NEC®* has been adopted. Compliance within and proper maintenance will result in an installation essentially free from hazard but not necessarily efficient, convenient, or adequate for good service or future expansion of electrical use. The *NEC®* is not an instruction manual for untrained persons, nor is it a design specification. However, it does offer design guidelines.

NEC® 90.2 the scope of this document, it is intended to cover all electrical conductors and equipment within or on public and private buildings or other structures, including mobile homes, recreational vehicles, and floating buildings, and other premises, such as yards, carnivals, parking and other lots, and industrial substations. Installations of conductors and equipment that connect to the supply of electricity, other installations of outside conductors on the premises, and installations of communications and optical fiber cables are clearly covered by the *NEC®*. The *NEC®* does not cover installations on ships, watercraft, railway rolling stock, automotive vehicles, underground mines, and surface mobile mining equipment. The *NEC®* does not cover installations governed by the utilities, such as communication equipment, transmission, generation, and distribution installations on right-of-way.

> *Note:* For the complete list of the exemptions and coverage, see 2011 *NEC®* 90.2. Mandatory rules are characterized by the word "shall." Explanatory rules are in the form of informational notes. All tables and footnotes are a part of the mandatory language. Material text extracted from other NFPA standards and documents shall not be compromised or violated. Editing is permitted for adapting the extracted text into *NEC®* style. A complete list of all NFPA documents referenced can be found in Annex A. New revisions are identified by shading, where entire paragraphs are deleted and bullet can be found in the margin and where new articles are added, a vertical line will be in the margin.

To use the *NEC®*, one must first have a thorough understanding of Article 90—The Introduction; Article 100—Definitions; Article 110—Requirements for Electrical Installations; and Article 300—Wiring Methods. The rest of the *Code* can be referred to on an as-needed basis.

New for the 2011 *NEC®*

The 2011 *NEC®* has some notable changes, many of which are quite significant. Some major changes are noted here.

What's New for the 2011 *NEC*®

During the 2011 *NEC* development process, several thousand changes and public comments were considered. The 2011 *NEC*® has now been published and will be adopted and used as the premier electrical installation safety standard by more jurisdictions than any other document.

Fine Print Notes are now referred to as "Informational Notes." The term "fine print" does not describe the function of a sentence or provision; it simply refers to the size of the text. By changing the term to "informational note," the *Code* makes it quite clear that these notes are intended to provide information and nothing else. The same logic applies to the change to "informative annexes." The style and layout of these notes and annexes have not changed, nor has the intent of them.

A new Article 694 for small wind electric systems has been added. This article covers small wind (turbine) electrical systems up to and including 100 kW. These systems can contain generators, alternators, inverters, and controllers.

A new Article 840 has also been added, which covers Premises-Powered Broadband Communications Systems. This article is similar to Article 830, but is often times more applicable. Expect for rare instances, Article 830 installations are made by a communications utility. Article 840 is intended to apply more regularly.

Article 90 contains the Scope of the *NEC* and is an enforceable article. This article contains information that applies to every installation. Every electrician should read this information carefully and understand it.

Only a few changes that occurred in Article 90. 90.2(B)(5) were revised to fix an error that occurred in 2008. This mistake was the removal of utility installations on Native American reservations, federal lands, and similar areas from *Code* exemption. This change now gives the AHJ a means to exempt utility wiring in these areas.

Chapter 1—General

Chapter 1 includes Article 100—Definitions. These definitions are unique to this document and apply where the term is used in the *NEC*®. Article 110 covers the general requirements for all electrical installations (except those covered by Chapter 8) and it contains information that applies to every installation. Every electrician should read this information carefully and understand it completely. These requirements apply to every installation, regardless of how large or small it is.

The definition of Arc-Fault Circuit Interrupter (AFCI) has been relocated from 210.12 to Article 100 because the term is used in more than one Article.

The definition of "bathroom" has been revised to add urinals, bidets, and similar equipment to the list of items that may be found in a bathroom. This change will result in a more uniform understanding of the Code.

"Bonding Jumper" is now referred to as "Bonding Conductor or Jumper." The 2008 *NEC*® (and previous editions) referred only to "bonding jumpers," which seems to imply a very short length of conductor used to bond things together. Although this may often be the case, they also may be much longer.

The term "grounding conductor" has been removed from the Code in not only Article 100, but also all other articles of the Code. The definition of "grounding conductor" seemed to apply only to communications systems and auxiliary grounding electrodes, yet this term was often used where "grounding electrode conductor" should have been used. Changes to the Code this cycle make for a more consistent and technically accurate document.

All "service Conductor" related definitions have been revised, deleted, or added. A real effort has been made to clarify that service conductors are those that are downstream of the "service point." Conductors upstream of the service point are not service conductors; they are utility conductors (and are not covered by the Code).

110.3(A)(1) has been changed to reflect the fact that some pieces of electrical equipment have special requirements, such as limitations on elevation, ambient temperature correction, power quality requirements or, specific types of overcurrent devices. This information may be marked on equipment, or it may be in the product literature, or listing and labeling information.

Changes to 110.10 clarify that, in addition to the impedance of the circuit, the short-circuit current ratings of equipment are a vital part of determining whether or not a system or circuit can withstand the effects of a short circuit or ground fault. Specific examples of the types of things that warrant consideration are always better than referring to "other characteristics."

New provisions for finely stranded conductors have been added to 110.14, recognizing the fact that these conductors warrant special consideration.

A substantial change was made to Article 110 with the addition of 110.24. This new section requires that the available fault current must be marked at the service equipment of all installations, other than dwelling units. This change is intended to make sections 110.9 and 110.10 easier for the EHJ to validate.

The location of Table 110.28 (formerly 110.20) has been changed to Part II of the article, so that it applies only to equipment operating at 600 volts or less.

Chapter 2—Wiring and Protection

Chapter 2 covers Articles 200 through 285. These are general requirements that apply to all installations; from Article 200, the use and identification of the grounded or neutral conductor, through Article 285, which covers surge protective devices operating at 1 kV or less. Although the requirements in this chapter are applicable to all installations, the electrician should refer to this material because there are too many requirements to try to memorize.

200.4 has been added to the Code, and it prohibits a single neutral conductor from serving multiple circuits (not including multiwire branch circuits). Nothing in previous editions of the Code prohibited a properly sized neutral conductor from serving multiple circuits with ungrounded conductors of the same phase. This change eliminates this oddity from the Code.

As has been the case for the last several Code cycles, 210.8, covering GFCI protection, has been revised. The test and reset functions of the GFCI device must now be in a readily accessible location so that they can be tested monthly, as required by the product standard. A revision was made to the requirements of GFCI protected receptacles in patient care areas, and a new requirement for GFCI protection in nondwelling unit indoor wet locations has been added. Nondwelling locker rooms with associated showering facilities also require GFCI protection now, as do all 15A and 20A, 125V receptacles installed in service bays, garages and similar areas. The areas in which this applies are where electrical diagnostic equipment, electrical hand tools, or portable lighting equipment are to be used.

Clarifications to the AFCI requirements have been made, especially as it pertains to branch circuit extensions and modifications.

Changes to Part II of Article 225 will help Code users to understand the applicability of the requirements, and changes throughout Article 230 will help to delineate the difference between which conductors the Code covers and which conductors the Code doesn't cover.

A new term: "Bonding Jumper, Supply-Side" has been added to the definitions in 250.2. It is a new term, and can be found throughout Article 250.

Experienced Code users will recognize immediately that 250.30, covering separately derived systems, have been rewritten.

Experienced Code users may find themselves clicking their heels when they see that 250.56 is gone, but they may also see their smiles turn to frowns when they see that it was only relocated to 250.53(A)(2).

250.118 has been revised (again) to help clarify the types of equipment grounding conductors (EGCs) recognized by the Code.

Chapter 3—Wiring Methods

Chapter 3 covers Article 300 through 398.

Article 300 covers the general wiring methods that apply to all of the wiring methods discussed in Chapter 3. Every electrician needs to read this article very carefully and understand it thoroughly because these rules apply to every installation, regardless of size. The rest of this chapter is used where applicable. The electrician should use it as reference material because there are too many requirements for anyone to memorize.

Chapter 3 has many significant changes. 300.4(E) has been revised to clarify the requirements for wiring near roof decks which were added in the 2008 *NEC*®, and a new subsection, 300.4(H), was added to address structural expansion and deflection joints.

300.11(A)(2) has been revised to require that all suspended ceiling wires supporting electrical equipment be marked and distinguishable from the other ceiling wires. This provision previously only applied when the ceiling was fire-resistance rated.

300.22 has been extensively revised to provide consistency with the terms used in widely adopted mechanical codes, such as the International Mechanical Code.

Article 310 has been extensively revised. The experienced Code user will recognize a complete renumbering of not only the sections, but the tables as well. The ampacity adjustment provisions have been clarified, the correction factors for raceways on rooftops have been made more stringent, and the ampacities of some conductors have been reduced!

Changes to 312.8 now require a warning label on some cabinets, and the requirements for the weight ratings of boxes in 314.27 have been revised.

Changes to Article 334, Type NM Cable, clarify the permitted uses of Type NM Cable in dwelling unit accessory buildings, and the ampacity adjustment requirements of 334.80 have been clarified.

The controversial issue of unsupported raceways (added in the 2008 *NEC*) has been removed altogether, resulting in a different controversy now. Support requirements for Flexible Metal Conduit and Liquidtight Flexible Metal Conduit have been changed as well.

Article 392 experienced a much-needed face-lift this Code cycle. Changes to this article include a uses permitted and uses not permitted section, similar to the 3xx.10 and 3xx.12 sections found in other Articles of Chapter 3. The provisions for the grounding and bonding of cable tray systems have also been made clearer in 2011.

Chapter 4—Equipment for General Use

Chapter 4 covers Articles 400 through 490. This chapter covers the requirements for general equipment for all installations. Although Chapter 4 covers general equipment, such as appliances and motors that are used in nearly all installations, the electrician should use this chapter as reference material because, once again, there are far too many requirements for anyone to memorize.

Significant changes to Chapter 4 can be found throughout the chapter, beginning in Article 404, Switches. 404.2(C) contains a very substantial change—one that has many electricians grumbling. With this change, the days of 2 conductor switch loops and dead-end 3-way switches are in the past, except for raceway systems and some unfinished areas of buildings. This new requirement will mandate a grounded conductor at each switch location for line-to-neutral switch controlled loads.

Another change in this article, this time to 404.9, will have some installers and many inspectors smiling. An allowance for certain switches to be installed without the benefit of an equipment grounding conductor has been made. These switch assemblies contain no metal parts, and can only be connected to nonmetallic cover plates, so safety is not compromised. Previously, the AHJ was forced to use 90.4 and waive the requirement if this product was to be installed.

As wiring systems become older, the Code has added provisions for updating systems, and this edition of the Code is no different. Replacement of AFCI-protected circuits, tamper-resistant receptacles, and weather-resistant receptacles are all now addressed in the Code, and tamper-resistant receptacles are now required in guest rooms, guest suites, and child care facilities.

Many revisions have been made throughout Article 410 to address LED luminaires and their drivers.

Perhaps the most controversial change in this edition can be found in Article 445, which covers generators. 445.20 now requires that all 125V receptacles that are part of a generator, 15 kW or smaller, must have GFCI protection.

Last but certainly not least in Chapter 4, a new section 450.14 has been added, which requires a disconnecting means for transformers (other than Class 2 and Class 3 transformers). Although commonly thought to already be a requirement, it has never been found in the Code until now, and will require substantial consideration in the design of an electrical system.

Chapter 5—Special Occupancies

Chapter 5 covers Articles 500 through 590. This chapter, like Chapters 6 and 7 does not apply generally—it supplements or modifies Chapters 1 though 4 (90.3). It is only applicable in the types of occupancies or portions of buildings or structures that contain areas with special occupancies, such as an area within a building that contains hazardous materials.

Listing requirements for different types of equipment in hazardous (classified) locations have been added for Class I, Class II, and Class III locations.

Clarifications to the disconnect requirements for fuel dispensers can be found in 514.11 and 514.13, and the "redundant" equipment grounding provisions in 517.13 have also been clarified.

Installers, and particularly designers, might be shocked to find that 517.16 now prohibits isolated ground receptacles from being used in patient care areas.

The GFCI requirements for assembly occupancies (Article 518) and carnivals, circuses, fairs, and similar events have been clarified. And a new requirement for GFPE protection at marinas and boatyards should result in a much safer environment in these areas.

Chapter 6—Special Equipment

Chapter 6 covers Articles 600–695. Like Chapters 5 and 7, this chapter supplements or modifies the first four chapters of the Code. The provisions in this chapter apply only where special equipment is involved, such as a sign on a building, a swimming pool, or a fire pump.

Article 645, covering information technology equipment and rooms, has been revised with new definitions, revised requirements, and new provisions.

Article 680 has been changed to incorporate a new concept, that of the "low voltage contact limit." Perhaps most substantially in Article 680, the rules on equipotential bonding have been revised . . . again, and the bonding of hydromassage bathtubs has been expanded to provide for replacement motors.

Because solar photovoltaic systems are such an expanding technology, it should come as no surprise that the article experienced an incredible amount of changes. One need only glance at the Code book to recognize this fact. Of particular interest to those involved in these systems, 690.47(D) has been deleted, removing the requirement for an additional grounding electrode for a ground or pole mounted PV array.

As mentioned earlier, a new Article 694 covering small wind systems has been added, and there are some changes in Article 695 for fire pumps, mostly having to do with the routing of fire pump circuitry.

Chapter 7—Special Conditions

Chapter 7 covers Articles 700 through 770. These articles do not apply generally, like Chapters 1 through 4 do. They supplement or modify the general rules found in Chapters 1 through 4. The Chapter 7 rules apply only where the special conditions covered exist, such as where an emergency system is installed, or where limited energy circuits (such as signaling circuits) are encountered. The electrician should use Chapter 7 as reference material, because there are too many requirements for anyone to try to memorize.

Articles 701 and 702 have been reorganized in an effort to harmonize the numbering system for emergency systems, legally required standby systems, and optional standby systems. Article 725 and 760 saw little action in this Code revision cycle. 725.3 and 760.3 saw most of the changes, which inform the Code user about which Article 300 provisions apply to these installations.

As is often the case, Article 770, covering optical fiber cables, saw extensive revisions, but most of them were editorial in nature. New to the 2011 *NEC* is the concept of "cable routing assemblies," a wiring trough for limited energy circuits. The definition for this system can be found in 770.2

Chapter 8—Communications Systems

This chapter covers Articles 800 through 840. Chapter 8 stands alone in the Code. It is not subject to the requirements of Chapters 1 through 7 unless the requirements are specifically referenced in Chapter 8. These articles in Chapter 8 do contain many references, such as the requirements for grounding and bonding, and the provisions for raceway fill. The electrician should use Chapter 8 only as reference material when installing communications systems, as there are too many requirements for anyone to memorize.

Like the past few Code cycles, Chapter 8 saw a lot of revisions in 2011, most of which were editorial in nature. With the removal of the term "grounding conductor" from the Code, many changes to the text were made. Removal of the requirement of insulated grounding electrode conductors makes for consistency throughout the Chapter 8 articles, and new tables added to these articles should make for easier navigation through these articles.

As discussed earlier in this book, a new Article 840 has been added, which covers premises-powered broadband communications systems.

Chapter 9—Tables

These tables are applicable where referenced elsewhere in the Code.

A new Table 10 has been added, which addresses finely stranded conductor. Most of the information in this table is borrowed from UL 486A-B, Table 14.

Annexes A through I

Annexes, formerly referred to as "appendixes," contain informational material that is not mandatory and not enforceable.

A new Annex, Annex I, has been added to provide tightening torques for terminations. This new annex consists of 2 tables, each of which is borrowed from UL 486A-B.

2011 NEC®	Title
300	General Wiring Methods
310	Conductors for General Wiring
312	Cabinets, Cutout Boxes and Meter Socket Enclosures
314	Outlet, Device, Pull and Junction Boxes, Conduit Bodies and Fittings
320	Armored Cable: Type AC
322	Flat Cable Assemblies: Type FC
324	Flat Conductor Cable: Type FCC
326	Integrated Gas Spacer Cable: Type IGS
328	Medium Voltage Cable: Type MV
330	Metal-Clad Cable: Type MC
332	Mineral-Insulated Metal-Sheathed Cable: Type MI

(Continued)

2011 NEC®	Title
334	Non-metallic-Sheathed Cable: Types NM, NMC, and NMS
336	Power and Control Tray Cable: Type TC
338	Service-Entrance Cable: Types SE and USE
340	Underground Feeder and Branch-Circuit Cable: Type UF
342	Intermediate Metal Conduit: Type IMC
344	Rigid Metal Conduit: Type RMC
348	Flexible Metal Conduit: Type FMC
350	Liquidtight Flexible Metal Conduit: Type LFMC
352	Rigid Polyvinyl Chloride Conduit Type PVC
353	High Density Polyethylene Conduit: Type HDPE
354	Nonmetallic Underground Conduit with Conductors: Type NUCC
355	Reinforced thermosetting resin conduit Type RTRC
356	Liquidtight Flexible Nonmetallic Conduit: Type LFNC
358	Electrical Metallic Tubing: Type EMT
360	Flexible Metallic Tubing: Type FMT
362	Electrical Nonmetallic Tubing: Type ENT
366	Auxiliary Gutters
368	Busways
370	Cable bus
372	Cellular Concrete Floor Raceways
374	Cellular Metal Floor Raceways
376	Metal Wireways
378	Nonmetallic Wireways
380	Multioutlet Assembly
382	Nonmetallic Extensions
384	Strut-Type Channel Raceway
386	Surface Metal Raceways
388	Surface Nonmetallic Raceways
390	Underfloor Raceways
392	Cable Trays
394	Concealed Knob-and-Tube Wiring
396	Messenger Supported Wiring
398	Open Wiring on Insulators
399	Outdoor, Overhead Conductors, Over 600 volts
404	Switches
406	Receptacles, Cord Connectors, and Attachment Plugs (Caps)
408	Switchboards and Panelboards
409	Industrial Control Panels

Example

Chapter: **Chapter 2—Wiring and Protection**

Article: **Article 250—Grounding**

Part: **IV. Conductors**

Section: **250.119** Identification of Equipment Grounding Conductors. Unless required.

Level 1: **(A)** Conductors larger than 6 AWG.

List item: (1) Stripping the insulation or covering from the entire exposed length...

List item: (2) Coloring the exposed insulation or covering green...

List item: (3) Marking the exposed insulation or covering...

Level 2: **(B)** Multiconductor Cable.

List item: (1) Stripping the insulation...

List item: (2) Coloring the exposed insulation...

List item: (3) Marking the exposed insulation...

LOCAL CODES AND REQUIREMENTS

Although most municipalities, countries, and states adopt the *NEC®*, they may not have adopted the latest edition, and in some cases, the AHJ may be using an older edition. Most jurisdictions make amendments to the *NEC®* or add local requirements. These may be based on environmental conditions, fire safety concerns, or other local experience. An example of one common local amendment is that all commercial buildings be wired in metal raceways (Rigid, IMC, or EMT). The *NEC®* generally does not differentiate between wiring methods in residential, commercial, or industrial installations; however, many local jurisdictions do. Metal wiring methods, especially in fire zones, are another common amendment. Some major cities have developed their own electrical code—for example, Los Angeles, New York, Chicago, and several metropolitan areas. In addition to the *NEC®* and all local amendments, the designer and installer must also comply with the local electrical utility rules. Most utilities have specific requirements for installing the service to the structure. There have been many unhappy designers and installers who have learned about special jurisdictional requirements after making the installation, thus incurring costly corrections at their own expense.

If you are preparing for a state or city examination, you must check with the AHJ and see exactly what the examination content is based on. If it is a locally developed examination, the content may vary widely. If it is a nationally developed examination, it will generally be based on the latest edition of the *NEC®*. However, some jurisdictions have supplemental examinations that cover their unique amendments, and in some jurisdictions a practical hands-on examination is given in addition to the written portion. (See Chapter 1 of this book for information related to some of these unique requirements.)

TESTING LABORATORIES

Underwriters Laboratories (UL) has long been the major product testing laboratory in the electrical industry. In addition to the testing, UL is a developer of product standards. By producing these standards and contracting for follow-up service after undergoing a listing procedure, a manufacturer is authorized to apply the UL label or to mark their product. (See Figure 2–1, Figure 2–2, and Figure 2–3.) Underwriters Laboratories is not the only testing laboratory evaluating electrical products. Over the years, numerous other electrical testing laboratories have arrived and are being officially recognized; for example, Electrical Testing Laboratories (ETL), Applied Research Laboratories (APL), and Canadian Standards Association (CSA). Some jurisdictions evaluate and approve laboratories; others accept them based on reputation. OSHA is now evaluating and approving testing laboratories. It is the responsibility of the entity responsible for specifying the materials to verify that the product has been evaluated by a testing laboratory acceptable by the AHJ where the installation is being made. It is the installer's responsibility that the product is installed in accordance with the product's listing (*NEC®* 110.3(B)).

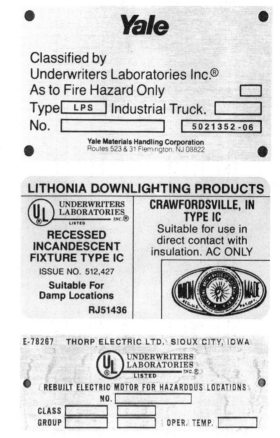

Figure 2–2(A) Examples of testing laboratories' labels as they appear on equipment. (*Reprinted with permission from Underwriters Laboratories, Inc.®*)

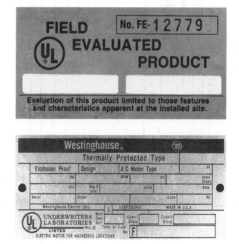

Figure 2–2(B) Examples of testing laboratories' labels as they appear on equipment. (*Reprinted with permission from Underwriters Laboratories, Inc.®*)

Figure 2–1 Common testing laboratory trademarks. (*Reprinted with permission from Underwriters Laboratories, Inc.®*)

IDENTIFICATION OF UL LISTED PRODUCTS THAT ARE ALSO UL CLASSIFIED IN ACCORDANCE WITH INTERNATIONAL PUBLICATIONS

Underwriters Laboratories Inc. (UL) provides a service for the Classification of products that not only meet the appropriate requirements of UL but also have been determined to meet appropriate requirements of the applicable international publication(s). For those products which comply with both the UL requirements and those of an international publication(s), the traditional UL Listing Mark and a UL Classification Marking, as described below, may appear on the product as a combination Listing and Classification Marking.

The combination of the UL Listing Mark and Classification Marking may appear as authorized by Underwriters Laboratories Inc.

LISTED
(Product Identity)
(Control Number)

ALSO CLASSIFIED BY UNDERWRITERS LABORATORIES INC.® IN ACCORDANCE WITH IEC PUBLICATION _____

LISTED
(Product Identity)
(Control Number)

ALSO CLASSIFIED BY UNDERWRITERS LABORATORIES INC.® IN ACCORDANCE WITH IEC PUBLICATION _____

UND. LAB. INC. ®   LISTED
(Product Identity)
(Control Number)

ALSO CLASSIFIED BY UNDERWRITERS LABORATORIES INC.® IN ACCORDANCE WITH IEC PUBLICATION _____

Underwriters   LISTED
Lab. Inc. ®   (Product Identity)
(Control Number)

ALSO CLASSIFIED BY UNDERWRITERS LABORATORIES INC.® IN ACCORDANCE WITH IEC PUBLICATION _____

(Product Identity)
CLASSIFIED BY
UNDERWRITERS LABORATORIES INC.®
IN ACCORDANCE WITH
IEC PUBLICATION _____
(Control Number)

OR

(Product Identity)
CLASSIFIED BY
UNDERWRITERS LABORATORIES INC.®
IN ACCORDANCE WITH
IEC PUBLICATION _____
(Control Number)

LOOK FOR THE CLASSIFICATION MARKING

(Product Identity)
(Control Number)

ALSO CLASSIFIED BY UNDERWRITERS LABORATORIES INC.® IN ACCORDANCE WITH IEC PUBLICATION _____

Underwriters
Laboratories Inc.®
LISTED
(Product Identity)
(Control Number)

ALSO CLASSIFIED BY UNDERWRITERS LABORATORIES INC.® IN ACCORDANCE WITH IEC PUBLICATION _____

UNDERWRITERS LABORATORIES INC. ®   (Product Identity)
LISTED   (Control Number)

ALSO CLASSIFIED BY UNDERWRITERS LABORATORIES INC.® IN ACCORDANCE WITH IEC PUBLICATION _____

LOOK FOR THE COMBINATION LISTING MARK AND CLASSIFICATION MARKING

IDENTIFICATION OF PRODUCTS CLASSIFIED TO INTERNATIONAL PUBLICATIONS ONLY

Underwriters Laboratories Inc. (UL) provides a service for the classification of products that have been determined to meet the appropriate requirements of the applicable international publication(s). For those products which comply with the requirements of an international publication(s), the Classification Marking may appear in various forms as authorized by Underwriters Laboratories Inc. Typical forms which may be authorized are shown below, and include one of the forms illustrated, the word "Classified", and a control number assigned by UL. The product name as indicated in this Directory under each of the product categories is generally included as part of the Classification Marking text, but may be omitted when in UL's opinion, the use of the name is superfluous and the Classification Marking is directly and permanently applied to the product by stamping, molding, ink-stamping, silk screening or similar processes.

Separable Classification Markings (not part of a name plate and in the form of decals, stickers or labels) will always include the four elements: UL's name and/or symbol, the word "Classified", the product category name, and a control number.

The complete Classification Marking will appear on the smallest unit container in which the product is packaged when the product is of such a size that the complete Classification Marking cannot be applied to the product or when the product size, shape, material or surface texture makes it impossible to apply any legible marking to the product. When the complete Classification Marking cannot be applied to the product, no reference to Underwriters Laboratories Inc. on the product is permitted.

Figure 2–3 Identification of marks listed and classified products in accordance with national and international publications. (*Reprinted with permission from Underwriters Laboratories, Inc.®*)

QUESTION REVIEW

> Each lesson is purposely designed to require the student to apply the entire *NEC®* text and not specific chapters or articles. It has been found that when studying for a timed, open-book examination, the student must gain proficiency in the Table of Contents, the Index, and the ability to move quickly from cover to cover to find the correct answer to each question in a timely fashion.

1. Is liquidtight flexible nonmetallic conduit permitted to be used in circuits in excess of 600 volts?

   Answer: _____

   _____

   _____

   _____

   Reference: _____

2. Who is the authority having jurisdiction (AHJ)?

   Answer: _____

   _____

   _____

   _____

   Reference: _____

3. Name four types of installations not covered by the *NEC®*.

   Answer: _____

   _____

   _____

   _____

   Reference: _____

4. How is explanatory information characterized in the *NEC®*?

   Answer: _____

   _____

   _____

   _____

   Reference: _____

5. Can conductors ever be spliced inside a raceway?

   Answer: _____

   _____

   _____

   _____

   Reference: _____

6. Are reduced neutrals allowed for manufactured home service conductors?

   Answer: _____

   _____

   _____

   _____

   Reference: _____

7. Are all installations by utilities exempt from the *NEC*®? Explain.

   Answer: _____

   _____

   _____

   _____

   Reference: _____

8. What does the gray highlighted text in the *NEC*® indicate?

   Answer: _____

   _____

   _____

   _____

   Reference: _____

9. IS Arc-fault circuit protection required for the lighting in the kitchen and the bathroom?

   Answer: _____

   _____

   _____

   _____

   Reference: _____

10. Who set up the first meeting that resulted in the *NEC*®? When and where was it held?

    Answer: _____

    _____

    _____

    _____

    Reference: _____

11. Can single insulated conductors be used for a 120/240-volt branch circuit as temporary wiring in a building under construction where the conductors are supported on insulators at intervals of not more than 10 feet (3 m)?

    Answer: _____

    _____

    _____

    _____

    Reference: _____

12. Which chapter of the *NEC*® is independent of all other chapters?

    Answer: _____

    _____

    _____

    _____

    Reference: _____

13. Is the *NEC*® considered a training manual?

    Answer: _____

    _____

    _____

    _____

    Reference: _____

14. Are all mining facilities exempt from the *NEC*®?

    Answer: _____

    _____

    _____

    _____

    Reference: _____

15. A comment often made by those involved in electrical design or installation states that the *NEC®* is a minimum standard. Where in the *NEC®* does it state that it is the true minimum permitted for electrical installations?

Answer: _____

_____

_____

_____

Reference: _____

16. As the inspector was making an inspection and looked at the size of overcurrent device and conductors supplying a motor, the inspector asked, "Is this motor circuit rated for continuous duty?" What is a continuous load?

Answer: _____

_____

_____

_____

Reference: _____

17. When you arrive in a nearby city to make an installation, the electrical inspector informs you that the city has not adopted the last 2 editions of the *NEC®*. Which edition would that city be enforcing?

Answer: _____

_____

_____

_____

Reference: _____

18. In wiring a small residence, the owners inform you that they wanted a doorbell mounted on the front and back of the house. What class wiring would this small 24-volt doorbell circuit be wired in and what article governs that wiring method?

Answer: _____

_____

_____

_____

Reference: _____

19. We have been asked to bid a nursing home where patients will have varied degrees of mobility. Some will come and go freely, cook their own meals, and do their own housekeeping; others may require meals to be prepared and minor medical attention, such as someone ensuring that they take their medicine regularly. Other patients may be bedridden and require oxygen or a doctor's care. Which article of the *NEC®* covers a nursing home facility of this type?

Answer: _____

_____

_____

_____

Reference: _____

20. In establishing a grounding electrode on a new installation, the owner says that although an underground metal water pipe exists, the owner prefers that it not be used and that a standard 8-feet (2.44 m) ground rod be used instead. Which section of the *NEC®* governs grounding electrodes?

Answer: _____

_____

_____

_____

Reference: _____

21. In wiring the kitchen of a residential home, the inspector informs you that the cord on the disposal is too long. Where are the requirements for the cord and attachment plug for a home disposal?

Answer: _____

_____

_____

_____

Reference: _____

22. In working in an industrial facility, you have been informed that some control circuitry will be installed to cut down on noise interference. An optical fiber cable will be used. Are optical fiber cables covered in the *NEC®*?

Answer: _____

_____

_____

_____

Reference: _____

23. You have recently been asked to wire an irrigation pump on a nearby farm. On arriving, you find that the farmer has a 3-phase irrigation pump; however, the utility can only provide single-phase power. The farmer tells you he has heard about a phase converter that can be purchased at a relatively low cost and can convert single-phase power to 3-phase power. Which section of the *NEC®* would cover an installation of this type?

Answer: _____

_____

_____

_____

Reference: _____

24. In making an installation, you are asked to connect de-icing equipment to the rain gutters on a house. Which article of the *NEC®* covers these requirements?

Answer: _____

_____

_____

_____

Reference: _____

25. In making an installation at a new golf course clubhouse, several decorative pools are to be installed in and around the golf course with electrical lighting in and around the pools. Which article of the *NEC®* covers this installation? What part of that article covers this area?

Answer: _____

_____

_____

_____

Reference: _____

CHAPTER 3

OBJECTIVES

After completing this chapter, you should be able to:

■ Understand the concepts of basic mathematics and mathematical terms

■ Apply basic mathematics to electricity

■ Explain the basic authority of the *NEC*®

■ Perform simple calculations relating to mathematical formulae

■ Perform simple calculations relating to Ohm's law

BASIC ELECTRICAL MATHEMATICS REVIEW

In this chapter, we refresh our knowledge of the basic mathematical calculations needed to perform the tasks required in our work as electrical contractors or electricians. Many tasks reviewed in this chapter may appear simple. However, if you have difficulty in solving any of these problems, then it may be necessary for you to obtain a more complete study guide on basic electrical mathematics. This chapter is not only a mathematical teaching aid but also a refresher to see that your mathematical skills are sufficient to solve the problems that you will encounter in most electrical examinations.

CALCULATOR MATH

Many of us were taught and still believe that mathematical study should be done manually. But in today's world, with the introduction of handheld calculators into the industrialized world, it would be somewhat silly to depend on all hand calculations. Most testing jurisdictions permit the use of handheld, silent, battery-powered, nonprogrammable calculators to use during the test. Therefore, it is advantageous to work the problems in this chapter using the same calculator that you will take to the examination. Many calculators are on the market, priced from free advertising giveaways to expensive complex scientific notation calculators. However, for your purposes, any quality handheld calculator will do. It should have the standard engineering functions, but it is not necessary to have one with scientific capabilities. The button arrangement should be good quality. It should be battery-powered, not solar-powered, because the room lighting might be inadequate for solar power. It should have a memory function and most modern mathematical functions, including the basic keys, input, error correction, combining operations, calculator hierarchy (calculations with a constant), roots, powers, reciprocal, factorial, percents, natural logarithm and natural antilogarithm, trigonometric functions, and error indication, accuracy and rounding, good memory usage for storing, recalling, and memory exchange, conversion factors from English to metric, and temperature conversions. A calculator of this quality is available in most discount stores for under $20.00. Get the calculator that you plan to use and familiarize yourself with it so that you are familiar with its operations before the examination.

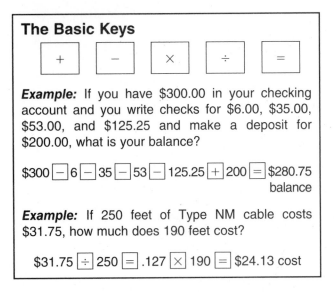

The Basic Keys

$$\boxed{+} \quad \boxed{-} \quad \boxed{\times} \quad \boxed{\div} \quad \boxed{=}$$

Example: If you have $300.00 in your checking account and you write checks for $6.00, $35.00, $53.00, and $125.25 and make a deposit for $200.00, what is your balance?

$300 $\boxed{-}$ 6 $\boxed{-}$ 35 $\boxed{-}$ 53 $\boxed{-}$ 125.25 $\boxed{+}$ 200 $\boxed{=}$ $280.75 balance

Example: If 250 feet of Type NM cable costs $31.75, how much does 190 feet cost?

$31.75 $\boxed{\div}$ 250 $\boxed{=}$ .127 $\boxed{\times}$ 190 $\boxed{=}$ $24.13 cost

Caution: Do not rely on programmable calculators to verify calculations or for preparation for the examination. They are unacceptable and are not permitted.

Although programmable calculators are not permitted in most testing organizations, there are good programmable handheld calculators on the market, as shown in Figure 3–1 and Figure 3–2. Although these are unacceptable in an examination, they can provide a convenient tool for the contractor or electrician or those studying to improve their skills in the electrical industry.

WORKING WITH FRACTIONS

Assuming that your basic addition, subtraction, multiplication, and division skills are still sharp, we begin with a brief review of fractions. A fraction is a part of a number. For instance, if we have a pie, we have one whole pie. If we cut that pie into four pieces, then each piece is one fourth of a pie. The four pieces together equal four fourths or one. If we cut that whole pie into five pieces, then each piece would be one fifth of the pie. The five pieces together would be one pie.

If we have two pieces of a pie that has been cut into four pieces, we would have two fourths. This fraction can be expressed in two different ways—by ²⁄₄ or ½. To reduce a fraction is to change it into another equal fraction. To do this, you divide the numerator (top number) and the denominator (bottom number) by the same number. For instance, to reduce ²⁄₆, divide the 2 and the 6 by 2. Then, the numerator would be 2 divided by 2, or 1. The denominator would be 6 divided

Figure 3–1 An example of a handheld calculator suitable for quick field calculation. (*Courtesy of Calculated Industries, Inc.*)

by 2, which equals 3. Therefore, the reduced fraction is $\frac{1}{3}$; $\frac{2}{6} = \frac{1}{3}$.

$$\textit{Example: } \frac{2}{6} = \frac{2 / 2}{6 / 2} = \frac{1}{3}$$

To reduce larger fractions, such as $\frac{24}{96}$, to the lowest form, we can divide both the 24 and the 96 by 24 (to get $\frac{1}{4}$) or we can divide the 24 and the 96 by 8 (which gives $\frac{3}{12}$) and then divide the 3 and the 12 by 3, which gives $\frac{1}{4}$, the answer.

$$\textit{Example: } \frac{24}{96} = \frac{24 / 24}{96 / 24} = \frac{1}{4}$$

Reducing Mixed Numbers to Improper Fractions

In a proper fraction, the numerator is always smaller than the denominator, such as $\frac{1}{3}$, $\frac{1}{6}$, $\frac{1}{12}$. Mixed numbers are made of two numbers: a whole number and a fraction; for example, $2\frac{1}{2}$, $5\frac{1}{2}$, $3\frac{9}{16}$—numbers that we use every day in our work. To change these to improper fractions, the numerator will always be larger than the denominator, such as $\frac{12}{3}$ or $\frac{9}{6}$ or $\frac{24}{10}$. For example, to change $2\frac{1}{2}$ to an improper fraction, the 2 will be a certain number of halves ($2 = \frac{4}{2}$), which we then add to the fraction we are given ($\frac{1}{2}$). Thus, $2\frac{1}{2} = \frac{4}{2} + \frac{1}{2} = \frac{5}{2}$.

$$\textit{Example: } 2\frac{1}{2} = \frac{2 \times 2 + 1}{2} = \frac{5}{2}$$

To change a mixed number to an improper fraction, multiply the denominator of the fraction by the whole number and add the numerator of the fraction. Place this answer over the denominator to make an improper fraction.

$$\textit{Example: } 5\frac{1}{2} = \frac{2 \times 5 + 1}{2} = \frac{11}{2}$$

Changing Improper Fractions to Mixed Numbers

To change an improper fraction, such as $\frac{13}{3}$, to a mixed number, divide the numerator by the denominator (13 divided by 3). Any remainder is placed over the denominator (13 divided by 3 = 4 with a remainder of 1). The resulting whole number and proper fraction form a mixed number. Thus, $\frac{13}{3} = 4 + \frac{1}{3} = 4\frac{1}{3}$.

$$\textit{Example: } \frac{13}{3} = 4 + \frac{1}{3} = 4\frac{1}{3}$$

Multiplication of Fractions

To multiply fractions, place the multiplication of the numerators over the multiplication of the denominators and reduce to lowest terms. For example: $\frac{1}{2} \times \frac{3}{4}$. Answer: $\frac{1}{2} \times \frac{3}{4}$ means that $1 \times 3 = 3$ and $2 \times 4 = 8$; thus, the answer is $\frac{3}{8}$. Example: $\frac{1}{2} \times 5$. Answer: $\frac{1}{2} \times 5$ means that $1 \times 5 = 5$ and $2 \times 1 = 2$, giving an improper fraction of $\frac{5}{2}$, which then has to be reduced to $2\frac{1}{2}$.

$$\textit{Example: } \frac{1}{2} \times \frac{3}{4} = \frac{1 \times 3}{2 \times 4} = \frac{3}{8}$$

$$\frac{1}{2} \times 5 = \frac{1 \times 5}{2 \times 1} = \frac{5}{2} = 2\frac{1}{2}$$

To cancel the numerator and denominator means to divide both numerator and denominator by the same

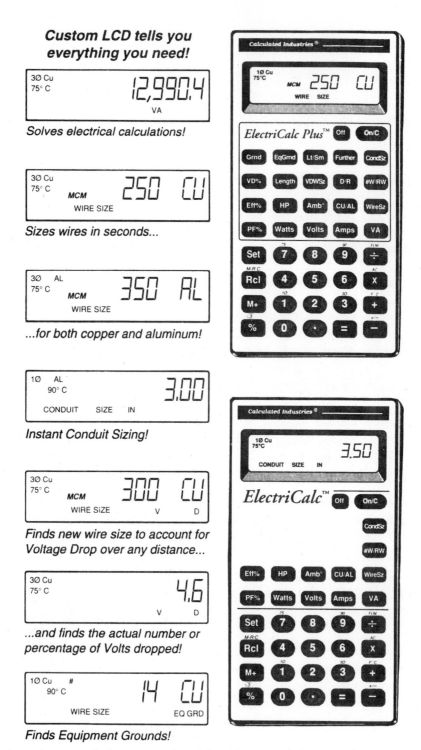

Custom LCD tells you everything you need!

Solves electrical calculations!

Sizes wires in seconds...

...for both copper and aluminum!

Instant Conduit Sizing!

Finds new wire size to account for Voltage Drop over any distance...

...and finds the actual number or percentage of Volts dropped!

Finds Equipment Grounds!

Figure 3–2 Examples of some of the types of multifunction tasks that can be quickly solved. (*Courtesy of Calculated Industries, Inc.*)

number. As an example, when multiplying ⅜ × ⁴⁄₉, notice that the 3 on top and the 9 on the bottom can both be divided by 3. Cross out the 3 and write a 1 over it. Cross out the 9 and write a 3 under it. Also, the 4 on the top and the 8 on the bottom can

both be divided by 4. Cross out the 4 and write a 1 over it. Cross out the 8 and write a 2 under it. The solution is 3 reduced to 1 × 4 reduced to 1 (1 × 1) over 8 reduced to 2 × 9 reduced to 3 (2 × 3), thus giving ⅙.

Example: $\dfrac{3}{8} \times \dfrac{4}{9} = \dfrac{\cancel{3}^{1} \times \cancel{4}^{1}}{\cancel{8}_{2} \times \cancel{9}_{3}} = \dfrac{1}{6}$

WORKING WITH DECIMALS

A decimal is a fraction in which the denominator is not written. The denominator is 1. Many answers to electrical problems will be fractions, such as ¼ of an ampere, ½ of an ampere, ½ of a volt, ⅓ of a volt, and the like. These will be correct mathematical answers, but they will be worthless to an electrician. The electrical measuring instruments give values expressed in decimals, not in fractions. In addition, the manufacturers of components give the values of parts in terms of decimals, such as 3.4 amperes, and the like on the nameplate of a motor. Suppose we worked out a problem and found that the current in the circuit should be ⅛ of an ampere, and using the ammeter, we test the circuit and find that .125 ampere flows. Is our circuit correct? How will we know? How can we compare ⅛ to .125? The easiest way is to change the fraction ⅛ to its equivalent decimal and then compare the decimals. To change a fraction into a decimal, divide the numerator by the denominator (1 divided by 8 = .125).

Example: $\dfrac{1}{8} = \dfrac{1}{8.000} = .125$

Converting Fractions to Decimals
Divide the numerator (top number) by the denominator (bottom number).

$$\tfrac{1}{2} = 1 \div 2 = 0.5$$
$$\tfrac{3}{4} = 3 \div 4 = 0.75$$
$$\tfrac{7}{8} = 7 \div 8 = 0.875$$

$$9\tfrac{3}{8} = \dfrac{9 \times 8 + 3}{8} = \dfrac{75}{8} = 75 \div 8 = 9.375$$

Converting Percentages to Decimals
Move decimal two places to the left.

$$100\% = 1.00$$
$$50\% = 0.50$$
$$37\tfrac{1}{2}\% = 0.375$$

WORKING WITH PERCENTAGES

To apply the rules of the *NEC®*, you will need to know how to work with percentages—for example, **430.110 (A) General.** *The disconnecting means for motor circuits rated 600 volts, nominal, or less shall have an ampere rating of at least 115% of the full-load current rating of the motor.* [Reprinted with permission from NFPA 70-2011.]

Question: A 480-volt motor has a full-load current of 28 amperes. What size disconnecting means is required?

A. 20 amperes
B. 25 amperes
C. 30 amperes
D. 40 amperes

Because the motor FLC is 28 amperes, you can quickly eliminate answers A and B.

To solve: 28 × 115% = 28 × 1.15 = 32.2 = 40 amperes disconnecting means required.

220.42 General Lighting. *The demand factors specified in Table 220.42 shall apply to that portion of the total branch-circuit load calculated for general illumination. They shall not be applied in determining the number of branch circuits for general illumination.* [Reprinted with permission from NFPA 70-2011.]

Question: A custom home (single-family dwelling) has a branch-circuit general illumination load of 200,000 VA (volt-amperes). What demand factor must be applied to this general illumination load?

A. 200,000
B. 43,950
C. 105,000
D. 63,950

Answer: D

To solve:

First 3000 at 100% = 3000 × 1	= 3000 VA
3001 to 120,000 = 117,000 at 35%	
= 117,000 × .35	= 40,950 VA
Over 120,000 = 80,000 at 25%	
= 80,000 × .25	= 20,000 VA
Total 200,000 VA lighting load demand	63,950 VA

POWERS

A power is a product of a number multiplied by itself one or more times. For an example, what is the value of 3^2? The exponent is 2; write 3 two times, or $3^2 = 3 \times 3$; $3^2 = 9$. What is the value of 5^3? The exponent is 3; write 5 three times; in other words, $5^3 = 5 \times 5 \times 5 = 125$. These powers are terms often used in the electrical field and should be known.

WORKING WITH SQUARE ROOTS

Roots are the opposite of powers. A square root is the opposite of the number to the second power. The sign of the square root is $\sqrt{\ }$. To find a square root, find a number that when multiplied by itself is the number inside the square root sign. For example, to find the value of the square root of 36, ask yourself what number times itself is 36. The answer is 6, so 6 is the square root of 36.

If you can find the square root of a number with the method that uses averages, suppose that you did not know that the square root of 144 is 12. When you divide the number by its square root, the answer is the square root. If you cannot find this answer, then

guess as close as possible. A good guess for 144 would be 10, because $10 \times 10 = 100$, which is close to 144. Divide 144 by 10 and the answer is 14 + a remainder. Average the guess, 10, and the answer to the division problem, 14. $10 + 14 = 24$, $^{24}/_2 = 12$, which is the correct answer.

Follow these steps to find the square root of the larger number: Guess the answer; divide the guess into the large number; average the guess and the answer to the division problem; and check. For example, find the value of the square root of 1024.

Step 1. Guess: In the list of square roots, $30 \times 30 = 900$. This is too small, but it is easy to divide by.

Step 2. Divide 1024 by 30. The answer is 34, with a remainder.

Step 3. Find the average of 30 and 34. $30 + 34 = 64$; $^{64}/_2 = 32$.

Step 4. Check: Multiply 32×32. The answer is 1024. Thus, 32 is the square root of 1024. When you use this method to find square roots, always guess a number that ends in 0. It is easier and faster to divide by these numbers. If the average is not the square root of the number, use the average as a new guess and try again.

Finding the Reciprocal of a Number

There are instances in the NEC® where it is easier to find the reciprocal of a number and then make the calculation rather than use the number in the calculation. For example, the reciprocal of 125% is 80%, and the reciprocal of 1.73 is .578, which may be rounded off to .58 for quick calculations.

To find a reciprocal of a number, divide the number into 1.

$$1 \div 1.73 = .5780 \text{ or } .58$$

If you are finding the reciprocal of a percentage, you must first convert the percent to a decimal.

$$125\% = 1 \div 1.25 = 0.8 \text{ or } 80\%$$

422.11 (E) Single Non–motor-Operated Appliance.
If the branch-circuit supplies a single non–motor-operated appliance, the rating of overcurrent protection shall
(1) Not exceed that marked on the appliance;
(2) Not exceed 20 amperes if the overcurrent protection rating is not marked and the appliance is rated 13.3 amperes or less; or

Table 220.42 Lighting Load Demand Factors		
Type of Occupancy	Portion of Lighting Load to Which Demand Factor Applies (Volt-Amperes)	Demand Factor (Percent)
Dwelling units	First 3000 or less at	100
	From 3001 to 120,000 at	35
	Remainder over 120,000 at	25
Hospitals*	First 50,000 or less at	40
	Remainder over 50,000 at	20
Hotels and motels, including apartment houses without provision for cooking by tenants*	First 20,000 or less at	50
	From 20,001 to 100,000 at	40
	Remainder over 100,000 at	30
Warehouses (storage)	First 12,500 or less at	100
	Remainder over 12,500 at	50
All others	Total volt-amperes	100

*The demand factors of this table shall not apply to the calculated load of feeders or services supplying areas in hospitals, hotels, and motels where the entire lighting is likely to be used at one time, as in operating rooms, ballrooms, or dining rooms.

Reprinted with permission from NFPA 70-2011.

(3) Not exceed 150% of the appliance rated current if the overcurrent protection rating is not marked and the appliance is rated over 13.3 amperes. Where 150% of the appliance rating does not correspond to a standard overcurrent device ampere rating, the next higher standard rating shall be permitted. [Reprinted with permission from NFPA 70-2011.]

Question: A quick recovery electrical water heater is rated at 4500 watts 240 volts. Is a 30-ampere circuit breaker permitted to protect this appliance?

Answer: Yes. $4500 \div 240 = 18.75 \times 150\% = 28.1 =$ the next higher standard circuit breaker rating is 30 amperes (240.6(A)).

You may first think a 20-ampere circuit breaker would be acceptable for an 18.75-ampere load. However, 422.10(A) requires the branch-circuit rating to be at least 125% of the marked rating. A 20-ampere circuit breaker is permitted to protect the reciprocal of 125% or 80% of the circuit breaker rating. Thus, $80\% \times 20 = 16$ amperes.

A 25-ampere circuit breaker would be acceptable. However, a 25-ampere circuit breaker is not a common size and may not be readily available.

OHM'S LAW

E = voltage; I = current (amperes); R = resistance

$$E = IR; \quad I = \frac{E}{R}; \quad R = \frac{E}{I}$$

All numbers are relative. For example, if $E = 12$, $I = 3$, and $R = 4$, then $3 \times 4 = 12$, $12/3 = 4$, and $12/4 = 3$. Remember, in any electrical circuit, the voltage force is the current through the conductor against its resistance. The resistance tries to stop the current from flowing. The current that flows in the circuit depends on the voltage and the resistance. The relationship among these three quantities is described by Ohm's law. Ohm's law applies to an entire circuit or any component part of a circuit.

Voltage Drop Calculations

We have discussed the methods for finding the resistance by using Ohm's law. Now, we can find the voltage drop for circuit loads using this same formula.

$$E = I \times R$$

Voltage Drop Formula

Single-phase voltage drop = amperes × length × resistance × (2)

Voltage drop % = voltage drop × 100/volts

Three-phase voltage drop—calculate as a single-phase circuit and multiply resultant by .866.

$$V_{(D)} = I \times 2L \times R$$

I = Amperes
L = Length of circuit
R = Resistance values from *NEC*® Chapter 9, Table 8
Resistance = R × [1 + a (temperature − 75)]
a = 0.00323 for copper, 0.000330 for aluminum
Temperature = ambient temperature

Voltage drop can be calculated with the known elements of the power circuit. The resistance of conductors commonly used for electrical power conductors can be found in *NEC*® Chapter 9, Table 8.

Example: What is the voltage drop for a 120-volt, 20-ampere, single-phase circuit supplying a 12-ampere load located 125 feet (37.5 m) away from the source of supply? The circuit conductors have a combined resistance of 0.35 ohm.

A. 21 volts
B. 42 volts
C. 2.1 volts
D. 4.2 volts

Answer: D

What is the question? What is the voltage drop? **Ohm's law works!**

$E = IR$ $\quad I = 12$ amperes $\quad R = 0.35$ ohm
$E = 12 \times 0.35$
$E = 4.2$ volts

Note that this is a 3½% voltage drop. 210.19 Informational note No. 4 recommends that branch circuits do not exceed a 3% drop. For maximum efficiency, the circuit conductors should be increased in size.

NEC® 2011 210.19(A)(1) Informational note No. 4: Conductors for branch circuits as defined in Article 100, sized to prevent a voltage drop exceeding 3% at the farthest outlet of power, heating, and lighting loads, or combinations of such loads, and where the maximum

(continued)

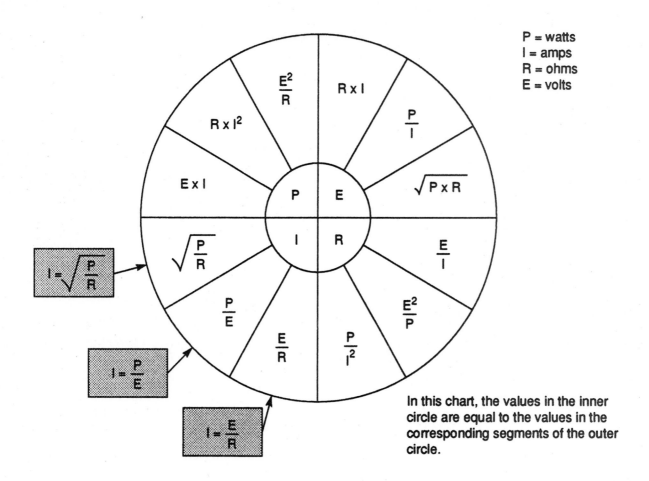

P = watts
I = amps
R = ohms
E = volts

In this chart, the values in the inner circle are equal to the values in the corresponding segments of the outer circle.

Total voltage drop on both feeders and branch circuits to the farthest outlet does not exceed 5%, provides reasonable efficiency of operation. See Informational note No. 2 of 215.2(A)(3) for voltage drop on feeder conductors.*

The *NEC®* references prescribe voltage drop values or percentages for feeder conductors and branch-circuit conductors; however, it should be noted these appear in an Informational Note and are only suggested and not mandatory requirements. *NEC®* 215.2(A)(3) Informational Note 2 recommends that the voltage drop should not exceed 3% of the farthest outlet of power, heating, or lighting loads or combinations of such loads, and the maximum voltage drop for both feeders and branch circuits to the farthest outlet is not to exceed 5% and will provide reasonable efficiency of operation.

Example: A dc circuit has 2 amperes of current flowing in it. A voltmeter reads 10 volts line to line. How much resistance is in the circuit? Answer: 5 ohms.

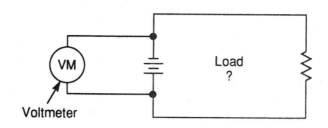

Voltmeter

$$R = \frac{E}{I} \quad R = \frac{10}{2} \quad R = 5 \text{ ohms}$$

Example: If the voltage is 100 volts and the resistance is 25 ohms, what amount of current will flow in the circuit? Answer: 4 amperes.

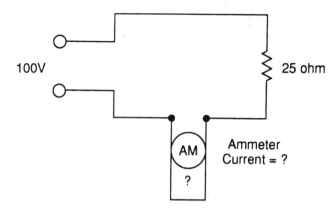

$$I = \frac{E}{R} \quad I = \frac{100}{25} \quad I = 4 \, \text{amperes}$$

Example: If the potential across a circuit is 120 volts and the current is 6 amperes, what is the resistance? Answer: 20 ohms.

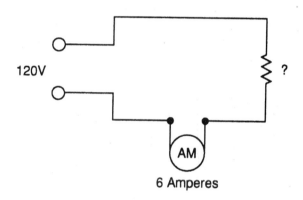

$$R = \frac{E}{I} \quad R = \frac{120}{6} \quad R = 20 \, \text{ohms}$$

A series circuit may be defined as a circuit in which the resistive elements are connected in a continuous run (end to end). It is evident that because the circuit has no branches, the same current flows in each resistance. The total potential across the circuit equals the sum of potential drops across each resistance.

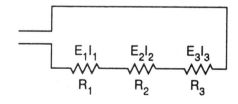

Example: $E_1 = IR_1$
$E_2 = IR_2$
$E_3 = IR_3$
$E = E_1 + E_2 + E_3$
$R = R_1 + R_2 + R_3$

The total potential of the circuit is:

$$E = IR_1 + IR_2 + IR_3 = I(R_1 + R_2 + R_3)$$

$$I = \frac{E}{R_1 + R_2 + R_3}$$

$R_1 = 10$ ohms, $R_2 = 5$ ohms, and $R_3 = 15$ ohms.

What amount of voltage must flow to force 0.5 amperes through the circuit?

$$R_T = 5 + 10 + 15 = 30 \, \text{ohms}$$

Hence, $E_T = 0.5 \times 30 = 15$ volts

What is the voltage drop across each resistance?

$E_1 = 0.5 \times 10 = 5.0$ volts
$E_2 = 0.5 \times 5 = 2.5$ volts
$E_3 = 0.5 \times 15 = \underline{7.5 \text{ volts}}$
$E_T = 15$ volts

When the resistance and current are known, what is the formula to determine the voltage?

The current is 2 amperes and the resistance is 15 ohms, 10 ohms, and 30 ohms.

$E = IR \qquad E = 2(15 + 10 + 30)$
$E = 2 \times 55 \quad E = 110$ volts

In the industry today, most circuits are connected in parallel. In a parallel circuit, the voltage is the same across each resistance (load).

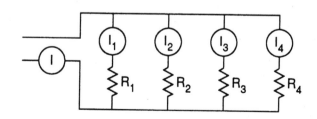

The total current is equal to the sum of all the currents.

$$E = I_1R_1 = I_2R_2 = I_3R_3 = I_4R_4$$

and

$$I = I_1 + I_2 + I_3$$

Individual Resistances =

$$I_1 = \frac{E}{R_1} \quad I_2 = \frac{E}{R_2} \quad I_3 = \frac{E}{R_3} \quad I_4 = \frac{E}{R_4}$$

Hence: $I_1 = \dfrac{E}{R_1} + I_2 = \dfrac{E}{R_2} + I_3 = \dfrac{E}{R_3} + I_4 = \dfrac{E}{R_4}$

or $I = E\left(\dfrac{1}{R_1} + \dfrac{1}{R_2} + \dfrac{1}{R_3} + \dfrac{1}{R_4}\right)$

Several resistances in parallel is

$$\frac{1}{R} = \frac{1}{R_1} + \frac{1}{R_2} + \frac{1}{R_3} + \frac{1}{R_4}$$

The sum of the resistances in parallel will always be smaller than the smallest resistance in the circuit.

$$R = \frac{R_1 \times R_2}{R_1 + R_2} \text{ ohms}$$

Power Factor

Power factor is a phase displacement of current and voltage in an ac circuit. The cosine of the phase angle of displacement is the power factor. The cosine is multiplied by 100 and is expressed as a percentage. A cosine of 90° is 0; therefore, the power factor is 0%. If the angle of displacement were 60°, the cosine of which is .5, the power factor would be 50%. This is true whether the current leads or lags the voltage. Power is expressed in dc circuits and ac circuits that are purely resistive in nature. Where these circuits contain only resistance, P (watts) = E × I. In ac circuits that contain inductive or capacitive reactances,

VA (volt-ampere) = E × I. In a 60-cycle ac circuit, if the voltage is 120 volts, the current is 12 amperes and the current lags the voltage by 60°. Find (a) the power factor, (b) the power and voltage-amperes (VA), and (c) the power in watts. The cosine of 60 is .5; therefore, the power factor is 50%. 120 × 12 = 1440 volt-amperes, which is called the apparent power. 120 × 12 × .5 = 720 watts, which is called the true power. Power factor is important. There is an apparent power of 1440 VA and a true power of 720 watts. There are also 12 amperes of line current and 6 amperes of in-phase or effective current. This means that all equipment from the source of supply to the power consumption device must be capable of handling a current of 12 amperes, when actually the device is only using the current of 6 amperes. A 50% power factor was used intentionally to make the results more pronounced. The I2R loss is based on the 12-ampere current, while only 6 amperes are effective. Power factor can be measured by a combined use of a voltmeter, ammeter, and wattmeter or by the use of a power factor meter. When using the three meters all connected properly in the circuit, the readings of the three meters are taken simultaneously under the same load conditions and calculated as follows: power factor = true power (watts)/apparent power (which is volt-amperes). Power factor = W/EI.

1000 W = 1 kW
1000 W operating for 1 hour = 1 KWH used
3413 BTU = 1 kW
3.413 BTU = 1 W
Fahrenheit = (9/5 × C) + 32
Celsius = 5/9 × (F − 32)
P. F. = kW / KVA
KVA = kW / P. F.

PRACTICE EXAM

1. 0.875 is the equivalent of which of the following?
    A. 9/32
    B. 15/16
    C. 7/8
    D. 5/8

2. How much current will a 240-volt circuit with 20-ohm load draw?
    A. 24 amperes
    B. 2.4 amperes
    C. 1.2 amperes
    D. 12 amperes

3. 4.75 is the equivalent of which of the following improper fractions?
    A. 23/4
    B. 17/5
    C. 15/3
    D. 38/8

4. A 10 kW load is increased by 12% every summer. What is the VA in the summer?
    A. 11.2
    B. 1120
    C. 11,200
    D. 112,000

5. What is the circumference of a circle with a 1-inch diameter?
    A. 1 inch
    B. 1.2 inches
    C. 3.14 inches
    D. 2½ inches

6. What is the radius of a circle with a 1-inch diameter?
    A. ½ inch
    B. 2½ inches
    C. 1 inch
    D. 1½ inches

7. What is the resistance of a 20-ampere, 120-volt circuit?
    A. 5 ohms
    B. .5 ohm
    C. .6 ohm
    D. 6 ohms

8. One meter equals how many feet?
    A. 32 inches
    B. 3.28 feet
    C. 2.88 feet
    D. 1½ feet

9. One meter equals how many millimeters?
   A. 100
   B. 10
   C. 1000
   D. 10,000

10. According to the *NEC*®, 10 feet converted to metric (soft conversion) equals
    A. 3 m.
    B. 3.05 m.
    C. 30 m.
    D. 300 m.

11. According to the *NEC*®, 3 feet converted to metric (hard conversion) equals
    A. .3 m.
    B. 300 mm.
    C. 900 mm.
    D. 9 m.

12. How many amperes does a 3-phase, 480-volt, 24 kW load draw? Round off to a whole number.
    A. 29
    B. 50
    C. 290
    D. 5

13. ¾ can be also shown as what percent?
    A. .75
    B. 75
    C. ¾
    D. none of the above

14. Unless specifically permitted in the *NEC*®, 12 AWG conductors with Type THWN overcurrent protection shall not exceed
    A. 16 amperes.
    B. 25 amperes.
    C. 20 amperes.
    D. 30 amperes.

15. The reciprocal of 125% is _____
    A. 0.8.
    B. .58.
    C. 1.73.
    D. 1.25.

QUESTION REVIEW

Directions: Reduce the following fractions to their lowest terms.

1. $3/6$ = _____

2. $4/16$ = _____

3. $4/40$ = _____

4. $40/100$ = _____

5. $6/8$ = _____

6. $15/20$ = _____

7. $18/36$ = _____

8. $9/15$ = _____

9. $30/50$ = _____

10. $4/20$ = _____

11. $5/20$ = _____

12. $6/24$ = _____

Directions: Change the following mixed numbers to improper fractions.

1. $1\frac{1}{2}$ = _____

2. $3\frac{1}{4}$ = _____

3. $5\frac{3}{8}$ = _____

4. $4\frac{1}{4}$ = _____

5. $7\frac{1}{8}$ = _____

6. $2\frac{1}{6}$ = _____

7. $5\frac{3}{4}$ = _____

8. $8\frac{1}{3}$ = _____

9. $6\frac{4}{7}$ = _____

10. $3\frac{5}{8}$ = _____

Directions: Change the following improper fractions to mixed numbers.

1. $9/2$ = _____

2. $12/5$ = _____

3. $64/5$ = _____

4. $26/8$ = _____

5. $29/3$ = _____

6. $31/2$ = _____

7. $5/3$ = _____

8. $13/3$ = _____

Directions: Multiply the following whole numbers and fractions. Give the answer in proper reduced form.

1. $8 \times 1/2$ = _____

2. $1/2 \times 3/4$ = _____

3. $1/5 \times 1/8$ = _____

4. $9 \times 1/2$ = _____

5. $3/4 \times 7$ = _____

6. $1/2 \times 5$ = _____

7. $3/4 \times 12$ = _____

8. $1/7 \times 15$ = _____

Directions: Convert the following fractions to decimal equivalents.

1. $1/4$ = _____

2. $5/8$ = _____

3. $3/4$ = _____

4. $13/15$ = _____

5. $3/8$ = _____

Directions: Convert to whole numbers.

1. 2^4 = _____

2. 9^2 = _____

3. 50^2 = _____

4. 1^5 = _____

5. 10^3 = _____

Directions: Find the square root of the following numbers.

1. Square root of 25 = _____

2. Square root of 81 = _____

3. Square root of 49 = _____

4. Square root of 3136 = _____

5. Square root of 10,201 = _____

6. Square root of 841 = _____

7. Square root of 6064 = _____

Directions: Solve the following problems.

1. A 20-ampere load is fed with two conductors that have a *combined* resistance of 0.3 ohm. If the source voltage is 120 volts dc, the calculated voltage drop on this circuit is _____%.

   Answer: _____

   _____

   _____

   _____

   Reference: _____

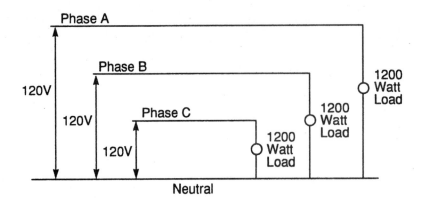

Phase A

Phase B

120V

120V

Phase C

120V

1200 Watt Load

1200 Watt Load

1200 Watt Load

Neutral

2. Refer to the above drawing.

I(LINE A) = I(LINE B) = I(LINE C) = 10 amperes.

The power factor in the circuit in the figure above is _____

Answer: _____

_____

_____

_____

Reference: _____

3. A 120-volt branch circuit has only six 100-watt, 120-volt incandescent luminaires (fixtures) connected to it. The current in the home run of this circuit is _____ amperes.

Answer: _____

_____

_____

_____

Reference: _____

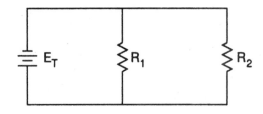

E_T R_1 R_2

4. Refer to the above drawing.

With a resistance at R_1 of 4 ohms and a resistance at R_2 of 2 ohms, the total resistance for the above circuit is _____ ohms.

Answer: _____

_____

_____

_____

Reference: _____

5. A 1000-watt, 120-volt lamp uses electrical energy at the same rate as a/an _____ ohm resistor.

Answer: _____

_____

_____

_____

Reference: _____

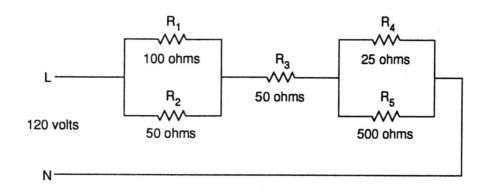

6. In the above diagram, eight equal resistors are connected in series to a 48-volt source. The voltage drop across each resistor is _____ volts.

Answer: _____

_____

_____

_____

Reference: _____

The above diagram applies to Questions 7 through 14.

7. What is the resistance of R_1 and R_2?

Answer: _____

_____

_____

_____

Reference: _____

8. What is the resistance of $R_3 + R_1$ and R_2?

   Answer: _____

   _____

   _____

   _____

   Reference: _____

9. What is the resistance of R_4 and R_5?

   Answer: _____

   _____

   _____

   _____

   Reference: _____

10. What is the total resistance in the circuit?

    Answer: _____

    _____

    _____

    _____

    Reference: _____

11. What is the voltage drop at R_1 and R_2?

    Answer: _____

    _____

    _____

    _____

    Reference: _____

12. What is the voltage drop at R_3?

    Answer: _____

    _____

    _____

    _____

    Reference: _____

13. What is the voltage drop at R_4 and R_5?

    Answer: _____

    _____

    _____

    _____

    Reference: _____

14. What is the total current in this circuit?

    Answer: _____

    _____

    _____

    _____

    Reference: _____

CHAPTER 4

OBJECTIVES

Studying this chapter along with the 2011 *NEC®* will give the reader a basic introduction to the *Code* and an understanding of how to start using the 2011 *NEC®* efficiently.

After completing this chapter, you should know:

■ The unique style the *NEC®* is written in

■ The types of installations covered and exempted

■ The application of Chapters 1–4 generally

■ The application of Chapters 5–7 to supplement and modify the general requirements

■ The application of Chapter 8 independently

■ The application of the tables in Chapter 9

■ The Annexes for information and examples

■ The importance of thoroughly understanding the contents of

   1. Article 90—Introduction

   2. Article 100, definitions of words and terms unique to the *NEC®*

   3. Article 110, general requirements that apply throughout

   4. Article 300, general wiring applications that apply to all wiring methods

INTRODUCTION TO THE *NEC*®

The *NEC*® is a reasonable document developed and written in a reasonable fashion. It is a design manual, but it is not a design specification. It is not the minimum, but there are many conveniences that we take for granted that are not required by this document. There are also many safety considerations that are not required by this text, such as smoke alarms and other safety and detection devices that are being developed every day. See NFPA 72 the National Fire Alarm Code for requirements related to smoke alarms and fire alarm equipment.

The first page in Article 90, 90.1(C), says it is not an instruction manual for untrained persons. You are going to find that that statement explains much of the confusion about this document. Most engineering schools, vocational schools, and apprentice programs do not have training programs on the proper use of this document as an installation and design standard. Open your 2011 *NEC*® and look up *NEC*® 90.1(C). You will need the 2011 *NEC*® throughout your study of this manual. There are several *NEC*® handbooks available. The NFPA *National Electrical Code*® *Handbook* contains the exact *NEC*® text and commentary. If you are using that text, remember that the commentary is only the author's opinion and may not be totally correct. Therefore, when studying in a classroom or for an examination, the *NEC*® text will be that from which questions are developed.

KNOW YOUR *CODE* BOOK

To use the *NEC*®, we must first have a thorough understanding of

- the Table of Contents
- the Index
- Article 90—Introduction
- Article 100—Definitions
- Article 110—Requirements for Electrical Installations
- Article 300—Wiring Methods

We will refer to the rest of the *Code* on an as-needed basis; but these are general articles, and we use them continually as applicable.

You must study these portions of the *Code* thoroughly several times until you have a complete understanding of the content in these articles if you expect to successfully pass your examination.

Introduction

Let us turn to Article 90. We often pass over the introduction of many books that we read or study. The introduction is an important part of any book. It is the application guidelines. Did you read the Introduction to this manual? I repeat! Read Article 90 several times. It is short, only about four pages, but it is a very important part of the document, especially for someone about to embark on the study of the *NEC*®.

Article 90 covers the foundation requirements of the *NEC*®, the purpose of which is the safeguarding of people and property from the hazards that can occur while using electricity. The *Code* contains rules that when complied with, along with proper maintenance, will result in an installation that is essentially free from hazards. But it may not be efficient, convenient, or adequate for good service. The requirements are minimal and generally do not provide capacity for future expansion of the electrical system.

> **90.1 Informational Note:** Hazards often occur because of overloading of wiring systems by methods or usage not in conformity with this *Code*. This occurs because initial wiring did not provide for increases in the use of electricity. An initial adequate installation and reasonable provisions for system changes provide for future increases in the use of electricity.
>
> Reprinted with permission from NFPA 70-2011.

The *NEC*® is not a design specification, but it is a design manual. For example, specific manufacturers' products are not specified; however, no responsible designer would begin to design an electrical system without considering minimal *Code* requirements such as load calculations, service size and location, or grounding.

> The *NEC*® is not an instruction manual for untrained persons.

Contained in the Introduction is the scope, and it is imperative that the user know the bounds of the *NEC*® (90.2(A)).

> **90.2 Scope. (A) Covered.** This *Code* covers the installation of electrical conductors, equipment, and raceways; signaling and communications conductors, cables, equipment, and raceways; and optical fiber cables and raceways for the following:
>
> *(continued)*

1. Public and private premises, including buildings, structures, mobile homes, recreational vehicles, and floating buildings
2. Yards, lots, parking lots, carnivals, and industrial substations
3. Installations of conductors and equipment that connect to the supply of electricity
4. Installations used by the electric utility, such as office buildings, warehouses, garages, machine shops, and recreational buildings, that are not an integral part of a generating plant, substation, or control center.

Reprinted with permission from NFPA 70-2011.

The *Code* does not cover (90.2(B)):

(B) Not Covered. This *Code* does not cover the following:

1. Installations in ships, watercraft other than floating buildings, railway rolling stock, aircraft, or automotive vehicles other than mobile homes and recreational vehicles

    FPN: Although the scope of this *Code* indicates that the *Code* does not cover installations in ships, portions of this *Code* are incorporated by reference into Title 46, *Code of Federal Regulations,* Parts 110–113.

2. Installations underground in mines and self-propelled mobile surface mining machinery and its attendant electrical trailing cable
3. Installations of railways for generation, transformation, transmission, or distribution of power used exclusively for operation of rolling stock or installations used exclusively for signaling and communications purposes
4. Installations of communications equipment under the exclusive control of communications utilities located outdoors or in building spaces used exclusively for such installations
5. Installations under the exclusive control of an electric utility where such installations
    (a) Consist of service drops or service laterals, and associated metering, or
    (b) Are on property owned or leased by the electric utility for the purpose of communications, metering, generation, control, transformation, transmission, or distribution of electric energy, or
    (c) Are located in legally established easements or rights-of-way, or

(d) Are located by other written agreements either designated by or recognized by public service commissions, utility commissions, or other regulatory agencies having jurisdiction for such installations. These written agreements shall be limited to installations for the purpose of communications, metering, generation, control, transformation, transmission, or distribution of electric energy where legally established easements or rights-of-way cannot be obtained. These installations shall be limited to federal lands, native American reservations through the U.S. Department of the Interior Bureau of Indian Affairs, military bases, lands controlled by port authorities and state agencies and departments, and lands owned by railroads.

Informational Note to (4) and (5): Examples of utilities may include those entities that are typically designated or recognized by governmental law or regulation by public service/utility commissions and that install, operate, and maintain electric supply (such as generation, transmission, or distribution systems) or communications systems (such as telephone, CATV, Internet, satellite, or data services). Utilities may be subject to compliance with codes and standards covering their regulated activities as adopted under governmental law or regulation. Additional information can be found through consultation with the appropriate governmental bodies, such as state regulatory commissions, the Federal Energy Regulatory Commission, and the Federal Communications Commission.

(C) Special Permission. The authority having jurisdiction for enforcing this *Code* may grant exception for the installation of conductors and equipment that are not under the exclusive control of the electric utilities and are used to connect the electric utility supply system to the service conductors of the premises served, provided such installations are outside a building or structure, or terminate inside nearest the point of entrance of the service conductors.

Reprinted with permission from NFPA 70-2011.

Included in 90.3 is the arrangement of the *Code* and hierarchy. The *Code* is divided into an Introduction, nine chapters, and annexes A–I.

• Chapters 1–4 apply generally and are applicable to all installations—branch circuitry, motors, appliances, lighting, etc.

- Chapters 5–7 apply to special occupancies—places of assembly, hazardous locations, agricultural buildings, mobile homes, health care facilities; special equipment—swimming pools, signs, elevators, welders; or other special conditions—emergency systems, fire alarms, power-limited circuits. Although the first four still apply, these chapters augment or modify the general rules for the particular conditions.

90.3 Code Arrangement. This *Code* is divided into the introduction and nine chapters, as shown in *[the figure make sure this remains correct on page 60]*. Chapters 1, 2, 3, and 4 apply generally; Chapters 5, 6, and 7 apply to special occupancies, special equipment, or other special conditions. These latter chapters supplement or modify the general rules. Chapters 1, 2, 3, and 4 apply except as amended by Chapters 5, 6, and 7 for the particular conditions.

Chapter 8 covers communications systems and is not subject to the requirements of Chapters 1 through 7 except where the requirements are specifically referenced in Chapter 8.

Chapter 9 consists of tables that are applicable as referenced.

Annexes are not part of the requirements of this *Code,* but are included for informational purposes only.

Reprinted with permission from NFPA 70-2011.

Example: Article 514 covers the classified hazardous requirements for wiring a service station. It does not cover other areas within that service station facility. It does not cover other electrical needs, such as service, the grounding, the branch circuits, or the feeders, and, more importantly, it does not cover other environmental concerns, such as corrosion or wet locations. The designer must take all these concerns and any others that apply and design the installation so that it will provide good service, have adequate capacity, and provide all the safety needed for the normal operation throughout the life of the facility.

- Chapter 8 is a stand-alone chapter and covers communications systems; it is independent of the other chapters except where they are specifically referenced.

- Chapter 9 consists of table.

- *Annex A through I are informational and not enforceable*

- Annex A covers Product Standards

- Annex B covers the application information for ampacity calculations

- Annex C contains conduit and tubing fill tables for conductors and fixture wire s of the same size

- Annex D contains Examples

- Annex F Availability and reliability for critical operations systems and development and implementation of functional performance tests (FPPTs) for critical operations power systems

- Annex G supervisory control and data acquisition (SCADA)

- Annex I Recommended tightening torque tables from the UL Standard 486A-B

The introduction lets us know that the *Code* is intended as suitable for mandatory application by those adopting it—that is, governmental bodies exercising legal jurisdiction over electrical installations and for use by insurance inspectors. It also places the responsibility for enforcement on the authority having jurisdiction (AHJ).

NEC® 90.5 gives the style of the document and covers the mandatory rules, which are characterized by the word "**shall**" and explains that explanatory material appears in the form of informational notes.

90.6 Formal Interpretations. To promote uniformity of interpretation and application of the provisions of this *Code,* formal interpretation procedures have been established and are found in the NFPA Regulations Governing Committee Projects.

Reprinted with permission from NFPA 70-2011.

Warning: The commentary found in the NFPA *National Electrical Code Handbook* is **not** a formal interpretation of the *NEC®*.

NEC® 90.7 states that factory-installed internal wiring or the construction of equipment need not be inspected at the time of installation of the equipment, except to detect alterations or damage, if the equipment has been listed by a qualified electrical testing laboratory that is recognized and that requires suitability for installation in accordance with this *Code.*

See Examination, Identification, Installation, and Use of Equipment, 110.3.
See definition of "Listed," Article 100.

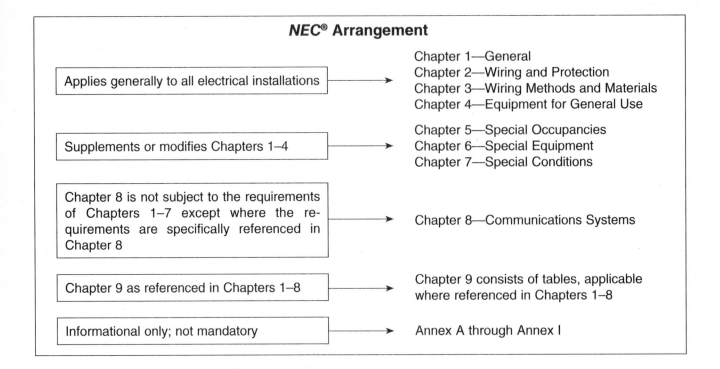

NEC® Arrangement

Applies generally to all electrical installations →
Chapter 1—General
Chapter 2—Wiring and Protection
Chapter 3—Wiring Methods and Materials
Chapter 4—Equipment for General Use

Supplements or modifies Chapters 1–4 →
Chapter 5—Special Occupancies
Chapter 6—Special Equipment
Chapter 7—Special Conditions

Chapter 8 is not subject to the requirements of Chapters 1–7 except where the requirements are specifically referenced in Chapter 8 →
Chapter 8—Communications Systems

Chapter 9 as referenced in Chapters 1–8 →
Chapter 9 consists of tables, applicable where referenced in Chapters 1–8

Informational only; not mandatory →
Annex A through Annex I

90.9(A) Units of Measurement. For the purpose of this *Code,* metric units of measurement are in accordance with the modernized metric system known as the International System of Units (SI).

Reprinted with permission from NFPA 70-2011.

Values of measurement in the *Code* text will be followed by an approximate equivalent value in metric units. Tables will have a footnote for metric conversion units used in the table.

Conduit size, wire size, horsepower designation for motors, and trade sizes that do not reflect actual measurements—for example, box sizes—will not be assigned dual designation metric units.

NEC® Units of Measurement (90.9)

- In the *NEC®*, metric units of measurement are in accordance with the modernized metric system known as the International System of Units (SI).
- The SI units appear first in the *NEC®*, followed by the inch-pound units in parentheses. The conversion from the inch-pound units to SI units is based on hard conversion, except as provided here.
  - *(C)(1) Trade Sizes. Where the actual measured size of a product is not the same as the nominal size, trade size designators shall be used rather than dimensions. Trade practices shall be followed in all cases.*

Example: Table 300.1(C) Metric Designator and Trade Sizes

Metric Designator	Trade Size
12	⅜
16	½
21	¾
27	1
35	1¼
41	1½
53	2
63	2½
78	3
91	3½
103	4
129	5
155	6

Note: The metric designators and trade sizes are for identification purposes only and are not actual dimensions.

- *(C)(2) Extracted Material. Where material is extracted from another standard, the context of the original material shall not be compromised or violated. Any editing of the extracted text shall be confined to making the style consistent with that of the NEC®.*
- *(C)(3) Industry Practice. Where industry practice is to express units in inch-pound units, the inclusion of SI units shall not be required.*

Example: Extracted Text to Section 516.3(B) from NFPA 33 and NFPA 34

516.3(B)(4)   For open dipping and coating operations, all space within a 1.5-m (5-ft) radial distance from the vapor sources extending from these surfaces to the floor. The vapor source shall be the liquid exposed in the process and the drainboard, and any dipped or coated object from which it is possible to measure vapor concentrations exceeding 25% of the lower flammable limit at a distance of 305 mm (1 ft), in any direction, from the object.

516.3(B)(5)   Sumps, pits, or belowgrade channels within 7.5 m (25 ft) horizontally of a vapor source. If the sump, pit, or channel extends beyond 7.5 m (25 ft) from the vapor source, it shall be provided with a vapor stop or it shall be classified as Class I, Division 1 for its entire length.

Reprinted with permission from NFPA 70-2011.

(C)(4) Safety. Where a negative impact on safety would result, soft conversion shall be used.

Example: Table 110.31.
The distance from the fence to live parts shall be not less than the following:

601 13,799 V	10 ft (3.05 m)
13,800 230,000 V	15 ft (4.57 m)
Over 230,000 V	18 ft (5.49 m)

(D) Compliance. Conversion from inch-pound units to SI units shall be permitted to be an approximate conversion. Compliance with the numbers shown in either the SI system or the inch-pound system shall constitute compliance with this Code.*

*Reprinted with permission from NFPA 70-2011.

NEC® 90.9 Units of Measurement

Informational Note No. 1: Hard conversion is considered a change in dimensions or properties of an item into new sizes that might or might not be interchangeable with the sizes used in the original measurement. Soft conversion is considered a direct mathematical conversion and involves a change in the description of an existing measurement but not in the actual dimension.

Informational Note No. 2: SI conversions are based on IEEE/ ASTM SI 10-1997, Standard for the Use of the International System of Units (SI): The Modern Metric System.

Reprinted with permission from NFPA 70-2011.

As you can see, Article 90 is very important, and one could never apply the *Code* correctly without first having a thorough understanding of the introduction, Article 90.

General Requirements

Article 100 contains definitions unique to this document. You must continually refer to these definitions as you apply the requirements set out in each section, article, or table. These terms are unique and essential to this document; that is why they have been included.

Definition Hierarchy

- A word or term unique to this document (*NEC®*) used one time will appear in the article in which it is used.
- A word or term unique to this document (*NEC®*) used two or more times will appear in Article 100.
- A word or term that is unique to the electrical industry but not unique to the *NEC®*, such as diode or transistor, will be defined in the *IEEE Dictionary.*
- All other words are defined by *Webster's Dictionary.*

Article 110 covers the general requirements. *NEC®* 110.3 tells us that we must select equipment and material suitable for the installation, environment, and/or application, and that it must be installed in accordance with any instructions included with its listing or labeling. *NEC®* 110.9 and 110.10 give us some very strong mandatory requirements on interrupting ratings and circuit impedance that we cannot ignore. If the equipment is intended to break current, it must be rated to interrupt the available current at fault conditions, and the system voltage and the circuit components including the impedance, shall be coordinated so that the circuit protective device will clear any fault without excessive damage to the equipment. *NEC®* 110.8 and 110.12 tell us that only recognized methods are permitted and that we are required to make our installation in a neat and professional manner. *NEC®*

110.14 gives us mandatory requirements for making our splices and terminations. *NEC®* 110.14(C) tells us that the terminations must be considered when determining the ampacity of a circuit. Read this *NEC®* section very carefully; similar requirements have appeared in the UL *Electrical Constructions Material Directory* (Green Book) and UL *Guide Information for Electrical Equipment Directory* (White Book).

NEC® 110.26 tells us that we are required to carefully select the proper location for our 600 volts or less equipment (for equipment rated at over 600 volts, the *Code* references are *NEC®* 110 Parts III and IV 110.30 through 110.59) so that proper work space can be maintained. This often requires us to check the building and mechanical plans so that encroachment can be avoided. To discover this encroachment in the last stages of the construction can be and often is very costly to one or more of the craft contractors. Other sections in Chapter 1 are equally important and must be referred to continually every time the *NEC®* is applied.

Article 110 Part V—Manholes and Other Electric Enclosures Intended for Personnel Entry, All Voltages—clarifies that all enclosures that are intended for personnel entry shall be of sufficient size to maintain safe work space and clearances from live parts.

Wiring and Protection

Now that we have thoroughly studied the general requirements, the next step is to take a look at the wiring and protection requirements. Let us turn to Chapter 2 of the *NEC®*, Article 210—Branch Circuits.

Note: Many of the requirements in this article apply only to dwellings; read the pertinent sections very carefully before applying the requirements.

Note: Although, generally, extra circuits are not necessary, careful arrangement of the required branch circuits is necessary for compliance with 210.11(A), (B), and (C).

You will also find specific branch-circuit requirements in other articles of the *NEC®*, such as Article 430 for Motors, Article 600 for Signs, and Article 517 for Health Care Facilities. *NEC®* 210.52 gives us the guidelines necessary to avoid the use of extension cords in dwellings. 210.52(A) requires outlets to be located so that no point along the floor line is more than 6 feet from an outlet and wall spaces 2 feet wide

have an outlet in them. This requirement ensures that it is not necessary to lay cords across door openings to reach an outlet. This section also points out that outlets that are a part of luminaires and appliances, located within cabinets, or above 5½ feet (1.7 m) cannot be counted as one of those required. 210.52(A) through (I) gives us the requirements for our outlets including the appliance outlets.

You will find as you study Article 210 and throughout the *Code* numerous references to 210.8(A) or 210.8(B). These sections cover requirements for Ground-Fault Circuit-Interrupter (GFCI) protection for personnel. 210.8(A) covers the GFCI requirements for dwelling units. 210.8(B) covers the GFCI requirements for other than dwelling units. These Class A GFCIs can be found in all supply warehouses and retail stores that carry electrical supplies. They are readily available as circuit breakers or GFCI-type receptacles. They are covered under UL Standard 943 and have a trip level of 4ma to 6ma and defined in Article 100.

Now let us look at *NEC®* 210.70(A) for the required lighting outlets and wall switches. Extra lighting can be added for the customer, but the required lighting will provide a safe illumination. *NEC®* 210.70(A) generally requires a switch-controlled light in each room. Stairways are required to be lighted with a switch at the top and bottom of stairs that have six or more risers. *NEC®* 210.70(A) also requires a switch-controlled lighting outlet in attics and under floor spaces where used for storage or equipment that requires servicing. This section also requires a switch-controlled light at outside exits or entrances. After we have established the branch-circuit requirements, we can do the general lighting load calculations in accordance with Table 220.12 and footnotes (see pages 64–65).

Because of the number of handheld appliances now being used in bathrooms, it is advisable that two bathrooms not share the same circuit; not a *Code* requirement but good design. The total calculated load cannot be ascertained until all of the equipment loads are known and added to the general purpose branch circuits. Chapter 2, Articles 215 and 225, will also guide us as we select and size our inside and outside feeders.

Article 225 for outside branch circuits and feeders has been revised in the past few *NEC®* editions, and Part II provides an almost mirror image of the service requirements for "buildings or other structures supplied by a feeder(s) or branch circuit(s)."

The requirements for a feeder supplying a building or structure from the service are almost identical to those for a service. The *NEC®* limits the building to be supplied by one feeder or branch circuit except for special conditions, such as fire pumps and emergency circuits and for special occupancies. These are requirements similar to those of *NEC®* 230.2. *NEC®* 225.31 through 225.39 cover the disconnect requirements. *NEC®* 225.40 requires ready access to the overcurrent devices. Part III covers the requirements for over-600-volt feeders branch circuits.

Article 230—Services is the logical next step and also the next article. We are beginning to see some rationale for the chapter and article arrangement and layout. The diagram 230.1 is an excellent aid for using this article as the sections apply to the installation. Although materials are not specified, the methods and requirements are quite specific (see page 68).

Note: It is very important when designing the service or solving questions related to the service that you properly define the components of the installation; see Article 100—Definitions. *NEC®* 230.42(A) requires the service entrance conductors to be of sufficient size to carry the loads as calculated in Article 220.

Article 240 covers the overcurrent protection for conductors equipment. This article of the *Code* has some very specific rules that will apply to all areas in the *Code*. For example, Parts I through VII of Article 240 cover the general requirements for overcurrent protection and overcurrent protective devices 600 volts or less. Part VIII covers overcurrent protection for those portions of supervised industrial installations operating at voltages of not more than 600 volts, nominal. Part IX covers overcurrent protection over 600 volts, nominal.

NEC® 240.2 contains special definitions that apply to this article of the *Code*.

220.10 Branch-circuit Load Calculations. Branch-circuit loads shall be calculated as shown in 220.12, 220.14, and 220.16.

220.12 Lighting Load for Specified Occupancies. A unit load of not less than that specified in Table 220.12 for occupancies specified therein shall constitute the minimum lighting load. The floor area for each floor shall be calculated from the outside dimensions of the building, dwelling unit, or other area involved. For dwelling units, the calculated floor area shall not include open porches, garages, or unused or unfinished spaces not adaptable for future use.

> FPN: The unit values herein are based on minimum load conditions and 100% power factor and may not provide sufficient capacity for the installation contemplated.

220.14 Other Loads — All Occupancies. In all occupancies, the minimum load for each outlet for general-use receptacles and outlets not used for general illumination shall not be less than that calculated in 220.14(A) through (L), the loads shown being based on nominal branch-circuit voltages.

Exception: The loads of outlets serving switchboards and switching frames in telephone exchanges shall be waived from the calculations.

(A) Specific Appliances or Loads. An outlet for a specific appliance or other load not covered in 220.14(B) through (L) shall be calculated based on the ampere rating of the appliance or load served.

(B) Electric Dryers and Electric Cooking Appliances in Dwelling Units. Load calculations shall be permitted as specified in 220.54 for electric dryers and in 220.55 for electric ranges and other cooking appliances.

(C) Motor Loads. Outlets for motor loads shall be calculated in accordance with the requirements in 430.22, 430.24, and 440.6.

(D) Luminaires. An outlet supplying luminaire(s) shall be calculated based on the maximum volt-ampere rating of the equipment and lamps for which the luminaire(s) [fixture(s)] is rated.

(E) Heavy-Duty Lampholders. Outlets for heavy-duty lampholders shall be calculated at a minimum of 600 VA.

(F) Sign and Outline Lighting. Sign and outline lighting outlets shall be calculated at a minimum of 1200 VA for each required branch circuit specified in 600.5(A).

(continued)

(Reprinted with permission from NFPA 70-2011, the *National Electrical Code®*, National Fire Protection Association, Quincy, MA 02269. This reprinted material is not the complete and official position of the National Fire Protection Association on the referenced subject, which is represented only by the standard in its entirety.)

Table 220.12   General Lighting Loads by Occupancy

Type of Occupancy	Unit Load	
	Volt-Amperes per Square Meter	Volt-Amperes per Square Foot
Armories and auditoriums	11	1
Banks	39[b]	3½[b]
Barber shops and beauty parlors	33	3
Churches	11	1
Clubs	22	2
Court rooms	22	2
Dwelling units[a]	33	3
Garages— commercial (storage)	6	½
Hospitals	22	2
Hotels and motels, including apartment houses without provision for cooking by tenants[a]	22	2
Industrial commercial (loft) buildings	22	2
Lodge rooms	17	1½
Office buildings	39	3½[b]
Restaurants	22	2
Schools	33	3
Stores	33	3
Warehouses (storage)	3	¼
In any of the preceding occupancies except one-family dwellings and individual dwelling units of two-family and multifamily dwellings:		
Assembly halls and auditoriums	11	1
Halls, corridors, closets, stairways	6	½
Storage spaces	3	¼

[a]See 220.12(J)
[b]See 220.14(K)

(G) Show Windows. Show windows shall be calculated in accordance with either of the following:

(1) The unit load per outlet as required in other provisions of this section
(2) At 200 volt-amperes per 300 mm (1 ft) of show window

(H) Fixed Multioutlet Assemblies. Fixed multioutlet assemblies used in other than dwelling units or the guest rooms or guest suites of hotels or motels shall be calculated in accordance with H(1) or H(2). For the purposes of this section, the calculation shall be permitted to be based on the portion that contains receptacle outlets.

(1) Where appliances are unlikely to be used simultaneously, each 1.5 m (5 ft) or fraction thereof of each separate and continuous length shall be considered as one outlet of not less than 180 VA .
(2) Where appliances are likely to be used simultaneously, each 300 mm (1 ft) or fraction thereof shall be considered as an outlet of not less than 180 VA.

(I) Receptacle Outlets. Except as covered in 220.14(J) and (K), receptacle outlets shall be calculated at not less than 180 VA for each single or for each multiple receptacle on one yoke. A single piece of equipment consisting of a multiple receptacle comprised of four or more receptacles shall be calculated at not less than 90 VA per receptacle. This provision shall not be applicable to the receptacle outlets specified in 210.11 (C)(1) and (2).

(J) Dwelling Occupancies. In one-family, two-family, and multifamily dwellings and in guest rooms or guest suites of hotels and motels, the outlets specified in (J)(1), (J)(2), and (J)(3) are included in the general lighting load calculations of 220.12. No additional load calculations shall be required for such outlets.

(1) All general-use receptacle outlets of 20-A rating or less, including receptacles connected to the circuits in 210.11(C)(3)
(2) The receptacle outlets specified in 210.52(E) and (G)
(3) The lighting outlets specified in 210.70(A) and (B)

(K) Banks and Office Buildings. In banks or office buildings, the receptacle loads shall be calculated to be the larger of (1) or (2):

(1) The calculated load from 220.14
(2) 11 VA/m^2 or 1 VA /ft^2

(L) Other Outlets. Other outlets not covered in 220.14 (A) through (K) shall be calculated based on 180 VA per outlet.

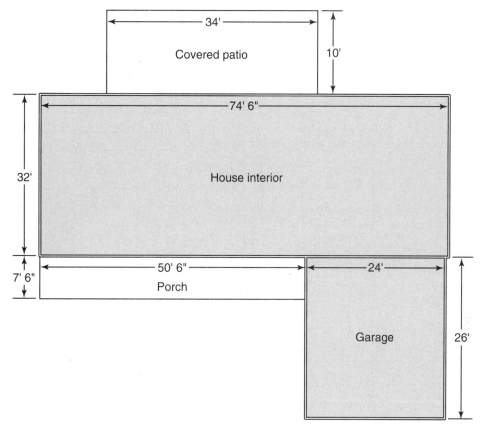

Figure 4–1 A typical single-family dwelling with an attached garage and covered porch on the front with a covered patio in the back. Neither the porch nor the patio is enclosed.

Example: In Figure 4–1, what is the general lighting load for this house?
A. 11,181 VA
B. 11,180.25 VA
C. 7152 VA
D. 9024 VA

Answer: C. NEC® 220.12 states that the computed area does not include porches and garages in NEC® Table 220.12. The general lighting load for a dwelling unita is 3 VA per square foot.

The superscript a directs us to a note at the bottom of Table 220.12 that references NEC® 220.14(J), which states that no additional load calculations are required.

240.1 Informational Note: Overcurrent protection for conductors and equipment is provided to open the circuit if the current reaches a value that will cause an excessive or dangerous temperature in conductors or conductor insulation. See also 110.9 for requirements for interrupting ratings and 110.10 for requirements for protection against fault currents.

Reprinted with permission from NFPA 70-2011.

230.6 Conductors Considered Outside the Building. Conductors shall be considered outside of a building or other structure under any of the following conditions:

(1) Where installed under not less than 50 mm (2 in.) of concrete beneath a building or other structure
(2) Where installed within a building or other structure in a raceway that is encased in concrete or brick not less than 50 mm (2 in.) thick
(3) Where installed in any vault that meets the construction requirements of Article 450, Part III
(4) Where installed in conduit and under not less than 450 mm (18 in.) of earth beneath a building or other structure.
(5) Where installed in overhead service masts on the outside surface of the building travelling through the cave of that building to meet the requirements of 230.29

Coordination (Selective). Localization of an overcurrent condition to restrict outages to the circuit or equipment affected, accomplished by the choice of overcurrent protective devices and their ratings or settings. (Article 100)

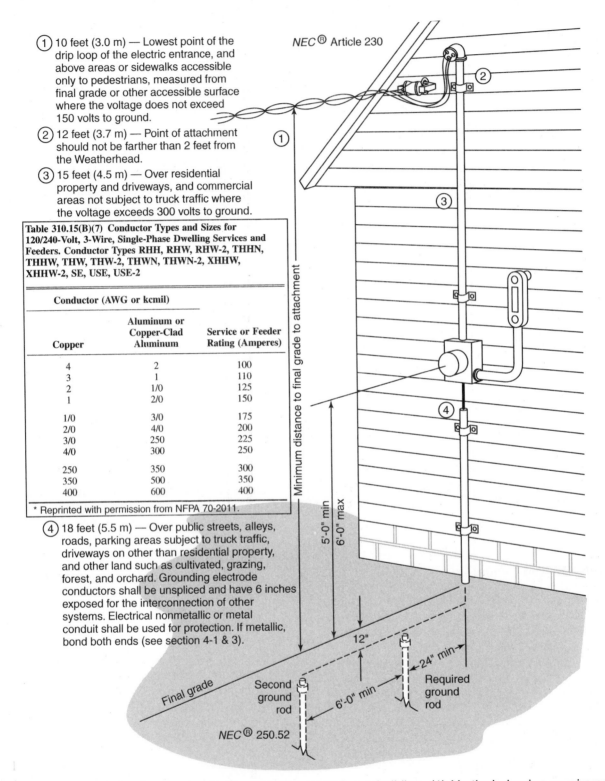

① 10 feet (3.0 m) — Lowest point of the drip loop of the electric entrance, and above areas or sidewalks accessible only to pedestrians, measured from final grade or other accessible surface where the voltage does not exceed 150 volts to ground.

② 12 feet (3.7 m) — Point of attachment should not be farther than 2 feet from the Weatherhead.

③ 15 feet (4.5 m) — Over residential property and driveways, and commercial areas not subject to truck traffic where the voltage exceeds 300 volts to ground.

Table 310.15(B)(7) Conductor Types and Sizes for 120/240-Volt, 3-Wire, Single-Phase Dwelling Services and Feeders. Conductor Types RHH, RHW, RHW-2, THHN, THHW, THW, THW-2, THWN, THWN-2, XHHW, XHHW-2, SE, USE, USE-2

Conductor (AWG or kcmil)		
Copper	Aluminum or Copper-Clad Aluminum	Service or Feeder Rating (Amperes)
4	2	100
3	1	110
2	1/0	125
1	2/0	150
1/0	3/0	175
2/0	4/0	200
3/0	250	225
4/0	300	250
250	350	300
350	500	350
400	600	400

* Reprinted with permission from NFPA 70-2011.

④ 18 feet (5.5 m) — Over public streets, alleys, roads, parking areas subject to truck traffic, driveways on other than residential property, and other land such as cultivated, grazing, forest, and orchard. Grounding electrode conductors shall be unspliced and have 6 inches exposed for the interconnection of other systems. Electrical nonmetallic or metal conduit shall be used for protection. If metallic, bond both ends (see section 4-1 & 3).

NEC® Article 230

Minimum distance to final grade to attachment

5'-0" min
6'-0" max

12"
24" min
Final grade
Second ground rod
6'-0" min
Required ground rod

NEC® 250.52

Figure 4–2 An example of a simple electrical service supplying a building. (1) Vertical clearing requirements; (2) Point of attachment should be located to remove stress on the mast; (3) NEC® Table 310.15(B)(6) [reprinted with permission from NFPA 70-2011]); (4) Example of grounding electrodes as required by NEC® 250.50 and 250.52.

General	Part I
Overhead Service-Drop Conductors	Part II
Underground Service-Lateral Conductors	Part III
Service-Entrance Conductors	Part IV
Service Equipment—General	Part V
Service Equipment—Disconnecting Means	Part VI
Service Equipment—Overcurrent Protection	Part VII
Services Exceeding 600 Volts, Nominal	Part VIII

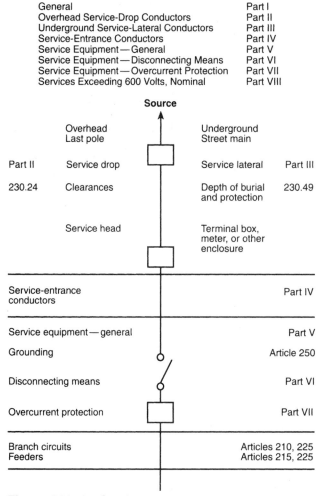

Figure 230–1   Services.

240.2 Current-Limiting Overcurrent Protective Device. A device that, when interrupting currents in its current-limiting range, reduces the current flowing in the faulted circuit to a magnitude substantially less than that obtainable in the same circuit if the device were replaced with a solid conductor having comparable impedance.

Supervised Industrial Installation. For the purposes of Part VIII, the industrial portions of a facility where all of the following conditions are met:

(1) Conditions of maintenance and engineering supervision ensure that only qualified persons will monitor and service the system.
(2) The premises wiring system has 2500 kVA or greater of load used in industrial process(es), manufacturing activities, or both, as calculated in accordance with Article 220.
(3) The premises has at least one service that is more than 150 volts to ground and more than 300 volts phase-to-phase.

This definition excludes installations in buildings used by the industrial facility for offices, warehouses, garages, machine shops, and recreational facilities that are not an integral part of the industrial plant, substation, or control center.

Tap Conductors. As used in this article, a tap conductor is defined as a conductor, other than a service conductor, that has overcurrent protection ahead of its point of supply that exceeds the value permitted for similar conductors that are protected as described elsewhere in 240.4.

240.4(E) Tap Conductors. *Tap conductors shall be permitted to be protected against overcurrent in accordance with the following:*

(1) 210.19(A)(3) and (A)(4) Household Ranges and Cooking Appliances and Other Loads
(2) 240.5(B)(2) Fixture Wire
(3) 240.21 Location in Circuit
(4) 368.17(B) Reduction in Ampacity Size of Busway
(5) 368.17(C) Feeder or Branch Circuits (busway taps)
(6) 430.53(D) Single Motor Taps

240.21 Location in Circuit. *Overcurrent protection shall be provided in each ungrounded circuit conductor and shall be located at the point where the conductors receive their supply except as specified in 240.21(A) through (H). Conductor supplied under the provisions of 240.21(A) through (H) shall not supply another conductor except through an*

overcurrent protective device meeting the requirements of 240.4.*

Note: This text in *NEC*® 240.21 clarifies that you can never tap a tap! (See Figure 4–3.)

NEC® 240.3 contains a list of about 40 *NEC*® articles that contains specific overcurrent requirements for systems and equipment, such as emergency systems, motors generators, and the like, that are in addition to Article 240.

NEC® 240.4 covers the protection of conductors and states that other than flexible cords and fixture wires, all conductors shall be protected against overcurrent in accordance with their ampacities specified in *NEC*® 310.15, unless otherwise permitted or required in 240.4(A) through 240.4(G).

*Reprinted with permission from NFPA 70-2011.

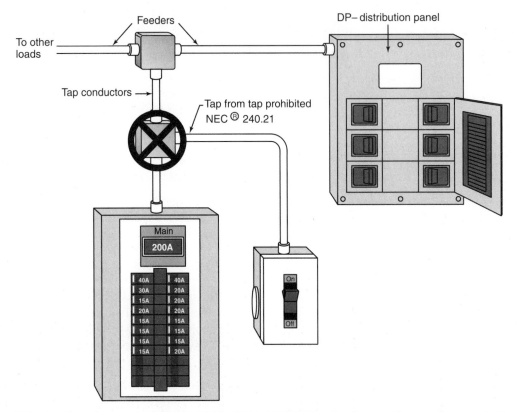

Figure 4–3 This drawing depicts a violation of the 2011 *NEC*®. The *Code* permits smaller conductor taps from a feeder to supply more than one piece of equipment. However, the tap from the tap conductors feeding the panelboard to supply the switch is not allowed. You are not permitted to "tap a tap." (See 240.2 Definition of a Tap, 240.4 Protection of Conductors, and 240.21.)

Example: Conductor overload protection is not required for which of the following where the interruption of the circuit would create a hazard?

    A. Material-handling magnet
    B. Motors
    C. Lighting
    D. All of these

Answer: A. The *NEC*® covers power loss hazards, such as a material-handling magnet. If there was a power loss while the magnet was moving a heavy object, the object would fall and possibly cause a severe injury.

Example: For overcurrent devices 800 amperes or less, the next higher overcurrent device is always permitted. True or False?

Answer: False. Although, this is permitted where all the conditions in *NEC*® 240.4(B) are met. It is not always permitted.

Example: For overcurrent devices 800 A or more and less than 600 volts, the next higher overcurrent device is never permitted. True or False?

Answer: True. *NEC*® 240.4(C) requires the overcurrent rating to be greater than the ampacity of the conductors it protects.

Example: A 120-volt, single-phase circulating hot water pump in a multifamily dwelling is supplied by a 12-2 W/Grd Type NM cable. What is the maximum size overcurrent circuit breaker or fuse protection for this circuit?

    A. 10 A
    B. 20 A
    C. 25 A
    D. Insufficient information

Answer: D. There is not enough information to answer this question. *NEC*® 240.3 states that motor overcurrent protection is governed by Article 430 Part III.

Small Size Conductor Overcurrent Protection Rules (240.4(D))

The overcurrent protection ratings (circuit breaker or fuse) cannot exceed 15 A for 14 AWG copper conductor, 20 A for 12 AWG copper conductor, and 30 A for 10 AWG copper conductor; or 15 A for 12 AWG and 25 A for 10 AWG aluminum and copper-clad aluminum conductor after any correction factors for ambient temperature and number of conductors have been applied except as permitted in *NEC*® 240.4(E) or 240.4(G). Overcurrent ratings for 16 AWG and 18 AWG copper conductors ratings are also included.

NEC® 240.4(E) permits tap conductors to be protected in accordance with the applicable *Code* section. For example, the rules for 25-foot tap conductors per 240.21(B)(2) would require the tap conductors to be protected from physical damage and have an ampacity of not less than one-third the rating of the overcurrent device protecting the feeder conductors terminating in a single circuit breaker or a single set of fuses that limit the load to the ampacity of the tap conductors. The device can supply any number of additional overcurrent devices on its load side.

Example: A 3 AWG THWN feeder protected by a 100-ampere circuit breaker is tapped in a junction box with _ AWG THWN conductors 20 ft (6 m) long to feed a 30-A single set of fuses. What is the minimum size THWN conductor permitted to be used for making this tap?

    A. 3 AWG
    B. 6 AWG
    C. 8 AWG
    D. 10 AWG

Answer: D. *NEC*® 240.21(B)(2) would require 1/3 of 100 or 33.34-ampere wire.
Table 310.16
• 3 AWG has an allowable ampacity of100 A.
• 10 AWG has an allowable ampacity of 35 A.

Section 240.21 covers the tap rules as they are permitted throughout the *Code*.

240.21 Location in Circuit. Overcurrent protection shall be provided in each undergrounded circuit conductor and shall be located at the point where the conductors receive their supply except as specified in 240.21(A) through (H). Conductors supplied under the provisions of 240.21(A) through (H) shall not supply another conductor except through an overcurrent protective device meeting the requirements of 240.4.

(A) Branch-Circuit Conductors. Branch-circuit tap conductors meeting the requirements specified in 210.19 shall be permitted to have overcurrent protection as specified in 210.20.

(B) Feeder Taps. Conductors shall be permitted to be tapped, without overcurrent protection at the tap, to a feeder as specified in 240.21(B)(1) through (B)(5). The provisions of 240.4(B) shall not be permitted for tap conductors.

(1) Taps Not over 3 m (10 ft) Long. Where the length of the tap conductors does not exceed 3 m (10 ft) and the tap conductors comply with all the following:

(1) The ampacity of the tap conductors is
  a. Not less than the combined calculated loads on the circuits supplied by the tap conductors, and
  b. Not less than the rating of the device supplied by the tap conductors or not less than the rating of the overcurrent protective device at the termination of the tap conductors.
(2) The tap conductors do not extend beyond the switchboard, panelboard, disconnecting means, or control devices they supply.
(3) Except at the point of connection to the feeder, the tap conductors are enclosed in a raceway, which shall extend from the tap to the enclosure of an enclosed switchboard, panelboard, or control devices, or to the back of an open switchboard.
(4) For field installations if the tap conductors leave the enclosure or vault in which the tap is made, the ampacity of the tap conductors is not less than one-tenth of the rating of the overcurrent device protecting the feeder conductors.

FPN: For overcurrent protection requirements for panelboards, see 408.36.

(continued)

---

(2) Taps Not over 7.5 m (25 ft) Long. Where the length of the tap conductors does not exceed 7.5 m (25 ft) and the tap conductors comply with all the following:

(1) The ampacity of the tap conductors is not less than one-third of the rating of the overcurrent device protecting the feeder conductors.
(2) The tap conductors terminate in a single circuit breaker or a single set of fuses that limit the load to the ampacity of the tap conductors. This device shall be permitted to supply any number of additional overcurrent devices on its load side.
(3) The tap conductors are protected from physical damage by being enclosed in an approved raceway or by other approved means.

(3) Taps Supplying a Transformer [Primary Plus Secondary Not over 7.5 m (25 ft) Long]. Where the tap conductors supply a transformer and comply with all the following conditions:

(1) The conductors supplying the primary of a transformer have an ampacity at least one-third the rating of the overcurrent device protecting the feeder conductors.
(2) The conductors supplied by the secondary of the transformer shall have an ampacity that is not less than the value of the primary-to-secondary voltage ratio multiplied by one-third of the rating of the overcurrent device protecting the feeder conductors.
(3) The total length of the one primary plus one secondary conductor, excluding any portion of the primary conductor that is protected at its ampacity, is not over 7.5 m (25 ft).
(4) The primary and secondary conductors are protected from physical damage by being enclosed in an approved raceway or by other approved means.
(5) The secondary conductors terminate in a single circuit breaker or set of fuses that limit the load current to not more than the conductor ampacity that is permitted by 310.15.

(4) Taps over 7.5 m (25 ft) Long. Where the feeder is in a high bay manufacturing building over 11 m (35 ft) high at walls and the installation complies with all the following conditions:

(1) Conditions of maintenance and supervision ensure that only qualified persons service the systems.

(2) The tap conductors are not over 7.5 m (25 ft) long horizontally and not over 30 m (100 ft) total length.
(3) The ampacity of the tap conductors is not less than one-third the rating of the overcurrent device protecting the feeder conductors.
(4) The tap conductors terminate at a single circuit breaker or a single set of fuses that limit the load to the ampacity of the tap conductors. This single overcurrent device shall be permitted to supply any number of additional overcurrent devices on its load side.
(5) The tap conductors are protected from physical damage by being enclosed in an approved raceway or by other approved means.
(6) The tap conductors are continuous from end-to-end and contain no splices.
(7) The tap conductors are sized 6 AWG copper or 4 AWG aluminum or larger.
(8) The tap conductors do not penetrate walls, floors, or ceilings.
(9) The tap is made no less than 9 m (30 ft) from the floor.

(5) Outside Taps of Unlimited Length. Where the conductors are located outdoors of a building or structure, except at the point of load termination, and comply with all of the following conditions:

(1) The conductors are protected from physical damage in an approved manner.
(2) The conductors terminate at a single circuit breaker or a single set of fuses that limit the load to the ampacity of the conductors. This single overcurrent device shall be permitted to supply any number of additional overcurrent devices on its load side.
(3) The overcurrent device for the conductors is an integral part of a disconnecting means or shall be located immediately adjacent thereto.
(4) The disconnecting means for the conductors is installed at a readily accessible location complying with one of the following:
   a. Outside of a building or structure
   b. Inside, nearest the point of entrance of the conductors
   c. Where installed in accordance with 230.6, nearest the point of entrance of the conductors

(C) Transformer Secondary Conductors. A set of conductors feeding a single load, or each set of conductors feeding separate loads, shall be permitted to be connected to a transformer secondary, without overcurrent protection at the secondary, as specified in 240.21(C)(1) through (C)(6). The provisions of 240.4(B) shall not be permitted for transformer secondary conductors.

Informational Note: For overcurrent protection requirements for transformers, see 450.3.

(1) Protection by Primary Overcurrent Device. Conductors supplied by the secondary side of a single-phase transformer having a 2-wire (single-voltage) secondary, or a three-phase, delta-delta connected transformer having a 3-wire (single-voltage) secondary, shall be permitted to be protected by overcurrent protection provided on the primary (supply) side of the transformer, provided this protection is in accordance with 450.3 and does not exceed the value determined by multiplying the secondary conductor ampacity by the secondary-to-primary transformer voltage ratio.

Single-phase (other than 2-wire) and multiphase (other than delta-delta, 3-wire) transformer secondary conductors are not considered to be protected by the primary overcurrent protective device.

(2) Transformer Secondary Conductors Not over 3 m (10 ft) Long. If the length of secondary conductor does not exceed 3 m (10 ft) and complies with all of the following:

(1) The ampacity of the secondary conductors is
 a. Not less than the combined calculated loads on the circuits supplied by the secondary conductors, and
 b. Not less than the rating of the device supplied by the secondary conductors or not less than the rating of the overcurrent-protective device at the termination of the secondary conductors
(2) The secondary conductors do not extend beyond the switchboard, panelboard, disconnecting means, or control devices they supply.
(3) The secondary conductors are enclosed in a raceway, which shall extend from the transformer to the enclosure of an enclosed switchboard, panelboard, or control devices or to the back of an open switchboard.
(4) For field installations where the secondary conductors leave the enclosure or vault in which

the supply connection is made, the rating of the overcurrent device protecting the primary of the transformer, multiplied by the primary to secondary transformer voltage ratio, shall not exceed 10 times the ampacity of the secondary conductor.

Informational Note: For overcurrent protection requirements for panelboards, see 408.36.

(3) Industrial Installation Secondary Conductors Not over 7.5 m (25 ft) Long. For industrial installations only, where the length of the secondary conductors does not exceed 7.5 m (25 ft) and complies with all of the following:

(1) Conditions of maintenance and supervision ensure that only qualified persons service the systems.
(2) The ampacity of the secondary conductors is not less than the secondary current rating of the transformer, and the sum of the ratings of the overcurrent devices does not exceed the ampacity of the secondary conductors.
(3) All overcurrent devices are grouped.
(4) The secondary conductors are protected from physical damage by being enclosed in an approved raceway or by other approved means.

(4) Outside Secondary Conductors. Where the conductors are located outdoors of a building or structure, except at the point of load termination, and comply with all of the following conditions:

(1) The conductors are protected from physical damage in an approved manner.
(2) The conductors terminate at a single circuit breaker or a single set of fuses that limit the load to the ampacity of the conductors. This single overcurrent device shall be permitted to supply any number of additional overcurrent devices on its load side.
(3) The overcurrent device for the conductors is an integral part of a disconnecting means or shall be located immediately adjacent thereto.
(4) The disconnecting means for the conductors is installed at a readily accessible location complying with one of the following:
 a. Outside of a building or structure
 b. Inside, nearest the point of entrance of the conductors
 c. Where installed in accordance with 230.6, nearest the point of entrance of the conductors

(continued)

(5) Secondary Conductors from a Feeder Tapped Transformer. Transformer secondary conductors installed in accordance with 240.21(B)(3) shall be permitted to have overcurrent protection as specified in that section.

(6) Secondary Conductors Not over 7.5 m (25 ft) Long. Where the length of secondary conductor does not exceed 7.5 m (25 ft) and complies with all of the following:

(1) The secondary conductors shall have an ampacity that is not less than the value of the primary-to-secondary voltage ratio multiplied by one-third of the rating of the overcurrent device protecting the primary of the transformer.
(2) The secondary conductors terminate in a single circuit breaker or set of fuses that limit the load current to not more than the conductor ampacity that is permitted by 310.15.
(3) The secondary conductors are protected from physical damage by being enclosed in an approved raceway or by other approved means.

(D) Service Conductors. Service conductors shall be permitted to be protected by overcurrent devices in accordance with 230.91.

(E) Busway Taps. Busways and busway taps shall be permitted to be protected against overcurrent in accordance with 368.17.

(F) Motor Circuit Taps. Motor-feeder and branch-circuit conductors shall be permitted to be protected against overcurrent in accordance with 430.28 and 430.53, respectively.

(G) Conductors from Generator Terminals. Conductors from generator terminals that meet the size requirement in 445.13 shall be permitted to be protected against overload by the generator overload protective device(s) required by 445.12.

(H) Battery Conductors. Overcurrent protection shall be permitted to be installed as close as practicable to the storage battery terminals in an unclassified location. Installation of the overcurrent protection within the hazardous classified location shall also be permitted.

(Reprinted with permission from NFPA 70-2011, the *National Electrical Code®* Copyright © 2011–, National Fire Protection Association, Quincy, MA 02269. This reprinted material is not the complete and official position of the National Fire Protection Association on the referenced subject, which is represented only by the standard in its entirety.)

This article also contains guidance for fuses and circuit breakers, their application, and some manufacturing requirements. For instance, 240.60(B) regulates current-limiting fuses. They must be manufactured so that they are different from standard fuses: Equipment designed or modified to accept current-limiting fuses will reject standard fuses. Manufacturers accomplish this by placing a notch in the load side (bottom) of the fuse blade, and equipment designed to accept them has a pin in the load side fuse holder. So, current-limiting types can be installed, but standard types cannot. *NEC®* 240.83 gives similar specifics for circuit breakers.

Article 240 is a general use article and must be referred to on all installations. Part VIII covers those portions of supervised industrial installations operating at voltages not exceeding 600 volts nominal. *NEC®* 240 Part VIII applies only to large industrial facilities and only to those portions of the electrical system in the supervised industrial installation used exclusively for manufacturing or process control activities. *NEC®* 240.2 defines supervised installations as the industrial portions of a facility where all conditions are met. This definition makes it clear that it does not apply to buildings used as offices, warehouses, garages, and so on, that are not part of the industrial plant, substation, or control center.

Part IX covers overcurrent protection over 600 volts nominal.

NEC® Article 250—Grounding and Bonding

The last major article in Chapter 2 is Article 250. (I did not forget Article 280—Surge Arresters over 1 KV and Article 285—Surge-Protective Devices (SPDs) 1 KV or less. If you have a question related to surge arresters or SPDs, do not forget these articles. Grounding and bonding are an important parts of any electrical installation project and will be an important part of your studies. Article 250 is very specific. As you design the grounding system for an installation, you will find it to be concise and complete for most installations. However, specific grounding requirements are referenced in several other articles of the *NEC®*. If you are researching a question related to grounding, you may find the answer in the article specifically addressing the question. This is an article that you will become very familiar with during your studies of the *Code*; however, you may find it more efficient to use the index to see all references and where they are located in the *Code*.

250.4 General Requirements for Grounding and Bonding. The following general requirements identify what grounding and bonding of electrical systems are required to accomplish. The prescriptive methods contained in Article 250 shall be followed to comply with performance requirements of this section.

(A) Grounded Systems.

(1) Electrical System Grounding. Electrical systems that are grounded shall be connected to earth in a manner that will limit the voltage imposed by lightning, line surges, or unintentional contact with higher-voltage lines and that will stabilize the voltage to earth during normal operation.

> Informational Note: An important consideration for limiting the imposed voltage is the routing of bonding and grounding electrode conductors so that they are not any longer than necessary to complete the connection without disturbing the permanent parts of the installation and so that unnecessary bends and loops are avoided.

(2) Grounding of Electrical Equipment. Normally non–current-carrying conductive materials enclosing electrical conductors or equipment, or forming part of such equipment, shall be connected to earth so as to limit the voltage to ground on these materials.

(3) Bonding of Electrical Equipment. Normally non–current-carrying conductive materials enclosing electrical conductors or equipment, or forming part of such equipment, shall be connected together and to the electrical supply source in a manner that establishes an effective ground-fault current path.

(4) Bonding of Electrically Conductive Materials and Other Equipment. Normally non–current-carrying electrically conductive materials that are likely to become energized shall be connected together and to the electrical supply source in manner that establishes an effective ground-fault current path.

(5) Effective Ground-Fault Current Path. Electrical equipment and wiring and other electrically conductive material likely to become energized shall be installed in a manner that creates a low-impedance circuit facilitating the operation of the overcurrent device or ground detector for high-impedance grounded systems. It shall be capable of safely carrying the maximum ground-fault current likely to be imposed on it from any point on wiring system where a ground fault may occur to the electrical supply source. The earth shall not be considered as an effective ground-fault current path.

(B) Ungrounded Systems.

(1) Grounding Electrical Equipment. Non–current-carrying conductive materials enclosing electrical conductors or equipment, or forming part of such equipment, shall be connected to earth in a manner that will limit the voltage imposed by lightning or unintentional contact with higher-voltage lines and limit the voltage to ground on these materials.

(2) Bonding of Electrical Equipment. Non–current-carrying conductive materials enclosing electrical conductors or equipment, or forming part of such equipment, shall be connected together and to the supply system grounded equipment in a manner that creates a low-impedance path for ground-fault current that is capable of carrying the maximum fault current likely to be imposed on it.

(3) Bonding of Electrically Conductive Materials and Other Equipment. Electrically conductive materials that are likely to become energized shall be connected together and to the supply system grounded equipment in a manner that creates a low-impedance path for ground-fault current that is capable of carrying the maximum fault current likely to be imposed on it.

(4) Path for Fault Current. Electrical equipment, wiring, and other electrically conductive material likely to become energized shall be installed in a manner that creates a low-impedance circuit from any point on wiring system to the electrical supply source to facilitate the operation of the overcurrent devices should a second ground fault from a different phase occur on the wiring system. The earth shall not be considered as an effective fault-current path.

(Reprinted with permission from NFPA 70-2011, the *National Electrical Code®* Copyright ©, National Fire Protection Association, Quincy, MA 02269. This reprinted material is not the complete and official position of the National Fire Protection Association on the referenced subject, which is represented only by the standard in its entirety.)

NEC® 250.4(A)(5) makes it clear that in all cases, the fault-current path must be electrically continuous and capable of carrying the maximum fault likely to be imposed on it. It must have sufficiently low impedance to facilitate the operation of the overcurrent device(s) under fault conditions. This section goes hand in hand with *NEC®* 110.10, which covers all components of a system.

NEC® **110.10 Circuit Impedance and Other Characteristics.** The overcurrent protective devices, the total impedance, the component short-circuit current ratings, and other characteristics of the circuit to be protected shall be selected and

(continued)

coordinated to permit the circuit-protective devices used to clear a fault to do so without extensive damage to the electrical components of the circuit. This fault shall be assumed to be either between two or more of the circuit conductors or between any circuit conductor and the grounding conductor or enclosing metal raceway. Listed products applied in accordance with their listing shall be considered to meet the requirements of this section.

Reprinted with permission from NFPA 70-2011.

NEC® 250.24 clearly lays out the requirements for system grounding.

250.24(A)(1) General. *The grounding electrode conductor connection shall be made at any accessible point from the load end of the service drop or service lateral to and including the terminal or bus to which*

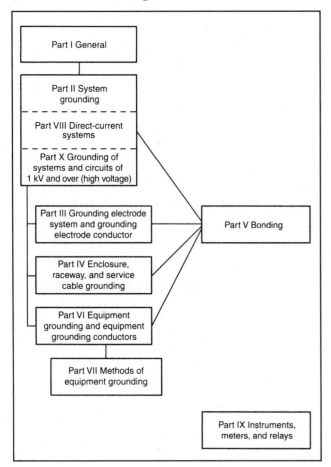

Figure 250–1   Grounding and Bonding

(Reprinted with permission from NFPA 70-2011, the *National Electrical Code®*, National Fire Protection Association, Quincy, MA 02269. This reprinted material is not the complete and official position of the National Fire Protection Association on the referenced subject, which is represented only by the standard in its entirety.)

the grounded service conductor is connected at the service disconnecting means.

250.24(C) Grounded Conductor Brought to Service Equipment. Where an ac system operating at less than 1000 volts is grounded at any point, the grounded conductor(s) shall be routed with the ungrounded conductors to each service disconnecting means and shall be connected to each disconnecting means grounded conductor(s) terminal or bus. A main bonding jumper shall connect the grounded conductor(s) to each service disconnecting means enclosure. The grounded conductor(s) shall be installed in accordance with **250.24(C)**(1) through (C)(4). *Exception: Where two or more service disconnecting means are located in a single assembly listed for use as service equipment, it shall be permitted to connect the grounded conductor(s) to the assembly common grounded conductor(s) terminal or bus. The assembly shall include a main bonding jumper for connecting the grounded conductor(s) to the assembly enclosure.*

(1) Sizing for a Single Raceway. The grounded conductor shall not be smaller than the required grounding electrode conductor specified in **Table 250.66** but shall not be required to be larger than the largest ungrounded service-entrance conductor(s). In addition, for sets of ungrounded service-entrance conductors larger than 1100 kcmil copper or 1750 kcmil aluminum, the grounded conductor shall not be smaller than 12½% of the circular mil area of the largest set of service-entrance ungrounded conductor(s).

NEC® 250.24(D) Grounding Electrode Conductor. A grounding electrode conductor shall be used to connect the equipment grounding conductors, the service-equipment enclosures, and, where the system is grounded, the grounded service conductor to the grounding electrode(s) required by Part III of this article. This conductor shall be sized in accordance with 250.66.

High-impedance grounded neutral system connections shall be made as covered in 250.36.

Reprinted with permission from NFPA 70-2011.

NEC® 250.32 makes it clear that the neutral shall not be regrounded at the second building or structure. This means that an equipment grounding conductor (250.118) must be run with the circuit conductors. There is an exception for existing installations. See *NEC®* Article 250 Part III for the accepted and required grounding electrode system rules.

NEC® 250.50 now requires that in each building or structure all the electrodes as described in 250.52 (A)(1) through 250.52(A)(7) present must be bonded together to form the grounding electrode system. The significance of this is that the concrete-encased (rebar) electrode is now required to be a part of the electrode system.

Where none of the electrodes covered in 250.52 (A)(1) through 250.52(6) exists, then one of the electrodes described in 250.52(A)(4) through 250.52 (A)(7) must be installed and used.

- *NEC®* 250.52(A) lists the types of electrodes permitted, such as underground water pipe in direct contact with earth 10 ft (3.0 m) or more, effectively grounded building steel, concrete encased electrode, and the like.

- *NEC®* 250.52(B) lists the two types of electrodes not permitted:

  1. Metal underground gas piping system
  2. Aluminum electrodes

- *NEC®* 250.53(A) through (H) covers the installation requirements for grounding electrodes.

    Note: Unless you test the resistance of the ground rod and it has 25 Ω or less, you must drive two electrodes to supplement the water pipe.

- *NEC®* 250.54 covers auxiliary grounding electrodes.

- *NEC®* 250.56 covers resistance requirements for rod, pipe, and plate-grounding electrodes.

- *NEC®* 250.58 covers "common" grounding electrodes.

- *NEC®* 250.60 covers the rules for the use of "air terminals."

- *NEC®* 250.62 covers the grounding electrode conductor material. *The material selected shall be resistant to any corrosive condition existing at the installation.*

- *NEC®* 250.64(A) through (F) covers the installation requirements for grounding electrode conductors.

- *NEC®* 250.66(A) through (C) and Table 250.66 cover the size of alternating current (ac) grounding electrode conductors.

- *NEC®* 250.68(A) and (B) covers the grounding electrode conductor and bonding jumper connection to the grounding electrode requirements.

- *NEC®* 250.70 covers the rules for making the connection of the grounding conductor to the electrode.

Now that we have our branch circuitry, the feeders, the service, the overcurrent protection, and the grounding designed into the installation, it is time to select the wiring methods and materials; those can all be found in Chapter 3 of the *NEC®*.

Wiring Methods

Chapter 3 Article Numbers and Acronyms

Article 300	Wiring Methods
Article 310	Conductors for General Wiring
Article 312	Cabinets, Cutout Boxes and Meter Socket Enclosures
Article 314	Outlet, Device, Pull and Junction Boxes, Conduit Bodies and Fittings
Article 320	Armored Cable **Type AC**
Article 322	Flat Cable Assemblies **Type FC**
Article 324	Flat Conductor Cable **Type FCC**
Article 326	Integrated Gas Spacer Cable **Type IGS**
Article 328	Medium Voltage Cable **Type MV**
Article 330	Metal-Clad Cable **Type MC**
Article 332	Mineral-Insulated Metal-Sheathed Cable **Type MI**
Article 334	Nonmetallic-Sheathed Cable **Types NM, NMC, and NMS**
Article 336	Power and Control Tray Cable **Type TC**
Article 338	Service Entrance Cable **Types SE and USE**
Article 340	Underground Feeder and Branch-Circuit Cable **Type UF**
Article 342	Intermediate Metal Conduit **Type IMC**
Article 344	Rigid Metal Conduit **Type RMC**
Article 348	Flexible Metal Conduit **Type FMC**
Article 350	Liquidtight Flexible Metal Conduit **Type LFMC**
Article 352	Rigid Polyvinyl Chloride Conduit **Type PVC**
Article 353	High Density Polyethylene Conduit **Type HDPE**
Article 354	Nonmetallic Underground Conduit with Conductors **Type NUCC**
Article 355	Reinforced Thermosetting Resin Conduit **Type RTRC**
Article 356	Liquidtight Flexible Nonmetallic Conduit **Type LFNC**
Article 358	Electrical Metallic Tubing **Type EMT**
Article 360	Flexible Metallic Tubing **Type FMT**
Article 362	Electrical Nonmetallic Tubing **Type ENT**
Article 366	Auxiliary Gutters
Article 368	Busways
Article 370	Cablebus
Article 372	Cellular Concrete Floor Raceways
Article 374	Cellular Metal Floor Raceways

Article 376	Metal Wireways
Article 378	Nonmetallic Wireways
Article 380	Multioutlet Assembly
Article 382	Nonmetallic Extensions
Article 384	Strut-Type Channel Raceway
Article 386	Surface Metal Raceways
Article 388	Surface Nonmetallic Raceways
Article 390	Underfloor Raceways
Article 392	Cable Trays
Article 394	Concealed Knob and Tube Wiring
Article 396	Messenger Supported Wiring
Article 398	Open Wiring on Insulators
Article 399	Outdoor Overhead Conductors over 600 volts

The individual article sections were renumbered using a common numbering system. For example, with only a few exceptions, all the wiring methods in Chapter 3 starts with a "Scope" numbered as 3XX.1. This common numbering system makes using the *Code* much quicker and easier. Once it is learned that "Securing and Supporting" requirements can be found in each Article 3XX.30 or that "Uses Not Permitted" can be found in 3XX.12, then the entire article will not need to be scanned to find these requirements.

Chapter 3 Wiring Methods Common Numbering	
Part I	General
3XX.1	Scope
3XX.2	Definitions
3XX.3	Other Articles
3XX.6	Listing Requirements
Part II	Installation
3XX.10	Uses Permitted
3XX.12	Uses Not Permitted
3XX.14	Dissimilar Metals
3XX.16	Temperature Limits
3XX.20	Size
3XX.22	Number of Conductors
3XX.24	Bends—How made
3XX.26	Bends—Number in one run
3XX.28	Reaming, Threading and Trimming
3XX.30	Securing and Supporting
3XX.40	Boxes and Fittings
3XX.42	Couplings and Connectors
3XX.44	Expansion Fittings
3XX.46	Bushings
3XX.48	Joints
3XX.50	Conductor Terminations
3XX.56	Splices and Taps
3XX.60	Grounding

Part III	Construction Specifications
3XX.100	Construction
3XX.110	Corrosion Protection
3XX.120	Marking
3XX.130	Standard Lengths

I hope you have studied Article 300. As I stated earlier, this article covers general requirements that must be applied to all the wiring methods. The wiring method selected must be appropriate for the environment, meet the physical protection requirements, temperature, and all other pertinent conditions that may affect the material selected.

Note: The uses permitted and not permitted appear in each wiring method article. Restrictions of use vary. The answer to your question may be in those *NEC®* sections.

The condition or question may address the wiring method directly or may require you to select an approved method based on these factors. Be sure to study all of the "uses permitted" and the "uses not permitted" in each specific article carefully.

For example, when selecting Rigid Metal Conduit (RMC), you only need to consider the corrosion elements (*NEC®* 300.6). However, when selecting Nonmetallic Sheath Cable (Type NM), many factors must be considered (300.4, and 336.10 and 336.12). Once this has been done, a wiring method may be selected from Chapter 3 based on all the pertinent factors.

Warning: You may have wiring methods that are not permitted or have additional restrictions in your jurisdiction. Check local ordinances that amend the *NEC®*.

The requirements of Article 590 permit the use of a less restrictive wiring for any temporary wiring needed during the construction phase or wiring used for short periods of time, such as holiday wiring. These requirements closely parallel OSHA requirements on construction sites. However, OSHA requirements may be much broader, and local experience with OSHA provides guidance in avoiding citations.

The conductor type is selected and sized from Article 310. The wiring method we select may need to be considered at the same time as the conductor type. Most installations require the use of several wiring methods to meet all the conditions, such as a method that is acceptable for dry, concealed locations. For this location, most of the methods in Chapter 3 are acceptable. We may also need to select a flexible wiring method for equipment connection and a material that can be used in wet locations and underground. There are methods that are acceptable in all locations: All are found in Chapter 3, including the junction, outlet, cabinets, and cutout boxes from Article 314 and Article 312. (To identify differences in these two types of enclosures, check Article 100—Definitions.) Once again, I remind you that you are required to comply with all the applicable provisions in Article 300 and the article that covers the specific wiring method.

> **Warning:** Some wiring methods are permitted in locations only when specific type conductors are installed. For example, 300.5(B) underground is a wet location. Therefore, conductors listed for wet locations must be used.

Equipment for General Use

Article 400 and Article 402 cover fixture wires and cords. These articles are used most in connection with equipment and, are not considered wring methods therefore, are placed in Chapter 4 rather than in Chapter 3.

The requirements for switches (Article 404) and switchboards and panelboards (Article 408) are now covered in Chapter 4. Read the scope of these articles. You see that Article 404 applies to all switches, switching devices, and circuit breakers where used as switches. Article 408's scope is not nearly as clear, and this is where the *NEC*® can be quite difficult to interpret. Motor control centers are not mentioned in Article 408; a informational note No. 1 to *NEC*® 430.1 states that motor control centers are covered in 110.26(E). You must study the *Code* carefully and thoroughly to successfully pass your examination the first time you take the test.

> **Example:** It is a common practice to use circuit breakers in large commercial installations or warehouses to regularly turn the lights on and off daily. Circuit breakers used as switches must be
>   A. hand-operable.
>   B. marked "ON" "OFF."
>   C. marked "SWD" or "HID."
>   D. all of these.
>
> **Answer:** D. 404.11, they must be of a "hand operable" type; 240.81, they must clearly indicate when they are "ON" and when they are "OFF" and where mounted in a vertical position the "ON" position must be up. *NEC*® 240.83(D): Circuit breakers used

as switches in 120-volt and 277-volt fluorescent lighting circuits are required to be listed and marked SWD or HID. Circuit breakers used as switches in high-intensity discharge lighting circuits are required to be listed and marked as HID.

Article 406 covers the rating, type, and installation of receptacles, cord connectors, and attachment plugs (cord caps).

Article 408 applies to all switchboards, panelboards, it does not apply to equipment over 600 volts unless specifically referenced elsewhere in the code.

Article 409 covers industrial control panels. These industrial control panels are intended for general use and operate from a voltage of 600 volts or less. These control panels are normally manufactured to the UL 508A Standard. They are defined in 409.2 as an assembly of two or more components such as motor controllers, overload relays, fused disconnect switches, and circuit breakers and related control devices but do not include the controlled equipment.

Article 410 covers the requirements for luminaires, lampholders, and lamps. The term "**luminaire**" replaced the term "lighting fixture." The *NEC*® is an international document used in many different countries. Design criteria can be found in this article. *NEC*® 410.4 covers wet locations and gives specific guideline rules for bathrooms where pendant-type luminaires are used. *NEC*® 410.8 covers all rules, allowances, and limitations for luminaires in clothes closets.

> **Note:** Electric discharge lighting includes fluorescent luminaires as well as other ballast-type fixtures.

As you look through Chapter 4, you will see articles to cover most utilization equipment found in all types of occupancies—that is, appliances, heating, air conditioning, refrigeration, motors, transformers, phase converters, and generators. Article 490 covers the general requirements for equipment operating at more than 600 volts and defines "high voltage" as more than 600 volts for the purpose of the article.

Special Equipment, Occupancies, and Systems

Chapters 5, 6, and 7 amend or modify the first four chapters. Chapter 8 stands alone, and earlier chapters can be applied only where specifically referenced (90.3). Chapter 9 contains tables and examples, which are used as applicable. These chapters are addressed in a later chapter of this book.

QUESTION REVIEW

Answer the following Question Review carefully. You will find it necessary to refer to each article in the *NEC®*, the Table of Contents, and the Index to solve these questions. You will find that answering these questions will prepare you as you proceed through the latter chapters of this book. The next eight chapters of the manual will help you gain that needed understanding of the *NEC®* and give you a wide variety of sample questions on each chapter of the *NEC®*.

> **Each lesson is purposely designed to require the student to apply the entire *NEC®* text and not specific chapters or articles. It has been found that when studying for a timed, open-book examination, the student must gain proficiency in the Table of Contents, the Index, and the ability to move quickly from cover to cover to find the correct answer to each question in a timely fashion.**

1. Would a floating restaurant located in a river or along a dock in a harbor be covered by the *NEC®*?

   Answer: _____

   _____

   _____

   _____

   Reference: _____

2. In which chapter in the *NEC®* would you find the requirements for grounding a community antenna television and radio distribution system?

   Answer: _____

   _____

   _____

   _____

   Reference: _____

3. In which article in the *NEC®* would you find the requirements for receptacles, cord connectors, and attachment plugs?

   Answer: _____

   _____

   _____

   _____

   Reference: _____

4. All requirements for the installation and calculation of branch circuits are found in Chapter 2 of the *NEC®*. True or False?

   Answer: _____

   _____

   _____

   _____

   Reference: _____

5. In what article would the requirements for installing luminaires in a paint spray booth be found?

Answer: _____

_____

_____

_____

Reference: _____

6. In what article would you find the requirements for the installation of central heating equipment, such as a natural gas furnace?

Answer: _____

_____

_____

_____

Reference: _____

7. Where in the *NEC®* will you find the allowable number of conductors in a conduit?

Answer: _____

_____

_____

_____

Reference: _____

8. Which article in the *NEC®* covers power and illumination in mine shafts?

Answer: _____

_____

_____

_____

Reference: _____

9. Where would you find the cover requirements for underground conductors running from building A to building B? (The conductors are direct burial conductors; the circuit is a 3-phase, 2300-volt, 4-wire system.)

Answer: _____

_____

_____

_____

Reference: _____

10. The conductors in Question 9 must terminate in a disconnecting means. What article and section of the *NEC®* cover the requirements for the disconnecting equipment at building B?

Answer: _____

_____

_____

Reference: _____

11. A trade size 4 (103) type PVC conduct is being installed on the outside of a building. The length of the conduit from the junction box to the point that it enters the building is 300 feet (91 m). Where in the *NEC®* is the amount of the expansion of the rigid nonmetallic conduit found?

Answer: _____

_____

_____

Reference: _____

12. You are going to wire a nightclub that will seat 300 people. Which article in the *NEC®* covers the wiring requirements for this nightclub?

Answer: _____

_____

_____

Reference: _____

13. Where in the *NEC®* would the requirements for grounding a portable generator be found?

Answer: _____

_____

_____

Reference: _____

14. In which section in the *NEC®* would the requirements for bonding the forming shell of an inground swimming pool be found?

Answer: _____

_____

_____

Reference: _____

15. In which section in the *NEC®* is material extracted from another NFPA standard explained?

Answer: _____

_____

_____

_____

Reference: _____

16. Where in the *NEC®* are the requirements for the location and installation of smoke detectors found?

Answer: _____

_____

_____

_____

Reference: _____

17. Under what article of the *NEC®* would the wiring requirements for fire alarms be found?

Answer: _____

_____

_____

_____

Reference: _____

18. An existing building is being remodeled. Permanent receptacles in the building are being used to supply temporary power for the construction. Must these permanent receptacles be GFCI protected?

Answer: _____

_____

_____

_____

Reference: _____

19. You have been given the responsibility for wiring a magnesium dust facility. What section of the *NEC®* covers this type of hazardous installation? (a) What class wiring is required? (b) If the hazardous dusts are not controlled and will be found on the day-to-day work schedule, what division must this area be wired in?

Answer: _____

_____

_____

_____

Reference: _____

20. You have been asked to wire a neon sign in the bedroom of a new home you are wiring. To which section of the *NEC®* would you refer to wire this neon sign for the bedroom?

    Answer: _____

    _____

    _____

    _____

    Reference: _____

21. You are asked to make an installation for a lawnmower repair shop where gasoline will be regularly used but not dispersed. The customer will also be repairing chainsaws and boat motors. To which article of the *NEC®* would you refer to wire this facility when gasoline is present in and around the equipment used for cleaning and testing these lawnmowers?

    Answer: _____

    _____

    _____

    _____

    Reference: _____

22. You have been asked to submit a bid for a new restaurant. After calculating the load requirements for the building, you determine that the requirements are too high. Is there an optional calculation you can use for restaurants?

    Answer: _____

    _____

    _____

    _____

    Reference: _____

23. You have been employed to wire a commercial building in electrical nonmetallic tubing. In which section in the *NEC®* are the requirements for running parallel to framing members with this raceway system?

    Answer: _____

    _____

    _____

    _____

    Reference: _____

24. You have been employed to install an optional standby generator system. To which article of the *NEC*® would you refer to make this installation?

    Answer: _____

    _____

    _____

    _____

    Reference: _____

25. A company that you have been working with has decided to burn their garbage, using the heat derived as an energy source to drive a generator and generate power to supplement the power being bought from the utility. To which article of the *NEC*® would you refer to make this installation?

    Answer: _____

    _____

    _____

    _____

    Reference: _____

CHAPTER 5

OBJECTIVES

Studying this chapter along with the 2008 *NEC®* gives the reader a basic introduction to the general requirements in the *Code* that apply to all installations. The introduction to *NEC®* Chapter 1 covers the general requirements that apply to all installations no matter how small or how large. The reader will study the application of installation rules for above and below 600 volts. The reader will learn how to apply all requirements and understand them.

After completing this chapter, you should know:

■ The definitions of words and terms that are unique to the 2011 *NEC®*

■ The application of Article 100—Definitions:

    1. Article 100, definitions of words and terms unique to the *NEC®*

    2. Why there are definitions elsewhere in the 2011 *NEC®*

    3. The importance of checking for a term as well as a word

■ The application of Article 110—General Requirements

■ The definitions of terms such as "listed", "identified", or "approved" as they apply to the *NEC®*

■ The minimal performance requirements such as "workmanship" and the "application of equipment"

■ The application of temperature requirements

■ The application of the complex working clearance requirements

GENERAL WIRING METHODS
Article 100

This article contains only definitions that are unique or essential to the application of the *Code*. It does not include commonly defined general terms or commonly defined technical terms from related codes and standards. In general, only those terms used in two or more articles are defined in Article 100. Other definitions are included in the article in which they are used, but may be referenced in Article 100.

Part I of Article 100 contains definitions that apply wherever the terms are used. Part II contains definitions of terms applicable only to those parts of the *Code* specifically covering installations and equipment operating at over 600 volts, nominal.

Electrical terms that are not specific to the *NEC®* may be defined in the *IEEE Dictionary*. Other words are defined in the standard *Webster's Dictionary*. An example of a definition with a specific meaning in the *NEC®* found in Article 100 would be "**accessible.**" Accessible is defined both for wiring methods and equipment. The definition found in Article 100 applies specifically to its use throughout the *NEC®*. Other terms or words found in Article 100 have meanings that generally differ from a standard *Webster's Dictionary*.

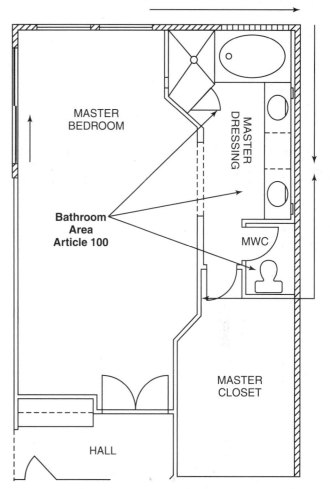

Figure 5–1 This is one example of a "bathroom area." It includes the separate water closet (toilet) room cut off by a door. Another example would be a typical Hotel-Motel "bathroom area" where the basin(s) are located outside the "bathroom" in the sleeping area.

Example: A bathroom is defined as
  A. a room with a basin, toilet and tub, or a shower.
  B. it is in a standard dictionary.
  C. an area including a basin, a toilet, and tub, or a shower.
  D. an area including a basin with one or more of the following: a toilet, urinal a tub, a shower, a bidet, or similar plumbing fixtures.

Answer: D. This definition, found in Article 100, is unique to the *NEC®*. (See Figure 5–1.)

This meaning is different than one would find in a standard dictionary, which defines "bathroom" as a room equipped for taking a bath or shower and usually includes a sink (basin) and toilet. This definition is needed since handheld grooming appliances have been the source of many serious and sometimes fatal accidents. These appliances are often used in and around the tub and basin areas. Basins are sometimes located outside the room commonly defined as the bathroom.

Thus, the *NEC®* defines the area that includes these items and not the room.

Article 100 has many new or revised definitions. Examples of some of the important new definitions and some definitions that are unique to the *NEC®*:

Bonded (Bonding). Connected to establish electrical continuity and conductivity.

Controller. A device or group of devices that serves to govern, in some predetermined manner, the electric power delivered to the apparatus to which it is connected.

Dusttight. Constructed so that dust will not enter the enclosing case under specified test conditions.

Duty, Continuous. Operation at a substantially constant load for an indefinitely long time.

*Reprinted with permission from NFPA 70-2011.

Intersystem Bonding Termination. A device that provides a means for connecting bonding conductors for communications systems to the grounding electrode system.

Kitchen. An area with a sink and permanent provisions for food preparation and cooking.

Switch, General-Use Snap. A form of general-use switch constructed so that it can be installed in device boxes or on box covers, or otherwise used in conjunction with wiring systems recognized by this *Code*.

Examples of words and terms defined in Article 100 are:

Energized. *Electrically connected to, or is, a source of voltage.*

Live Parts. *Energized conductive components.*

Damp Locations (under **Location, Damp**). *Locations protected from weather and not subject to saturation with water or other liquids but subject to moderate degrees of moisture. Examples of such locations include partially protected locations under canopies, marquees, roofed open porches, and like locations, and interior locations subject to moderate degrees of moisture, such as some basements, some barns, and some cold-storage warehouses.* (See Figure 5–2.)

Motor Control Center. *An assembly of one or more enclosed sections having a common power bus and principally containing motor control units.*

Nonlinear Load. *A load where the wave shape of the steady-state current does not follow the wave shape of the applied voltage. Informational Note: Electronic equipment, electronic/electric-discharge lighting, adjustable-speed drive systems, and similar equipment may be nonlinear loads.*

Qualified Person. *One who has the skills and knowledge related to the construction and operation of the electrical equipment and installations and has received safety training to recognize and avoid the hazards involved.* (See Figure 5–3.)

Revision of the definition of a **Receptacle.** *A receptacle is a contact device installed at the outlet for the connection of* [a single contact device]. *A single receptacle is a single contact device with no other contact device on the same yoke. A multiple receptacle is two or more contact devices on the same yoke.*

Figure 5–2 This is one example of a location that meets the Article 100 definition for "damp location." It is an industrial area protected on one side and open on the other three sides. It is covered with a canopy.

Figure 5–3 An example of "qualified person(s)" as defined in *NEC*® Article 100. Construction electricians are typically licensed based on their technical training and work experience. They typically receive regular safety and OSHA training at their workplace. (Refer to NFPA 70E–2009 Standard for Electrical Safety in the Workplace for Electrical Training Requirements.)

Service Conductors. *The conductors from the service point to the service disconnecting means.* (See Figure 5–4.)

Switchboard. *A large single panel, frame, or assembly of panels on which are mounted on the face, back, or both, switches, overcurrent and other protective devices, buses, and usually instruments. Switchboards are generally accessible from the rear as well as from the front and are not intended to be installed in cabinets.* (See Figure 5–5.)

Transfer Switch (under **Switch, Transfer**). *An automatic or nonautomatic device for transferring one or more load conductor connections from one power source to another.*

Electronically Actuated Fuse. *An overcurrent protective device that generally consists of a control module that provides current sensing, electronically derived time-current characteristics, energy to initiate tripping, and an interrupting module that interrupts current when an overcurrent occurs. Electronically actuated fuses may or may not operate in a current-limiting fashion, depending on the type of control selected.*

Article 110

Article 110 contains the general requirements for all electrical installations. *NEC*® 110.2 clarifies the term "approval." The word approved is defined in Article 100; however, "approval" that is required for installations, equipment, and instructions to judge it suitable is the responsibility of the authority having jurisdiction to approve or give approval. Many inspectors require listing and labeling by a recognized testing laboratory before they will give approval.

NEC® 110.3(A) and (B) requires the examination of equipment for identification and use. It also requires that the installation be made in accordance with its listing and or manufacturing instructions and in accordance with the provisions of the *Code*. *NEC*® 110.4 states that the voltages considered are those at which the circuit operates, and *NEC*® 110.6 points out that the *Code* sizes conductors in AWG (American Wire Gauge) and circular mils.

Workmanship

NEC® 110.7, 110.8, 110.11, 110.12, and 110.13 remind us that only wiring methods recognized as suitable are

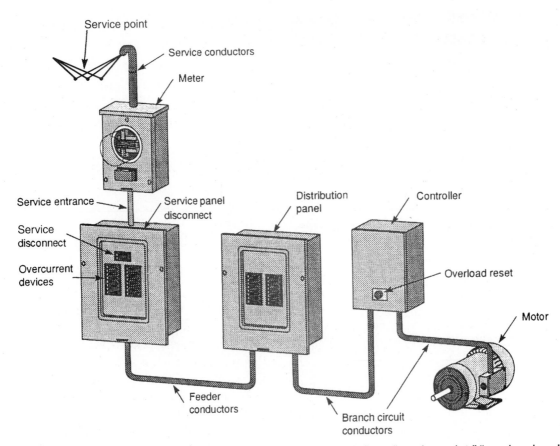

Figure 5–4 Specific system equipment and circuit types. Article 100 defines "service point," "service drop," "service conductors," "feeder," "branch circuit," and so on.

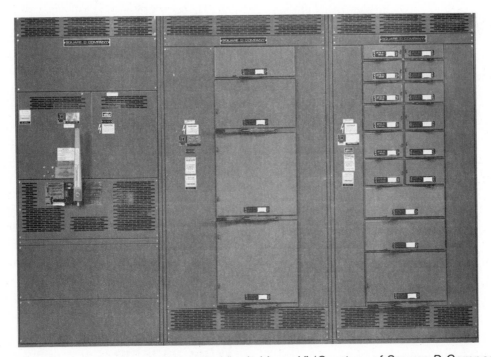

Figure 5–5 An example of an electrical "switchboard." (*Courtesy of Square D Company*)

included in the *Code* and that only recognized methods of wiring are permitted to be installed in any type of building or occupancy as provided in the *NEC®*. This is important because materials are often misused and wiring methods are devised in the field that are not in accordance with the manufacturer's instructions or design recommendations of the equipment being used.

All equipment and materials must be designed for the environment they will be subjected to over the life of the installation—that is, wet and damp locations, excessive temperatures, exposure to fumes, gases, vapors, liquids, or other elements that may have a deteriorating effect on them. It reminds us that all work must be installed in a neat and professional manner and that unused openings must be closed. Conductors and cables must be racked and supported neatly in manholes. All equipment must be installed carefully and protected throughout construction phases from contamination from paint, plaster, solvents, and cleaners that might harm contacts, bus bars, terminations, and insulators. All equipment must be firmly and securely mounted. Adequate clearances for all equipment must be mounted and located so that heat can dissipate. Equipment provided with ventilated openings must be mounted so that the openings are not obstructed.

> Although the *NEC®* is a minimal requirement, it still contains some very specific rules for good workmanship and to ensure a safe installation.

NEC® 110.9 and 110.10 require that all circuitry and equipment intended to break current must be properly sized and installed so that it will operate correctly and have an interrupting rating sufficient for the nominal circuit voltage and current available at the terminals of the equipment. *The overcurrent protective devices, the total impedance, the equipment short-circuit current ratings, and other characteristics of the circuit to be protected shall be selected and coordinated to permit the circuit-protective devices used to clear a fault to do so without extensive damage to the electrical equipment of the circuit.*

*Reprinted with permission from NFPA 70-2011.

Wire Temperature Termination Requirements

Lack of knowledge of wire temperature ratings and temperature termination requirements for equipment results in rejected installations or installations that fail long before they should need replacing. Information about this topic can be found in testing agency directories, product testing standards, and manufacturers' literature, but most do not consult these sources or read the information provided with the equipment until it is too late.

NEC® 110.14 warns that dissimilar metals, such as copper and aluminum, should not be mixed. Only terminals identified for this purpose are permitted to splice these dissimilar metals together. Although 14 AWG and 12 AWG aluminum building wire are no longer manufactured, there are still some installations that were originally wired with aluminum conductors. Until recently, it has been relatively difficult to splice these aluminum conductors to copper conductors. However, there are now several listed products on the market to accomplish this without difficulty. These methods vary from some requiring special tools to a simple twist-on connector. (Note that the twist-on device is only listed for copper-to-aluminum and not aluminum-to-aluminum terminations.)

This section also requires that materials, such as solder, flux, and inhibitors, be suitable for the use and not damage the insulation or conductors. Solder connections must be made so as not to depend on the solder for making a good connection.

New to the 2011 *NEC®* is a requirement that finely stranded conductor terminals and connectors be identified for the application.

NEC® 110.14(C) requires that terminations must be considered as a part of every circuit when calculating the allowable ampacity the circuit conductors can carry.

110.14(C) Temperature Limitations. The temperature rating associated with the ampacity of a conductor shall be selected and coordinated so as not to exceed the lowest temperature rating of any connected termination, conductor, or device. Conductors with temperature ratings higher than specified for terminations shall be permitted to be used for ampacity adjustment, correction, or both.
(1) Equipment Provisions. The determination of termination provisions of equipment shall be based on

---

110.14(C)(1)(a) or (C)(1)(b). Unless the equipment is listed and marked otherwise, conductor ampacities used in determining equipment termination provisions shall be based on Table 310.15(B)(16) as appropriately modified by 310.15(B)(7).

(a) Termination provisions of equipment for circuits rated 100 amperes or less, or marked for 14 AWG through 1 AWG conductors, shall be used only for one of the following:

(1) Conductors rated 60°C (140°F).

(2) Conductors with higher temperature ratings, provided the ampacity of such conductors is determined based on the 60°C (140°F) ampacity of the conductor size used.

(3) Conductors with higher temperature ratings if the equipment is listed and identified for use with such conductors.

(4) For motors marked with design letters B, C, or D, conductors having an insulation rating of 75°C (167°F) or higher shall be permitted to be used, provided the ampacity of such conductors does not exceed the 75°C (167°F) ampacity.

(b) Termination provisions of equipment for circuits rated over 100 A, or marked for conductors larger than 1 AWG, shall be used only for one of the following:

(1) Conductors rated 75°C (167°F)

(2) Conductors with higher temperature ratings, provided the ampacity of such conductors does not exceed the 75°C (167°F) ampacity of the conductor size used, or up to their ampacity if the equipment is listed and identified for use with such conductors

(2) Separate Connector Provisions. Separately installed pressure connectors shall be used with conductors at the ampacities not exceeding the ampacity at the listed and identified temperature rating of the connector.

Informational Note: With respect to 110.14(C)(1) and (C)(2), equipment markings or listing information may additionally restrict the sizing and temperature ratings of connected conductors.

Reprinted with permission from NFPA 70-2011.

Why Are Temperature Ratings Important?

Conductors carry a specific temperature rating based on the type of insulation employed on the conductor. Common insulation types can be found in

Table 310.13 of the *NEC*®, and corresponding ampacities can be found in Table 310.15(B)(16).

Example: A 1/0 AWG copper conductor ampacity based on different conductor insulation types:

Insulation Type	Temperature Rating	Ampacity
TW	60°C	125 A
THW	75°C	150 A
THHN	90°C	170 A

Even though the wire size has not changed (1/0 AWG Cu), the ampacity *has* changed due to the temperature rating of the insulation on the conductor. Higher-rated insulation allows a smaller conductor to be used at the same ampacity as a larger conductor with lower-rated insulation, and, as a result, the amount of copper and even the number of conduit runs needed for the job may be reduced.

One of the most common misapplications of conductor temperature ratings occurs when the rating of the equipment termination is ignored. Conductors must be sized by giving consideration to where they will terminate and how that termination is rated. If a termination is rated for 75°C, this means that the temperature at that termination may rise up to 75°C when the equipment is loaded to its ampacity. If 60°C insulated conductors were employed in this example, the additional heat at the connection above the 60°C conductor insulation rating could result in failure of the conductor insulation.

When a conductor is selected to carry a specific load, the user/installer or designer must know the termination ratings for the equipment involved in the circuit.

Example: Using a circuit breaker with 75°C termination and a 150-A load, if a THHN (90°C) conductor is selected for the job, from Table 310.15(B)(16), select a conductor that will carry the 150 A. Although Type THHN has a 90°C ampacity rating, the ampacity from the 75°C column must be selected because the circuit breaker termination is rated at 75°C. Looking at the table, a 1/0 AWG copper conductor is acceptable. The installation would be as shown in Figure 5–6, with proper heat dissipation at the termination as well as along the conductor length. Had the temperature

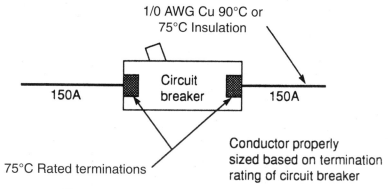

Figure 5–6   (*Courtesy of Square D Company*)

rating of the termination not been considered, a 1 AWG conductor, based on the 90°C ampacity, may have been selected, which may have led to overheating at the termination or premature opening of the overcurrent device due to the smaller conductor size. (See Figure 5–7.)

In this example, a conductor with a 75°C insulation type (THW, RHW, USE, etc.) would also be acceptable because the termination is rated at 75°C. A 60°C insulation type (TW or perhaps UF) is not acceptable because the temperature at the termination could rise to a value greater than the insulation rating.

General Rules for Application

When applying equipment with conductor terminations, the following basic rules apply:

- The termination and splicing provisions of equipment shall be based on 110.14(C)(1)(a) or (C)(1)(b).

- Conductor ampacities used in determining equipment termination provisions shall be based on Table 310.15(B)(16) as appropriately modified by 310.15(B)(7).

- **110.14(C)(1)(a).** *Termination provisions of equipment for circuits rated 100 A or less, or marked for 14 AWG through 1 AWG, shall be used only for one of the following:*
  (1) Conductors rated 60°C (140°F).
  (2) Conductors with higher temperature ratings, provided the ampacity of such conductors is determined based on the 60°C (140°F) ampacity of the conductor size used.
  (3) Conductors with higher temperature ratings if the equipment is listed and identified for use with such conductors.
  (4) For motors marked with design letters B, C, or D, conductors having an insulation rating of 75°C (167°F) or higher shall be permitted to be

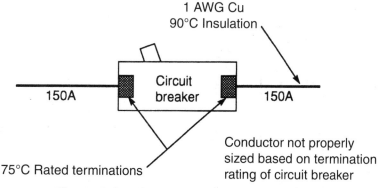

Figure 5–7   (*Courtesy of Square D Company*)

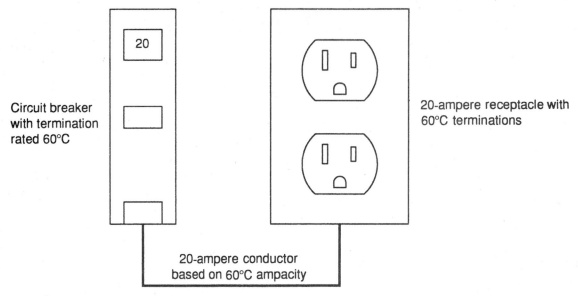

Figure 5–8 (*Courtesy of Square D Company*)

used, provided the ampacity of such conductors does not exceed the 75°C (167°F) ampacity. (See Figure 5–8.)

- **110.14(C)(1)(b).** *Termination provisions of equipment for circuits rated over 100 A, or marked for conductors larger than 1 AWG, shall be used only for one of the following:*
  (1) Conductors rated 75°C (167°F)
  (2) Conductors with higher temperature ratings, provided the ampacity of such conductors does not exceed the 75°C (167°F) ampacity of the conductor size used, or up to their ampacity if the equipment is listed and identified for use with such conductors. (See Figure 5–9.)

There are exceptions to the rules. Two of the most important exceptions are as follows:

- Conductors with higher temperature insulation may be terminated on lower temperature-rated terminations provided that the ampacity of the conductor is based on the lower rating. This is illustrated in the example, where the THHN (90°C) conductor ampacity is based on the 75°C rating to terminate in a 75°C termination. The following table provides a quick reference of how this exception would apply to common terminations.

*Reprinted with permission from NFPA 70-2011.

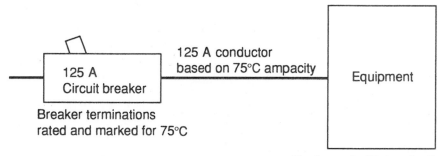

Figure 5–9 (*Courtesy of Square D Company*)

Conductor Insulation Versus Equipment Termination Ratings

Termination Rating	Conductor Insulation Rating		
	60°C	75°C	90°C
60°C	OK	OK (at 60°C ampacity)	OK (at 60°C ampacity)
75°C	No	OK	OK (at 75°C ampacity)
60/75°C	OK	OK (at 60°C or 75°C ampacity)	OK (at 60°C or 75°C ampacity)
90°C	No	No	OK*

*The equipment must have a 90°C rating to terminate 90°C wire at its 90°C ampacity.

- The termination may be rated for a value higher than the value permitted in the general rules if the equipment is listed and marked for the higher temperature rating.

   Example: A 30-ampere safety switch could have 75°C rated terminations if the equipment was listed and identified for use at this rating.

A Word of Caution

When terminations are inside equipment, such as panelboards, motor control centers, switchboards, enclosed circuit breakers, safety switches, and the like, it is important to note that the temperature rating identified on the equipment labeling should be followed—not the rating of the lug itself. It is common to use 90°C rated lugs (that is, marked AL9CU), but the equipment rating may be only 60°C or 75°C. The use of the 90°C rated lugs in this type of equipment does not give permission for the installer to use 90°C wire at the 90°C ampacity.

   The labeling of all devices and equipment should be reviewed for installation guidelines and possible restrictions.

Equipment Terminations Available Today

Remember, a conductor has two ends, and the termination on each end must be considered when applying the sizing rules.

   Example: Consider a conductor that will terminate in a 75°C rated termination on a circuit breaker at one end and a 60°C rated termination on a receptacle at the other end. This circuit must be wired with a conductor that has an insulation rating of at least 75°C (due to the circuit breaker) and sized based on the ampacity of 60°C (due to the receptacle).

   In the realm of electrical equipment, terminations are typically rated at 60°C, 75°C, or 60/75°C. There is no listed distribution or utilization equipment that is listed and identified for the use of 90°C wire at its 90°C ampacity. This includes distribution equipment, wiring devices, transformers, motor control devices, and even utilization equipment, such as HVAC, motors, and luminaires. Installers and designers who have not realized this fact have been faced with jobs that do not comply with the *NEC*® and jobs that have been turned down by the electrical inspectors.

   Example: 90°C wire may be used at its 90°C ampacity (see Figure 5–10). Note that the 2 AWG 90°C rated conductor does not terminate directly in the distribution equipment but in a terminal or tap box with 90°C rated terminations.

   Frequently, manufacturers are asked when distribution equipment will be available with terminations that will permit 90°C conductors at the 90°C ampacity. The answer is complex and requires not only significant equipment redesign (to handle the additional heat) but also coordination of the downstream equipment where the other end of the conductor will terminate. Significant changes in the product testing/listing standards would also have to occur.

   A final note about equipment: Generally, equipment requiring the conductors to be terminated in the equipment has an insulation rating of 90°C but has an ampacity based on 75°C or 60°C. This type of equipment might include 100% rated circuit breakers, fluorescent luminaires, and so on and will include a marking to indicate such a requirement. Check with the manufacturer of the equipment to see whether any special considerations need to be taken into account.

Higher-Rated Conductors and Adjustment or Correction

One advantage to conductors with higher insulation ratings is noted when ampacity adjustment or correction is required. Ampacity adjustment or correction may be required due to the number of conductors in a conduit, higher ambient temperatures, or possibly internal design requirements for a facility. By beginning the ampacity adjustment or correction process at the ampacity of the conductor based on the higher insulation value, upsizing the conductors to compensate

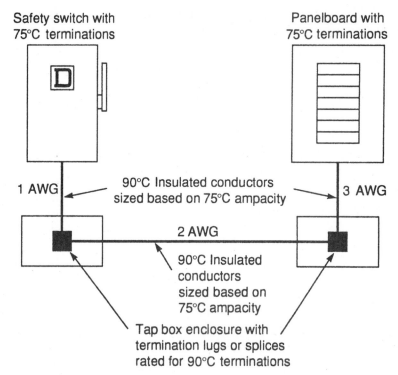

Safety switch with 75°C terminations

Panelboard with 75°C terminations

1 AWG

90°C Insulated conductors sized based on 75°C ampacity

3 AWG

2 AWG

90°C Insulated conductors sized based on 75°C ampacity

Tap box enclosure with termination lugs or splices rated for 90°C terminations

Figure 5–10 (*Courtesy of Square D Company*)

for the ampacity adjustment or correction may not be required.

For the following example of this ampacity adjustment or correction process, the following two points must be considered:

1. The ampacity determined after applying the ampacity adjustment or correction must be equal to or less than the ampacity of the conductor based on the temperature limitations at its terminations.

2. The adjusted or corrected ampacity becomes the allowable ampacity of the conductor, and the conductor must be protected against overcurrent in accordance with this ampacity.

Example: Assume a 480Y/277 Vac, 3Ø4W feeder circuit to a panelboard supplying 200 A of fluorescent luminaire load, and assume that the conductors will be in a 40°C ambient temperature. Also, assume that the conductors originate and terminate in equipment with 75°C terminations.

1. Because the phase and neutral conductors will all be in the same conduit, the issue of adjacent current carrying conductors must be considered. 310.15(B)(5) states that the neutral must be considered to be a current-carrying conductor

because it is supplying electric discharge lighting, a nonlinear load.

2. With four current-carrying conductors in the raceway, apply Table 310.15(B)(3)(a). This requires an 80 % reduction in the conductor ampacity based on four to six current-carrying conductors in the raceway.

3. The correction factors of Table 310.15(B)(2)(b) must also be applied. A correction factor of .88 for 75°C and .91 for 90°C is required where applicable.

Now for the calculations:

Using a 75°C conductor such as THWN:

300 kcmil copper has a 75°C ampacity of 285 A. Using the factors from the example, the calculations are as follows:

$$285 \times .80 \times .88 = 201 \text{ A}$$

Thus, 201 A is now the allowable ampacity of the 300 kcmil copper conductor for this circuit. Had the ampacity adjustment and correction factors for adjacent current-carrying conductors and ambient temperatures not been required, a 3/0 AWG copper conductor would have met these requirements.

Using a 90°C conductor such as THHN:

250 kcmil copper has a 90°C ampacity of 290 A. Using the factors from the example, the calculations are as follows:

$$290 \times .8 \times .91 = 211 \text{ A}$$

211 A is less than the 75°C ampacity of a 250 kcmil copper conductor (255 A), so the 211 A would now be the allowable ampacity of the 250 kcmil conductor. Had the calculation resulted in a number larger than the 75°C ampacity, the actual 75°C ampacity would have been required to be used as the allowable ampacity of the conductor. This is critical because the terminations are rated at 75°C.

> *Note:* The primary advantage to using 90°C conductors is exemplified here. The conductor is permitted to be reduced by one size (300 kcmil to 250 kcmil) and still accommodate all the required ampacity adjustment or correction factors for the circuit.

In summary, when using 90°C wire for ampacity adjustment and correction purposes, begin by using at the 90°C ampacity. Compare the result of the calculation to the ampacity of the conductor based on the termination rating (60°C or 75°C). The smaller of the two numbers then becomes the allowable ampacity of the conductor. Note that if the load dictates the size of the required overcurrent device (that is, continuous load × 125%), then the conductor allowable ampacity must be protected by the required overcurrent device. (See 240.4, which permits the conductor to be protected by the next higher standard size overcurrent device as per 240.6) This may require moving up to a larger conductor size and beginning the ampacity adjustment and corrections. In the example, if the 200 A of load were continuous, a 250-A overcurrent device would be required. The 201 or 211 A from our calculations would not be protected by the 250-A overcurrent device and, as such, we would be required to move to a larger conductor meeting these requirements to begin the ampacity adjustment and correction process.

Summary

There are a number of factors that affect how the allowable ampacity of a conductor is determined. The key is not to treat the wire as a system in itself, but as a component of the total electrical system. The terminations, the equipment ratings, and the environment all affect the ampacity that can be assigned to the conductor. If the designer and the installer keep each of the rules in mind, the installation will go much more smoothly from the beginning.

NEC® 110.14 is especially important in that the requirements for all electrical connections must be made in the proper manner. The *Code* text clarifies the termination requirements to meet the listing laboratories recommendations.

High-Leg Marking

NEC® 110.15 High-Leg Marking. The general requirements for marking the high leg has been appropriately located in Article 110. The rules for services remain in 230.56. For switchboards and panelboards see 408.3(F) and 408.3(G).

110.15 High-Leg Marking. On a 4-wire, delta-connected system where the midpoint of one phase winding is grounded, only the conductor or busbar having the higher phase voltage to ground shall be durably and permanently marked by an outer finish that is orange in color or by other effective means. Such identification shall be placed at each point on the system where a connection is made if the grounded conductor is also present.

Reprinted with permission from NFPA 70-2011.

Requirements for worker protection against electrical burns from "arc-flash" are found in *NEC®* 110.16. These requirements only require field marking warnings to persons performing maintenance or servicing certain electrical equipment in other than dwellings.

110.16 Arc-Flash Hazard Warning. Electrical equipment such as switchboards, panelboards, industrial control panels, meter socket enclosures, and motor control centers that are in other than dwelling units and are likely to require examination, adjustment, servicing, or maintenance while energized shall be field marked to warn qualified persons of potential electric arc flash hazards. The marking shall be located so as to be clearly visible to qualified persons before examination, adjustment, servicing, or maintenance of the equipment.

Informational Note No. 1: NFPA 70E-2009, *Standard for Electrical Safety in the Workplace,* provides assistance in determining severity of potential exposure, planning safe work practices, and selecting personal protective equipment.

Section 110.28 and Table 110.28 require enclosure types. This section was formerly located in Article 430 to provide guidance for selecting enclosures (other than surrounding fences or walls) of switchboards, panelboards, industrial control panels, motor control centers, meter sockets, and motor controllers rated not over 600 volts, nominal. They must be marked with an Enclosure Type number as shown in Table 110.28, which provides the basis for selecting the appropriate enclosures for use in specific locations other than hazardous (classified) locations.

110.22 Identification of Disconnecting Means

(A) General. Each disconnecting means shall be legibly marked to indicate its purpose unless located and arranged so the purpose is evident. The marking shall be of sufficient durability to withstand the environment involved.

(B) Engineered Series Combination Systems. Equipment enclosures for circuit breakers or fuses applied in compliance with series combination ratings selected under engineering supervision in accordance with 240.86(A) shall be legibly marked in the field as directed by the engineer to indicate the equipment has been applied with a series combination rating. The marking shall be readily visible and state the following:

CAUTION—ENGINEERED SERIES COMBINATION SYSTEM RATED ____ AMPERES. IDENTIFIED REPLACEMENT COMPONENTS REQUIRED.

(C) Tested Series Combination Systems. Equipment enclosures for circuit breakers or fuses applied in compliance with the series combination ratings marked on the equipment by the manufacturer in accordance with 240.86(B) shall be legibly marked in the field to indicate the equipment has been applied with a series combination rating. The marking shall be readily visible and state the following:

*Reprinted with permission from NFPA 70-2011.

CAUTION—SERIES COMBINATION SYSTEM RATED ____ AMPERES. IDENTIFIED REPLACEMENT COMPONENTS REQUIRED.

Working Clearances

NEC® 110.26 reminds us that we must provide adequate working clearances around electrical equipment of 600 volts nominal or less. This section contains requirements for providing access and working space about all electric equipment for the ready and safe operation and maintenance of this equipment (see Figure 5–5 and Figure 5–11).

Lockable enclosures housing electrical apparatus that are controlled by lock and key shall be considered accessible to qualified persons.

Working space for equipment operating at 600 volts, nominal, or less to ground and likely to require examination, adjustment, servicing, or maintenance while energized shall comply with the dimensions of 110.26(A)(1), (A)(2), and (A)(3) or as required or permitted elsewhere in this Code.* 110.26(B) reminds us that this space is not to be used for storage but must be kept clear for servicing and maintenance purposes and shall be suitably guarded when located in passage ways or clear spaces.

NEC® 110.26(C), titled "Entrance to Working Space," covers the minimum entrance space and the requirements for access to electrical rooms or spaces containing large electrical equipment.

Large equipment is considered as equipment rated 1200 A or more and over 6 feet (1.8 m) wide that contains overcurrent devices, switching devices, or control devices.

110.26(C) Entrance to and Egress from Working Space.
(1) Minimum Required. At least one entrance of sufficient area shall be provided to give access to and egress from working space about electrical equipment.
(2) Large Equipment. For equipment rated 1200 A or more and over 1.8 m (6 ft) wide that contains overcurrent devices, switching devices, or control devices, there shall be one entrance to the required working space not less than 610 mm (24 in.) wide and 2.0 m (6½ ft) high at each end of the working space. Where the entrance has a personnel door(s) that is less than 1.8 m (6 ft) from the working space, the door(s) shall open in the direction of egress and be equipped with panic bars, pressure

plates, or other devices that are normally latched but open under simple pressure. A single entrance to and egress from the required working space shall be permitted where either of the conditions in 110.26(C)(2)(a) or (C)(2)(b) is met.

(a) Unobstructed Exit. Where the location permits a continuous and unobstructed way of exit travel, a single entrance to the working space shall be permitted.

(b) Extra Working Space. Where the depth of the working space is twice that required by 110.26(A)(1), a single entrance shall be permitted. It shall be located so that the distance from the equipment to the nearest edge of the entrance is not less than the minimum clear distance specified in Table 110.26(A)(1) for equipment operating at that voltage and in that condition.

NEC® 110.26(A)(3) states that a minimum of 6½ feet (2 m) head room or to the top of the equipment, whichever is greater, must be maintained in the required working spaces to work on the equipment. This means that the area 30 inches (750 mm) wide or the width of the equipment, whichever is greater, by the depth required in Table 110.26(A)(1) must have a minimum headroom clearance of the height of the equipment or 6½ feet (2 m), whichever is greater, to provide a safe place for the worker to work on the equipment. Proper illumination must be provided in these working spaces and around the equipment. If an adequate lighting source is provided in these rooms or areas, additional specified lighting is not required.

Working Clearances Over 600 Volts

Over-600-volt requirements are found in Part III, which supplements or modifies the preceding sections. In no case shall the provisions of this part apply to equipment on the supply side of the service point. These requirements are very specific regarding entrance and access to working space, guarding barriers, enclosures for electrical installations, and separation from other circuitry (see Figure 5–11 and Figure 5–12).

Over-600-volt installations must be located in locked rooms or enclosures, except where under the observation of qualified persons at all times. Table 110.34(E) covers the elevation of unguarded live parts above working space for nominal voltages 601 through 35 kV.

Many of the requirements for over-600-volt installations are similar to those for 600 volts or less; however, when preparing for an examination or applying these sections, it is very important that they be studied carefully.

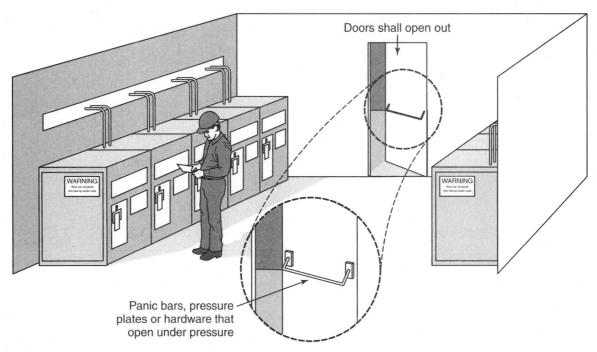

Figure 5–11 *NEC®* 110.26(C) (up to 600 volts) and 110.33(A)(1) (over 600 volts) defines large equipment and the required entrance and access to large electrical equipment. Where the qualified worker is required to exit through a door to get out of the working space, the door shall open out and be equipped with panic-type hardware.

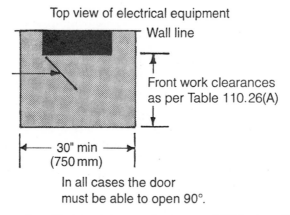

Figure 5–12 *NEC®* 110.26(A) requires that a minimum clearance of 30 inches (750 mm) wide be maintained in front of electrical equipment or the width of the equipment, whichever is greater; in addition, all doors must have clearance to open at least 90°. The distance in front of the equipment must be in accordance with Table 110.26(A). See Part III to Article 110 for required clearances over 600 volts and Table 110.34(A) for clearances.

QUESTION REVIEW

> **Each lesson is purposely designed to require the student to apply the entire NEC® text and not specific chapters or articles. It has been found that when studying for a timed, open-book examination, the student must gain proficiency in the Table of Contents, the Index, and the ability to move quickly from cover to cover to find the correct answer to each question in a timely fashion.**

The following drawing applies to Questions 1–5.

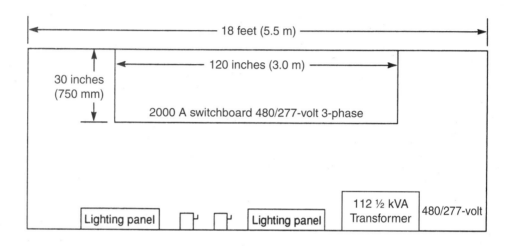

This proposed electrical equipment room is to be constructed of concrete floor and concrete. Block walls will be painted.

1. What is the space required between the switchboard and the 112½ kVA transformer?

   Answer: _____

   _____

   _____

   _____

   Reference: _____

2. How many doors are required for entering and leaving this room?

   Answer: _____

   _____

   _____

   _____

   Reference: _____

3. Is this electrical equipment room required to be illuminated?

Answer: _____

_____

_____

_____

Reference: _____

4. What is the required headroom in front of the electrical equipment needed for those performing tests or maintenance on this electrical equipment?

Answer: _____

_____

_____

_____

Reference: _____

5. If the proposed electrical room size was increased from 10 × 18 to 18 × 18, how many doors would be required, provided that the room had no additional electrical equipment added?

Answer: _____

_____

_____

_____

Reference: _____

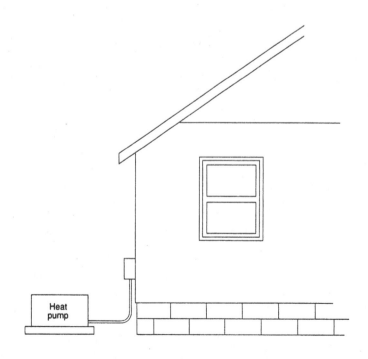

6. Does the heat pump installation shown comply with the *NEC®*? This is a single-family dwelling. The supplementary overcurrent device and required disconnecting means are located on the dwelling wall directly behind the heat pump for convenience.

   Answer: _____

   _____

   _____

   _____

   Reference: _____

7. The working space in front of enclosed electrical equipment operating at 480 volts must have a width when facing the equipment of at least _____ inches (_____ mm).

   Answer: _____

   _____

   _____

   _____

   Reference: _____

8. In an electrical system, an overcurrent device shall be placed in series with each _____ conductor.

   Answer: _____

   _____

   _____

   _____

   Reference: _____

9. A conduit nipple of 18 in (450 mm) in length is installed between two boxes. This nipple can be filled to a maximum of _____ % without applying ampacity adjustment to the conductors.

   Answer: _____

   _____

   _____

   _____

   Reference: _____

10. A panelboard with branch circuit overcurrent devices must *not* be installed in _____ conductor.

   Answer: _____

   _____

   _____

   _____

   Reference: _____

11. An unprotected cable is installed through bored holes in wood studs. The holes must be bored so that the edge of the hole is at least _____ inch(es) (_____ mm) from the nearest edge of the stud.

    Answer: _____

    _____

    _____

    _____

    Reference: _____

12. Circuit conductors run in electrical nonmetallic tubing (ENT) cannot exceed _____ volt(s).

    Answer: _____

    _____

    _____

    _____

    Reference: _____

13. Can liquidtight flexible nonmetallic conduit be used as a service raceway?

    Answer: _____

    _____

    _____

    _____

    Reference: _____

14. When a metal raceway is used as physical protection for a grounding electrode conductor, must this metallic raceway be bonded even though the grounding electrode conductor is bare copper? If the answer is yes, how must this metal raceway be bonded?

    Answer: _____

    _____

    _____

    _____

    Reference: _____

15. The size of the equipment grounding conductor routed with the feeder or branch-circuit conductors is determined by what?

    Answer: _____

    _____

    _____

    _____

    Reference: _____

16. How many outdoor receptacle outlets are required for a one-family dwelling?

    Answer: _____

    _____

    _____

    _____

    Reference: _____

17. Are the metal parts of electrical equipment associated with a hydromassage tub required to be bonded?

    Answer: _____

    _____

    _____

    _____

    Reference: _____

18. Where rooms within dwellings are separated by railings planters, free standing bar type counters and so forth, are receptacles required to be installed in or along these items?

    Answer: _____

    _____

    _____

    _____

    Reference: _____

19. Are the receptacle outlet spacings for room dividers, bar-type counters, and fixed room dividers the same for site-built dwellings as they are for mobile homes?

    Answer: _____

    _____

    _____

    _____

    Reference: _____

20. You have been asked to install a receptacle on a rooftop for the air-conditioning repair people to plug their instruments and portable tools into. In which section of the *NEC*® would this requirement be found?

    Answer: _____

    _____

    _____

    _____

    Reference: _____

21. You have encountered a problem in making a new service installation and need to know the requirements of the meter enclosure on the outside of the building. Which article in the *NEC*® covers meter enclosures?

Answer: _____

_____

_____

_____

Reference: _____

22. In making an installation in a home, the homeowners ask you to install a luminaire over their hydromassage bathtub. Which section of the *NEC*® governs luminaires over hydromassage tubs?

Answer: _____

_____

_____

_____

Reference: _____

23. As you are getting ready to install the conductors for the service drop on a large residence, you discover that all you have on the truck is a roll of 1 AWG THW copper. The service is 200 A. Realizing that this is not large enough, you parallel two 1 AWGs for each phase and two 1 AWGs for the neutral. Is this method acceptable for making this installation in accordance with the *NEC*®?

Answer: _____

_____

_____

_____

Reference: _____

24. In Question 23, what size single conductors would be required for a 200-A service?

Answer: _____

_____

_____

_____

Reference: _____

25. In making an installation of over 200 feet (61 m), a concern is voltage drop. Where does one obtain the resistance for conductors to be installed in a raceway or in direct burial for the purposes of calculating voltage drop?

Answer: _____

_____

_____

_____

Reference: _____

CHAPTER 6

OBJECTIVES

Studying this chapter along with the 2011 *NEC®* gives the reader a basic introduction to the branch-circuit and feeder requirements in the *Code* that apply to all installations above and below 600 volts. Voltage drop is also introduced in this chapter.

After studying this chapter, you should know:

- The definitions and the application of terms that are applicable and that are unique to the 2011 *NEC®*
- The many *Code* articles in which branch circuits and feeders are addressed as they apply to the *NEC®*
- Branch-circuit sizing and feeder-circuit sizing
- Branch-circuit conductor sizing and feeder conductor sizing
- Branch-circuit overcurrent device sizing and feeder overcurrent device sizing
- Branch-circuit raceway sizing and feeder raceway sizing
- Where GFCIs and AFCIs are required by the *NEC®*
- Where GFCIs and AFCIs are typically used
- Voltage drop
- The application of voltage drop recommendations and requirements

BRANCH CIRCUITS

Branch circuits are covered in *NEC®* Article 210. Part I covers the general provisions for branch circuits; Part II covers the branch-circuit ratings; and Part III gives the required outlets. Outdoor branch circuits and feeders are found in *NEC®* Article 225. Article 220 covers the calculations for branch circuits, feeders, and services as well as the optional calculations for calculating feeder and service loads. However, branch circuits are found throughout the *Code*. Air-conditioning branch circuits are found in *NEC®* Article 440, appliance branch circuits in *NEC®* Article 422, heating branch circuits in *NEC®* Article 424, and motor branch circuits in *NEC®* Article 430, and so on. Chapters 5–7 supplement or augment the first four chapters of the *Code* (90.3). Therefore, for branch-circuit studies, it is necessary that you be familiar with the entire *NEC®* and utilize the index to quickly find the specific type of branch circuits with which you are dealing.

Branch Circuit. The circuit conductors between the final overcurrent device protecting the circuit and the outlet(s).

Branch Circuit, Appliance. A branch circuit that supplies energy to one or more outlets to which appliances are to be connected and that has no permanently connected luminaries (lighting fixtures) that are not a part of an appliance.

Branch Circuit, General-Purpose. A branch circuit that supplies two or more receptacles or outlets for lighting and appliances.

Branch Circuit, Individual. A branch circuit that supplies only one utilization equipment.

Branch Circuit, Multiwire. A branch circuit that consists of two or more ungrounded conductors that have a voltage between them, and a grounded conductor that has equal voltage between it and each ungrounded conductor of the circuit and that is connected to the neutral or grounded conductor of the system.

Reprinted with permission from NFPA 70-2011.

Branch circuits are defined in *NEC®* Article 100 and are defined as a branch circuit, an appliance branch circuit, a general purpose branch circuit, an individual branch circuit, a multiwire branch circuit, and the branch-circuit selection current. It is very important that you read these definitions before applying

the requirements in the *NEC®*, so that you are properly applying the requirements. (See Figure 6–1.)

General Purpose Branch Circuits

The simplest forms of branch circuits are general purpose branch circuits, which supply a number of outlets for lighting and appliances, and these are in 15-, 20-, 30-, 40-, and 50-ampere sizes. Where conductors or higher ampacity are used for any reason, the ampere rating or the setting of the specified overcurrent device determines the circuit classification (210.3). Multiwire branch circuits greater than 50 amperes may be permitted for nonlighting outlet loads on industrial premises where maintenance and supervision indicate that a qualified person will service the equipment. A common multioutlet branch circuit greater than 50 amperes is often found in industrial buildings utilizing welding receptacles so that the maintenance personnel can move their welders throughout the plant on an as-needed basis. In these instances, you may find many receptacles on one 60-ampere or above rated circuit to supply these receptacles. There is generally no danger of overloading this circuit because the plant has a limited number of welding machines and personnel capable of utilizing those machines. There are no voltage limitations on branch circuits because branch circuits can supply the equipment and motors of many varying different voltages. However, branch circuits exceeding 600 volts, nominal, between conductors are only permitted to supply utilization equipment in establishments where the conditions of maintenance and supervision ensure that only qualified persons service the installation (210.6(E)).

Qualified Person. One who has skills and knowledge related to the construction and operation of the electrical equipment and installations and has received safety training to recognize and avoid the hazards involved.

Informational Note: Refer to NFPA 70E-2009, *Standard for Electrical Safety in the Workplace*, for electrical safety training requirements.

Reprinted with permission from NFPA 70-2011.

NEC® 210.6 lists the limitations for branch-circuit voltages. In occupancies such as dwelling units and guest rooms of hotels, motels, and similar

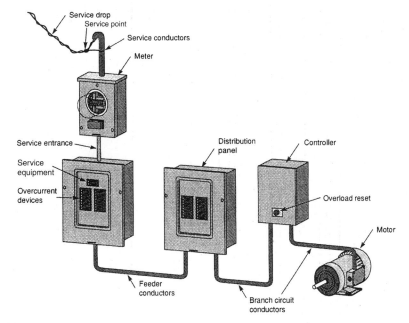

Figure 6–1 Specific conductor types as defined in *NEC®* Article 100.

occupancies, the voltage is not permitted to exceed 120 volts between conductors that supply terminals of luminaires and cord-and-plug-connected loads 1440 VA nominal or less, or less than ¼ horsepower, and permitted to supply terminals of medium-base, screw-shell lampholders, or lampholders of other types applied within their voltage ratings, auxiliary equipment of electric discharge lamps, and cord-and-plug-connected or permanently connected utilization equipment. Circuits exceeding 120 volts, nominal, and not exceeding 277 volts, nominal, to ground are permitted to supply listed electric-discharge luminaires equipped with medium-base screw-shell lampholder, luminaires with mogul-base screw-shell lampholders, and lampholders other than the screw-shell type applied within their voltage ratings, auxiliary equipment of electric discharge lamps, and cord-and-plug connected or permanently connected utilization equipment. Circuits exceeding 277 volts, nominal, to ground and not exceeding 600 volts, nominal, between conductors are permitted to supply auxiliary equipment of electric discharge lamps mounted in permanently installed luminaires where the luminaires are mounted in accordance with *NEC®* 210.6(D) (see Figure 6–2). Branch circuits supplying equipment over 600 volts are required to comply with the appropriate *Code* sections and articles related to that equipment.

Example: The circuit conductors supplying a 600-horsepower, 2300-volt 3-phase induction motor are described in the *NEC®* as a
A. feeder.
B. branch circuit.
C. service conductors.
D. none of the above.

Answer: **B. Branch circuit.** *The circuit conductors between the final overcurrent device protecting the circuit and the outlet(s) (Article 100).*

Example: What is the maximum circuit rating for a cord-and-plug-connected load that can be connected to a circuit that contains two or more 15-ampere receptacles? What is the maximum load that may be connected?

Step 1. Refer to the Index "Branch Circuits, Overcurrent Protection," which references 210.20.

Step 2. *NEC®* 210.20 gives the rules for 15- and 20-ampere branch circuits and references Table 210.21(B)(2).

Step 3. Table 210.21(B)(2).

Answer: A 15- or 20-ampere circuit with a maximum connected load of 12 amperes.

Ranges. For the computation of the range loads in these examples, Column C of Table 220.55 has been used. For optional methods, see Column A and B of Table 220.55. Except where the computation result in a major fraction of a kilowatt (0.5 or larger), such fractions are permitted to be dropped.

SI Units. For metric conversions, 0.093 m² = 1 f² and 0.3048 m = 1ft.

Example D6 Maximum Demand for Range Loads

Table 220.55, Column C, applies to ranges not over 12 kW. The application of Note 1 to ranges over 12 kW (and not over 27 kW) and Note 2 to ranges over 8¾ kW (and not over 27 kW) is illustrated in the following two examples.

A. Ranges All the Same Rating (*see Table 220.55, Note 1*)
Assume 24 ranges, each rated 16 kW.

From Table 220.55, Column C, the maximum demand for 24 ranges of 12 kW rating is 39 kW. 16 kW exceeds 12 kW by 4.
5% × 4 = 20% (5% increase for each kW in excess of 12)
39 kW × 20% = 7.8 kW increase

Solution: 39 kW + 7.8 kW = 46.8 kW (value to be used in selection of feeders)

B. Ranges of Unequal Rating (*see Table 220.55, Note 2*)
Assume 5 ranges, each rated 11 kW; 2 ranges, each rated 12 kW; 20 ranges, each rated 13.5 kW; 3 ranges, each rated 18 kW.

5 ranges × 12 kW =	60 kW (use 12 kW for range rated less than 12)	
2 ranges × 12 kW =	24 kW	
20 ranges × 13.5 kW =	270 kW	
3 ranges × 18 kW =	54 kW	
30 ranges Total kW =	408 kW	

408 kW ÷ 30 ranges = 13.6 kW (average to be used for calculation)

From Table 220.55, Column C, the demand for 30 ranges of 12 kW rating is 15 kW + 30 (1 kW × 30 ranges) = 45 kW. 13.6 kW exceeds 12 kW by 1.6 kW (use 2 kW).

5% × 2 = 10% (5% increase for each kW in excess of 12 kW)

45 kW × 10% = 4.5 kW increase

Solution: 45 kW + 4.5 kW = 49.5 kW (value to be used in the selection of feeders)

Figure 6–2 Branch circuit, feeder, and service-load calculations for a single-family dwelling. (Reprinted with permission from NFPA 70-2011, the *National Electrical Code*®, © 2011, National Fire Protection Association, Quincy, MA 02269. This reprinted material is not the complete and official position of the National Fire Protection Association on the referenced subject, which is represented only by the standard in its entirety.)

Example: How many general-use, duplex 20-ampere receptacles can be connected to one 120-volt, 20-ampere circuit breaker?

Step 1. This question is not complete enough to reach a single answer; therefore, we must look at both residential dwelling units and at other type installations.

Step 2. Table of Contents; branch-circuit calculations are found in Article 220.

Step 3. Calculation of branch circuits, 220.10.

Step 4. Table 220-12; from the table, we see "Dwelling Units."* See 220.14(J).

Step 5. Dwelling Occupancies. *In one-family, two-family, and multifamily dwellings and in guest rooms or guest suites of hotels and motels, the outlets specified in (J)(1), (J)(2), and (J)(3) are included in the general lighting load calculations of 220.12. No additional load calculations shall be required for such outlets.*

(1) All general-use receptacle outlets of 20-ampere rating or less, including receptacles connected to the circuits in 210.11(C)(3)
(2) The receptacle outlets specified in 210.52(E) and (G)
*(3) The lighting outlets specified in 210.70(A) and (B)**

Answer: For dwelling units, the number is not limited.

Step 6. For all other occupancies, we must go to 220.16(B) because we have no special instructions under Table 220.12.

Step 7. Other Outlets. Other than dwelling outlets shall be computed based on 180 volt-amperes per outlet.*

Step 8. I = VA ÷ E or I = 180 ÷ 120 or I = 1.5 amperes per 220.12 or 220.11 as applicable.

*Reprinted with permission from NFPA 70-2011.

Step 9. Table 210.24 references 210.23(A), which limits cord-and-plug-connected loads to 80% of the branch-circuit rating.

Step 10. 20-ampere circuit breaker multiplied by 80% equals 16 amperes.

Step 11. 16 ÷ 1.5 = 10.667.

Answer: For other occupancies, 10 duplex receptacles are permitted.

Ground-Fault Circuit Interruption Protection

The *Code* requires 125-volt 15- and 20-ampere receptacles to be protected by a ground-fault circuit-interrupter protection (GFCI) device in many locations where safety of the person using that circuit is of greater concern than in normal locations. In all but a few applications, this protection can be achieved by installing a GFCI circuit breaker or a GFCI receptacle. However, 620.85 requires that receptacles installed in hoistways, in elevator pits, on elevator car tops, and in escalator and moving walk wellways shall be of the GFCI type. But either type can protect receptacles installed in machine rooms and machinery spaces.

The GFCI rules for dwellings are found in 210.8(A), and 210.8(B) covers locations in other than dwelling units.

210.8 Ground-fault circuit-interruption for personnel shall be provided as required in **210.8(A)** through (C). The ground-fault circuit-interrupter shall be installed in a readily accessible location.

Informational Note: See **215.9** for ground-fault circuit-interrupter protection for personnel on feeders.

(A) Dwelling Units. All 125-volt, single-phase, 15- and 20-A receptacles installed in the locations specified in **210.8(A)**(1) through (8) shall have ground-fault circuit-interrupter protection for personnel.

(1) Bathrooms

(2) Garages, and also accessory buildings that have a floor located at or below grade level not intended as habitable rooms and limited to storage areas, work areas, and areas of similar use

(3) Outdoors

Exception to (3): Receptacles that are not readily accessible and are supplied by a branch circuit dedicated to electric snow-melting, deicing, or pipeline and vessel heating equipment shall be permitted to be installed in accordance with 426.28 or 427.22, as applicable

(4) Crawl spaces—at or below grade level

(5) Unfinished basements—for purposes of this section, unfinished basements are defined as portions or areas of the basement not intended as habitable rooms and limited to storage areas, work areas, and the like

Exception to (5): A receptacle supplying only a permanently installed fire alarm or burglar alarm system shall not be required to have ground-fault circuit-interrupter protection.

Informational Note: See **760.41(B)** and **760.121(B)** for power supply requirements for fire alarm systems.

Receptacles installed under the exception to **210.8(A)**(5) shall not be considered as meeting the requirements of **210.52(G)**.

(6) Kitchens—where the receptacles are installed to serve the countertop surfaces

(7) Sinks—located in areas other than kitchens where receptacles are installed within 1.8 m (6 ft) of the outside edge of the sink

(8) Boathouses

NEC® 210.8(B) applies to locations other than in dwelling units. This would include common areas in multifamily dwellings and commercial and industrial installations. The important thing to remember about these requirements is that normally in other than dwelling units, a receptacle is not required to be installed. If they are installed, then the requirement must be followed.

The requirement for placing a receptacle on a rooftop is found in 210.63. Where equipment is located on the rooftop, then this requirement must be followed.

All 15- and 20-ampere receptacles installed in kitchens must be GFCI protected. The term "kitchen" is defined, in Article 100. The AHJ will approve these locations and make the final determination which receptacles are covered.

(B) Other Than Dwelling Units. All 125-volt, single-phase, 15- and 20-ampere receptacles installed in the locations specified in **210.8(B)**(1) through (8) shall have ground-fault circuit-interrupter protection for personnel.

(1) Bathrooms
(2) Kitchens
(3) Rooftops
(4) Outdoors

*Exception No. 1 to (3) and (4):Receptacles that are not readily accessible and are supplied by a branch circuit dedicated to electric snow-melting, deicing, or pipeline and vessel heating equipment shall be permitted to be installed in accordance with **426.28** or **427.22**, as applicable.*

*Exception No. 2 to (4): In industrial establishments only, where the conditions of maintenance and supervision ensure that only qualified personnel are involved, an assured equipment grounding conductor program as specified in **590.6(B)**(2) shall be permitted for only those receptacle outlets used to supply equipment that would create a greater hazard if power is interrupted or having a design that is not compatible with GFCI protection.*

(5) Sinks—where receptacles are installed within 1.8 m (6 ft) of the outside edge of the sink.

Exception No. 1 to (5): In industrial laboratories, receptacles used to supply equipment where removal of power would introduce a greater hazard shall be permitted to be installed without GFCI protection.

*Exception No. 2 to (5): For receptacles located in patient bed locations of general care or critical care areas of health care facilities other than those covered under **210.8(B)**(1), GFCI protection shall not be required.*

(6) Indoor wet locations
(7) Locker rooms with associated showering facilities
(8) Garages, service bays, and similar areas where electrical diagnostic equipment, electrical hand tools, or portable lighting equipment are to be used

© Boat Hoists. GFCI protection shall be provided for outlets not exceeding 240 volts that supply boat hoists installed in dwelling unit locations.

406.4(D)(3) Ground-fault circuit-interrupter protected receptacles shall be provided where replacements are made at receptacle outlets that are required to be so protected elsewhere in this *Code.*

Reprinted with permission from NFPA 70-2011.

Other requirements for ground-fault circuit interrupters can be found in specific articles, such as Article 511 for commercial garages, Article 517 for health care facilities, and Article 680 for swimming pools, and so on. 210.52 covers the requirements and general provisions for installing receptacle outlets in dwellings and does not cover installations other than dwelling installations generally; there are some exceptions. There are provisions in the *Code,* in Article 220, that do cover these receptacle outlets in other than dwelling installations when installed. Lighting outlets for dwellings are covered in 210.70. Just as for receptacle outlets, the lighting requirements are generally not specified in the *NEC*® for other than dwellings. However, the installation procedures are specified when they are installed in these installations.

When calculating the loads for branch-circuit conductors, one must verify whether these loads are continuous or noncontinuous. Continuous loads is defined in Article 100. Most loads in dwellings are considered to be noncontinuous loads because they would be on for less than three hours at any given time. An example of a continuous load would be the lighting in a commercial or industrial establishment, which would generally be energized for more than three hours at a time. Most branch-circuit calculations are made in Article 220. However, specific branch-circuit requirements for branch circuits supplying specific utilization equipment are also found in the appropriate article for that equipment, such as Article 440 for air-conditioning equipment and Article 430 for motors. There are also many other specific requirements throughout the latter chapters in the *NEC*®. Several examples for calculating branch circuits are found in the latter sections of this chapter. Article 215 covers the installation requirements and minimum size and ampacity for conductors for feeders supplying branch-circuit loads, as calculated in Article 220.

Example: Is it permissible for the required single 20-A laundry circuit in a dwelling to supply both an automatic washer with a nameplate of 9.8 A and a gas dryer with a nameplate of 7.2 A?

The *Code* is not generally retroactive; however, the requirements in *NEC*® 406.4(D)(3) make the GFCI safety requirements retroactive.

Step 1. *NEC®* 210.52(F) requires at least one receptacle outlet.

Step 2. *NEC®* 210.11(C)(2) requires at least one 20-A circuit for laundry outlet(s) to supply the receptacle(s) required by 210.52(F); 422.10(B); also see references in 210.23.

> *Note:* This means all outlets for laundry purposes; that is, washer, 120-volt dryer, and ironing closet.

Step 3. *NEC®* 210.24 and Table 210.24 reference 210.23(A).

Step 4. *NEC®* 210.23(A); the rating of any one utilization equipment supplied from a cord and plug shall not exceed 80% of the branch-circuit rating.

Step 5. $20 \times 80\% = 16$ A. Therefore, that requirement is acceptable.

Step 6. $9.8 + 7.2 = 17$ A. The rating of the circuit is not exceeded.

> *Answer:* Yes. Reference: 422.10(B), 210.24, 210.23, 210.23(A)(1), and 210.11(C)(2), also see definition of "**continuous load.**" These would not be considered a continuous load.

Arc-Fault Circuit-Interrupters

The requirements for providing arc-fault circuit-interrupter (AFCI) protection for dwellings were first introduced in the 1999 *NEC®*. The initial rule was to require that all bedroom receptacle circuits be protected by AFCI. This requirement only applies to dwellings. The requirements in the 2011 *NEC®* are that all 15- and 20-A, 120-volt receptacles located in dwelling unit family rooms, dining rooms, living rooms, parlors, libraries, dens, bedrooms, sunrooms, recreation rooms, closets, hallways, or similar rooms or areas are required to be protected by a listed AFCI combination type and installed to protect the branch circuit.

110.12

(A) Dwelling Units. All 120-volt, single phase, 15- and 20-ampere branch circuits supplying outlets installed in dwelling unit family rooms, dining rooms, living rooms, parlors, libraries, dens, bedrooms, sunrooms, recreation rooms, closets, hallways, or similar rooms or areas shall be protected by a listed arc-fault circuit interrupter,

---

combination-type, installed to provide protection of the branch circuit.

Informational Note No. 1: For information on types of arc-fault circuit interrupters, see UL 1699-1999, *Standard for Arc-Fault Circuit Interrupters*.

Informational Note No. 2: See 11.6.3(5) of *NFPA 72-2010, National Fire Alarm and Signaling Code*, for information related to secondary power supply requirements for smoke alarms installed in dwelling units.

Informational Note No. 3: See **760.41(B)** and **760.121(B)** for power-supply requirements for fire alarm systems.

*Exception No. 1: If RMC, IMC, EMT, Type MC, or steel armored Type AC cables meeting the requirements of **250.118** and metal outlet and junction boxes are installed for the portion of the branch circuit between the branch-circuit overcurrent device and the first outlet, it shall be permitted to install an outlet branch-circuit type AFCI at the first outlet to provide protection for the remaining portion of the branch circuit.*

Exception No. 2: Where a listed metal or nonmetallic conduit or tubing is encased in not less than 50 mm (2 in.) of concrete for the portion of the branch circuit between the branch-circuit overcurrent device and the first outlet, it shall be permitted to install an outlet branch-circuit type AFCI at the first outlet to provide protection for the remaining portion of the branch circuit.

*Exception No. 3: Where an individual branch circuit to a fire alarm system installed in accordance with **760.41(B)** or **760.121(B)** is installed in RMC, IMC, EMT, or steel-sheathed cable, Type AC or Type MC, meeting the requirements of 250.118, with metal outlet and junction boxes, AFCI protection shall be permitted to be omitted.*

(B) Branch Circuit Extensions or Modifications — Dwelling Units. In any of the areas specified in **210.12(A)**, where branch-circuit wiring is modified, replaced, or extended, the branch circuit shall be protected by one of the following:

(1) A listed combination-type AFCI located at the origin of the branch circuit
(2) A listed outlet branch-circuit type AFCI located at the first receptacle outlet of the existing branch circuit

Outside Feeders and Branch Circuits

Article 225 covers outside branch circuits and feeders. Article 225 covers the electrical equipment and wire for the supply of utilization equipment located on or

Arc-fault Circuit-interrupter Protection

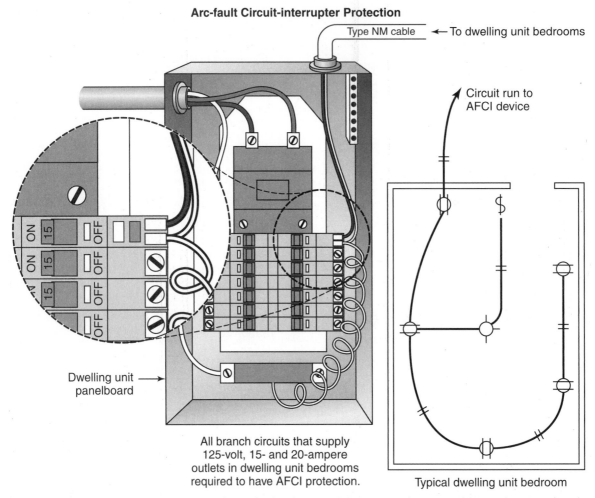

All branch circuits that supply
125-volt, 15- and 20-ampere
outlets in dwelling unit bedrooms
required to have AFCI protection.

Typical dwelling unit bedroom

Figure 6–3   The present rule requires that branch circuits supplying 120-volt, 15- and 20-ampere outlets be AFCI protected. (See definition of "dwelling unit" in Article 100.)

attached to the outside of public and private buildings or run between buildings, other structures, or poles on premises served. For clearances, see Figure 6–4 and Figure 6–5. An example of a feeder would be those conductors being fed from an overcurrent device in the service equipment to a subpanel in another part of the building or feeding a separate building on the premises, terminating in a panel or group of overcurrent devices within the second building or in another location. A feeder also serves separately derived systems in many cases, such as a transformer where the voltage is reduced to serve lighting and receptacle branch-circuit loads.

NEC® 225.32 provides that where more than one building or structure is on the same property and under single management, each building or other structure served shall be provided with a means of disconnecting all ungrounded conductors, which must

be installed either inside or outside the building or structure in a readily accessible location nearest the point of entrance of the supply feeder conductors. The disconnect disconnecting these feeder conductors must be installed in accordance with 230.70 and 230.72 and must be suitable as service equipment. However, it is required to have overcurrent protection in accordance with Article 220 for branch circuits and Article 215 for feeders. The reference to Article 220 is necessary because residential buildings and structures, such as garages and other outbuildings, are permitted to be supplied with branch circuits where the loads permit. For examples of branch-circuit and feeder calculations, see Figure 6–6.

Tap conductors are defined in 240.2. Tap rules are found in 240.21. The conductors from the secondary of an outside transformer to the building, are clearly defined as feeders, and the rules are found in 240.21(C).

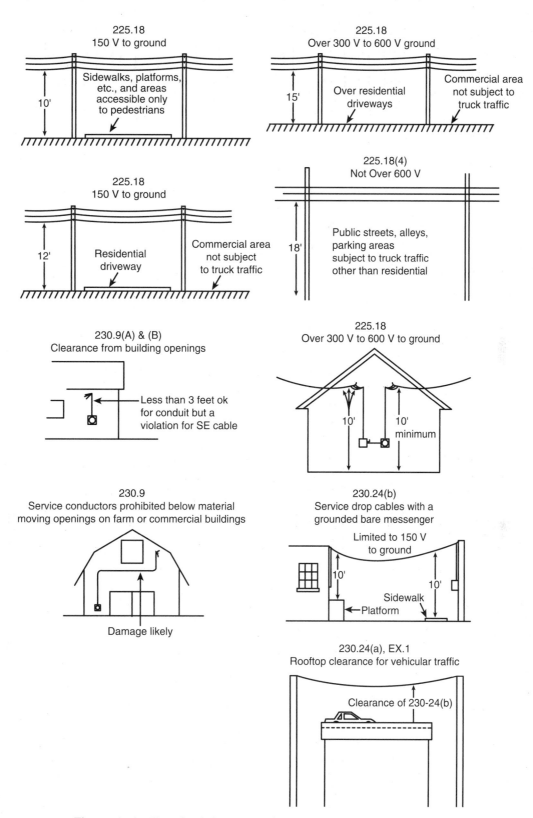

Figure 6–4 Required clearances for service and feeder conductors.

Example D1(a). One-Family Dwelling

The dwelling has a floor area of 1500 ft², exclusive of an unfinished cellar not adaptable for future use, unfinished attic, and open porches. Appliances are a 12-kW range and a 5.5-kW, 240-volt dryer. Assume range and dryer kW ratings equivalent to kVA ratings in accordance with 220.54 and 220.55.

Calculated Load [see 220.40]
General Lighting Load: 1500 ft² at 3 VA/ft² = 4500 VA

Minimum Number of Branch Circuits Required [see 210.11(A)]
General Lighting Load: 4500 VA ÷ 120 V = 37.5 A
   This requires three 15-A, 2-wire or two 20-A, 2-wire circuits
Small Appliance Load: Two 2-wire, 20-A circuits [see 210.11(C)(1)]
Laundry Load: One 2-wire, 20-A circuit [see 210.11(C)(2)]
Bathroom Branch Circuit: One 2-wire, 20-A circuit (no additional load calculation is required for this circuit) [see 210.11(C)(3)]

Minimum Size Feeder Required [see 220.40]

General Lighting	4500 VA
Small Appliance	3000 VA
Laundry	1500 VA
Total	9000 VA
3000 VA at 100%	3000 VA
9000 VA – 3000 VA = 6000 VA at 35%	2100 VA
Net Load	5100 VA
Range (see Table 220.19)	8000 VA
Dryer (see Table 220.54)	5500 VA
Net Calculated Load	18,600 VA

Net Calculated Load for 120/240-V, 3-wire, single-phase service or feeder

$$18,600 \text{ VA} \div 240 \text{ V} = 77.5 \text{ A}$$

Sections 230.42(B) and 230.79 require service conductors and disconnecting means rated not less than 100 amperes.

Calculation for Neutral for Feeder and Service

Lighting and Small Appliance Load	5100 VA
Range: 8000 VA at 70% (see 220.61)	5600 VA
Dryer: 5500 VA at 70% (see 220.61)	3850 VA
Total	14,550 VA

Calculated Load for Neutral

$$14,550 \text{ VA} \div 240 \text{ V} = 60.6 \text{ A}$$

Example D1(b). One-Family Dwelling

Assume same conditions as Example No. D1(a), plus addition of one 6-A, 230-volt, room air-conditioning unit and one 12-A, 115-volt, room air-conditioning unit,* one 8-A, 115-volt rated waste disposer, and one 10-A, 120-volt rated dishwasher.* See Article 430 for general motors and Article 440, Part VII, for air-conditioning equipment. Motors have nameplate ratings of 115 volts and 230 volts for use on 120-volt and 240-volt nominal voltage systems.

* (For feeder neutral, use larger of the two appliances for unbalance.)

   From Example No. D1(a), feeder current is 78 A (3-wire, 240 V).

	Line A	Neutral	Line B
Amperes from Example D1(a)	78	61	78
One 230-V air conditioner	6	—	6
One 115-V air conditioner and 120-V dishwasher	12	12	10
One 115-V disposer	—	8	8
25% of largest motor (see 430.24)	3	3	2
Total amperes per Conductor	99	84	104

   Therefore, the service would be rated 110 A.

Example D2(a). Optional Calculation for One-Family Dwelling, Heating Larger than Air-Conditioning [see 220.82]

The dwelling has a floor area of 1500 ft², exclusive of an unfinished cellar not adaptable for future use, unfinished attic, and open porches. It has a 12-kW range, a 2.5-kW water heater, a 1.2-kW dishwasher, 9 kW of electric space heating installed in five rooms, a 5-kW clothes dryer, and a 6-A, 230-volt, room air-conditioning unit. Assume range, water heater, dishwasher, space heating, and clothes dryer kW ratings equivalent to kVA.

Air Conditioner kVA Calculation

$$6 \text{ A} \times 230 \text{ V} \div 1000 = 1.38 \text{ kVA}$$

This 1.38 kVA [Item 1 from 220.82(C)] is less than 40%

Figure 6–5 Examples of branch-circuit, feeder, and service-load calculations for different types of occupancies. (Reprinted with permission from NFPA 70-2011, *the National Electrical Code*, © 2011, National Fire Protection Association, Quincy, MA 02269. This reprinted material is not the complete and official position of the National Fire Protection Association on the referenced subject, which is represented only by the standard in its entirety.)

of 9 kVA of separately controlled electric heat [Item 6 from 220.82(C)], so the 1.38 kVA need not be included in the service calculation.

General Load

1500 ft² at 3 VA	4500 VA
Two 20-A appliance outlet circuits at 1500 VA each	3000 VA
Laundry circuit	1500 VA
Range (at nameplate rating)	12,000 VA
Water heater	2500 VA
Dishwasher	1200 VA
Clothes dryer	5000 VA
Total	29,700 VA

Application of Demand Factor [See 220.82(B)]

First 10 kVA of general load at 100%	10,000 VA
Remainder of general load at 40% (19.7 kVA × 0.4)	7880 VA
Total of general load	17,880 VA
9 kVA of heat at 40% (9000 VA × 0.4) =	3600 VA
Total load	21,480 VA

Calculated Load for Service Size

$$21.48 \text{ kVA} = 21,480 \text{ VA}$$

$$21,480 \text{ VA} \div 240 \text{ V} = 89.5 \text{ A}$$

Therefore, the minimum service size would be 100 A in accordance with 230.42 and 230.79.

Feeder Neutral Load, in accordance with 220.61

1500 ft² at 3 VA	4500 VA
Three 20-A circuits at 1500 VA	4500 VA
Total	9000 VA
3000 VA at 100%	3000 VA
9000 VA – 3000 VA = 6000 VA at 35%	2100 VA
Subtotal	5100 VA
Range: 8 kVA at 70%	5600 VA
Clothes dryer: 5 kVA at 70%	3500 VA
Dishwasher	1200 VA
Total	15,400 VA

Calculated Load for Neutral

$$15,400 \text{ VA} \div 240 \text{ V} = 64.2 \text{ A}$$

Example D2(b). Optional Calculation for One-Family Dwelling, Air Conditioning Larger than Heating [see 220.82(A) and 220.82.(C)]

The dwelling has a floor area of 1500 ft², exclusive of an unfinished cellar not adaptable for future use, unfinished attic, and open porches. It has two 20-A small appliance circuits, one 20-A laundry circuit, two 4-kW wall-mounted ovens, one 5.1-kW counter-mounted cooking unit, a 4.5-kW water heater, a 1.2-kW dishwasher, a 5-kW combination clothes washer and dryer, six 7-A, 230-volt room air-conditioning units, and a 1.5-kW permanently installed bathroom space heater. Assume wall-mounted ovens, counter-mounted cooking unit, water heater, dishwasher, and combination clothes washer and dryer kW ratings equivalent to kVA.

Air Conditioning kVA Calculation

$$\text{Total amperes} = 6 \text{ units} \times 7 \text{ A} = 42 \text{ A}$$

$$42 \text{ A} \times 240 \text{ V} \div 1000 = 10.08 \text{ kVA (assume PF} = 1.0)$$

Load Included at 100%

Air-Conditioning: Included below [see item 1 in 220.82(C)]
Space Heater: Omit [see item 5 in 220.82(C)]

General Load

1500 ft² at 3 VA	4500 VA
Two 20-A small appliance circuits a at 1500 VA each	3000 VA
Laundry circuit	1500 VA
Two ovens	8000 VA
One cooking unit	5100 VA
Water heater	4500 VA
Dishwasher	1200 VA
Washer/dryer	5000 VA
Total general load	32,800 VA
First 10 kVA at 100%	10,000 VA
Remainder at 40% (22.8 kVA × 0.4 × 1000)	9120 VA
Subtotal general load	19,120 VA
Air conditioning	10,080 VA
Total	29,200 VA

Calculated Load for Service

$$29,200 \text{ VA} \div 240 \text{ V} = 122 \text{ A (service rating)}$$

Figure 6–5 (*continued*) (Reprinted with permission from NFPA 70-2011, the *National Electrical Code*, © 2011, National Fire Protection Association, Quincy, MA 02269. This reprinted material is not the complete and official position of the National Fire Protection Association on the referenced subject, which is represented only by the standard in its entirety.)

Feeder Neutral Load, in accordance with 220.61

Assume that the two 4-kVA wall-mounted ovens are supplied by one branch circuit, the 5.1-kVA counter-mounted cooking unit by a separate circuit.

1500 ft² at 3 VA		4500 VA
Three 20-A circuits at 1500 VA		4500 VA
	Subtotal	9000 VA
3000 VA at 100%		3000 VA
9000 VA − 3000 VA = 6000 VA		
at 35%		2100 VA
	Subtotal	5100 VA

Two 4-kVA ovens plus one 5.1-kVA cooking unit = 13.1 kVA. Table 220.55 permits 55% demand factor or, 13.1 kVA × 0.55 = 7.2 kVA feeder capacity.

Subtotal from above		5100 VA
Ovens and cooking unit:		
7200 VA × 70% for neutral load		5040 VA
Clothes washer/dryer:		
5 kVA × 70% for neutral load		3500 VA
Dishwasher		1200 VA
	Total	14,840 VA

Calculated Load for Neutral

$$14{,}840 \text{ VA} \div 240 \text{ V} = 61.83 \text{ A (use 62 A)}$$

Example D2(c). Optional Calculation for One-Family Dwelling with Heat Pump (Single-Phase, 240/120-Volt Service) (see 220.82)

The dwelling has a floor area of 2000 ft², exclusive of an unfinished cellar not adaptable for future use, unfinished attic, and open porches. It has a 12-kW range, a 4.5-kW water heater, a 1.2-kW dishwasher, a 5-kW clothes dryer, and a 2½-ton (24-A) heat pump with 15 kW of backup heat.

Heat Pump kVA Calculation

$$24 \text{ A} \times 240 \text{ V} \div 1000 = 5.76 \text{ kVA}$$

This 5.76 kVA is less than 15 kVA of the backup heat; therefore, the heat pump load need not be included in the service calculation [see 220.82(C)].

General Load

2000 ft² at 3 VA	6000 VA
Two 20-A appliance outlet circuits	
at 1500 VA each	3000 VA
Laundry circuit	1500 VA
Range (at nameplate rating)	12,000 VA

Water heater	4500 VA
Dishwasher	1200 VA
Clothes dryer	5000 VA
Subtotal general load	33,200 VA
First 10 kVA of general load at 100%	10,000 VA
Remainder of general load at 40%	
(23,200 VA × 0.4)	9280 VA
Total net general load	19,280 VA

Heat Pump and Supplementary Heat*

$$240 \text{ V} \times 24 \text{ A} = 5760 \text{ VA}$$

15-kW Electric Heat:

$$5760 \text{ VA} + (15{,}000 \text{ VA} \times 65\%) =$$
$$5.76 \text{ kVA} + 9.75 \text{ kVA} = 15.51 \text{ kVA}$$

*If supplementary heat is not on at same time as heat pump, heat pump kVA need not be added to total.

Totals

Net general load	19,280 VA
Heat pump and supplementary heat	15,510 VA
Total	34,790 VA

Calculated Load for Service

$$34.79 \text{ kVA} \times 1000 \div 240 \text{ V} = 144.96 \text{ A}$$

This dwelling unit would be permitted to be served by a 150-A service.

Example D3. Store Building

A store 50 ft by 60 ft, or 3000 ft², has 30 ft of show window. There are a total of 80 duplex receptacles. The service is a 120/240-volt, single-phase 3-wire service. Actual connected lighting load is 8500 VA.

Computed Load (see 220.40)

Noncontinuous Loads
Receptacle Load (see 220.44)

80 receptacles at 180 VA		14,400 VA
10,000 VA at 100%		10,000 VA
14,400 VA − 10,000 VA =		
4400 VA at 50%		2,200 VA
	Subtotal	12,200 VA

Figure 6–5 (continued) (Reprinted with permission from NFPA 70-2011, the *National Electrical Code®*, © 2011, National Fire Protection Association, Quincy, MA 02269. This reprinted material is not the complete and official position of the National Fire Protection Association on the referenced subject, which is represented only by the standard in its entirety.)

Continuous Loads
General Lighting*
3000 ft² at 3 VA/ft²   9000 VA
Show Window Lighting Load
   30 ft at 200 VA/ft   6000 VA
Outside Sign Circuit *[see 220.14(F)]*   1200 VA
           Subtotal   16,200 VA

Subtotal from noncontinuous   12,200 VA
Total noncontinuous +
     continuous loads =   28,400 VA

*In the example, 125% of the actual connected lighting load (8500 VA × 1.25 = 10,625 VA) is less than 125% of the load from Table 220.12, so the minimum lighting load from Table 220.12 is used in the calculation. Had the actual lighting load been greater than the value calculated from Table 220.12, 125% of the actual connected lighting load would have been used.

Minimum Number of Branch Circuits Required
General Lighting: Branch circuits need only be installed to supply the actual connected load *[see 210.11(B)]*.

$$8500 \text{ VA} \times 1.25 = 10,625 \text{ VA}$$

$$10,625 \text{ VA} \div 240 \text{ V} = 44 \text{ A for 3-wire, 120/240 V}$$

The lighting load would be permitted to be served by 2-wire or 3-wire, 15- or 20-A circuits with combined capacity equal to 44 A or greater for 3-wire circuits or 88 A or greater for 2-wire circuits. The feeder capacity as well as the number of branch-circuit positions available for lighting circuits in the panelboard must reflect the full calculated load of 9000 VA × 1.25 = 11,250 VA.

Show Window
$$6000 \text{ VA} \times 1.25 = 7500 \text{ VA}$$

$$7500 \text{ VA} \div 240 \text{ V} = 31 \text{ A for 3-wire, 120/240 V}$$

The show window lighting is permitted to be served by 2-wire or 3-wire circuits with a capacity equal to 31 A or greater for 3-wire circuits or 62 A or greater for 2-wire circuits.

Receptacles required by 210.62 are assumed to be included in the receptacle load above if these receptacles do not supply the show window lighting load.

Receptacles
Receptacle Load: 14,400 VA ÷ 240 V = 60 A for 3-wire, 120/240 V

The receptacle load would be permitted to be served by 2-wire or 3-wire circuits with a capacity equal to 60 A or greater for 3-wire circuits or 120 A or greater for 2-wire circuits.

Minimum Size Feeder (or Service) Overcurrent Protection *[see 215.3 or 230.90]*

Subtotal noncontinuous loads   12,200 VA
Subtotal continuous load at 125%
  (16,200 VA × 1.25)   20,250 VA
        Total   32,450 VA

$$32,450 \text{ VA} \div 240 \text{ V} = 135 \text{ A}$$

The next higher standard size is 150 A *(see 240.6)*.

Minimum Size Feeders (or Service Conductors) Required *[see 215.2 and 230.42(A)]*

For 120/240-V, 3-wire system,

$$32,450 \text{ VA} \div 240 \text{ V} = 135 \text{ A}$$

Service or feeder conductor is 1/0 Cu in accordance with 310.15(B)(16) (form 0514310.16) and Table 310.16 (with 75°C terminations).

Figure 6–5 *(continued)* (Reprinted with permission from NFPA 70-2011, the *National Electrical Code*®, © 2011, National Fire Protection Association, Quincy, MA 02269. This reprinted material is not the complete and official position of the National Fire Protection Association on the referenced subject, which is represented only by the standard in its entirety.)

Example D8. Motor Circuit Conductors, Overload Protection, and Short-Circuit and Ground-Fault Protection *(see 240.6, 430.6, 430.22, 430.23, 430.24, 430.32, 430.52, and 430.62, Table 430.52 and Table 430.250)*

Determine the minimum required conductor ampacity, the motor overload protection, the branch-circuit short-circuit and ground-fault protection, and the feeder protection, for three induction-type motors on a 480-volt, 3-phase feeder, as follows:

(a) One 25-hp, 460-volt, 3-phase, squirrel-cage motor, nameplate full-load current 32 A, Design B, Service Factor 1.15

(b) Two 30-hp, 460-volt, 3-phase, wound-rotor motors, nameplate primary full-load current 38 A, nameplate secondary full-load current 65 A, 40°C rise.

Conductor Ampacity

The full-load current value used to determine the minimum required conductor ampacity is obtained from Table 430.250 *[see 430.6(A)]* for the squirrel-cage motor and the primary of the wound-rotor motors. To obtain the minimum required conductor ampacity, the full-load current is multiplied by 1.25 *[see 430.22 and 430.23(A)]*.

For the 25-hp motor,

$$34 \text{ A} \times 1.25 = 42.5 \text{ A}$$

For the 30-hp motors,

$$40 \text{ A} \times 1.25 = 50 \text{ A}$$

$$65 \text{ A} \times 1.25 = 81.25 \text{ A}$$

Motor Overload Protection

Where protected by a separate overload device, the motors are required to have overload protection rated or set to trip at not more than 125% of the nameplate full-load current *[see 430.6(A) and 430.32(A)(1)]*.

For the 25-hp motor,

$$32 \text{ A} \times 1.25 = 40.0 \text{ A}$$

For the 30-hp motors,

$$38 \text{ A} \times 1.25 = 47.5 \text{ A}$$

Where the separate overload device is an overload relay (not a fuse or circuit breaker), and the overload device selected at 125% is not sufficient to start the motor or carry the load, the trip setting is permitted to be increased in accordance with 430.32(C).

Branch-Circuit Short-Circuit and Ground-Fault Protection

The selection of the rating of the protective device depends on the type of protective device selected, in accordance with 430.52 and Table 430.52. The following is for the 25-hp motor.

(a) Nontime-Delay Fuse: The fuse rating is 300% × 34 A = 102 A. The next larger standard fuse is 110 A *[see 240.6 and 430.52(C)(1), Exception No. 1]*. If the motor will not start with a 110-A nontime-delay fuse, the fuse rating is permitted to be increased to 125 A because this rating does not exceed 400% *[see 430.52(C)(1), Exception No. 2(a)]*.

(b) Time-Delay Fuse: The fuse rating is 175% × 34 A = 59.5 A. The next larger standard fuse is 60 A *[see 240.6 and 430.52(C)(1), Exception No. 1]*. If the motor will not start with a 60-A time-delay fuse, the fuse rating is permitted to be increased to 70 A because this rating does not exceed 225% *[see 430.52(C)(1), Exception No. 2(b)]*.

Feeder Short-Circuit and Ground-Fault Protection

The rating of the feeder protective device is based on the sum of the largest branch-circuit protective device (example is 110 A) plus the sum of the full-load currents of the other motors, or 110 A + 40 A + 40 A = 190 A. The nearest standard fuse that does not exceed this value is 175 A *[see 240.6 and 430.62(A)]*.

Figure 6–6 Examples of motor calculations. (Reprinted with permission from NFPA 70-2011, the *National Electrical Code*, © 2011, National Fire Protection Association, Quincy, MA 02269. This reprinted material is not the complete and official position of the National Fire Protection Association on the referenced subject, which is represented only by the standard in its entirety.)

FEEDERS

A feeder is defined in Article 100 as the circuit conductor between the service equipment or the source of a separately derived system and the final branch-circuit overcurrent device.

Caution: A feeder generally terminates in more than one overcurrent device.

Many luminaries and other equipment contain supplementary overcurrent devices at the luminaire or equipment. The conductors feeding that equipment are still be considered branch-circuit conductors and not feeder conductors. Feeders are generally found in *NEC®* Articles 215 and 225. Calculations for sizing feeders are found in *NEC®* Article 220 (see Figure 6–5). Other references to feeders may be found in *NEC®* Article 368 for busways, *NEC®* Article 430 for motors (see Figure 6–6), *NEC®* Article 550 for mobile homes, and *NEC®* Article 530 for motion picture studios. See the index for a complete listing of where feeders may be found in the *NEC®*.

Examples for calculating feeders and outside branch circuits are shown in this chapter.

VOLTAGE DROP

The resistance or impedance of conductors may cause a substantial difference between the voltage measured at the service entrance of a facility and the point of use at the equipment utilizing the said voltage. It is necessary to minimize the voltage drop in conductors to ensure good service. Excessive voltage drop impairs the operation of electrical equipment. Lower than rated voltage will result not only in decreased light output from fluorescent lighting but may also cause starting and operating difficulties. The total lumen output for a 2-lamp rapid-stop fluorescent luminaire in general will vary about 1% for each percent change in line voltage. A similar voltage drop of 1% causes about 3½% loss in lumen output for filament-type lamps.

Although motors are generally designed for a 10% overall voltage variation, they operate more efficiently at nameplate voltages. Excessive voltage drop (more than 5% of the line voltage) can cause serious motor overheating. If the overheating is serious enough, it could cause a fire, should the overcurrent or thermal protective devices not operate properly.

Although the *NEC®* does not make mandatory requirements to govern voltage drop generally, it would be a violation of *NEC®* 110.3(B) to supply voltage to equipment at less than the nameplate rating. *NEC®* 695.7 for Fire Pumps contains specific requirements for limiting the voltage drop to 5% of the voltage rating of the motor. *NEC®* 210.19 includes Informational note that includes guides to the allowable voltage drop in branch-circuit conductors; the *NEC®* includes a similar Informational Note for feeder conductors. Adherence to these Informational notes provides a reasonable efficiency of operation. As a matter of information, the *NEC®* states the following:

Conductors for branch circuits as defined in Article 100 size to prevent voltage drop exceeding 3% at the farthest outlet of power, heating, and lighting loads, or combinations of such loads, and where the maximum total voltage drop on both feeders and branch circuits to the farthest outlet does not exceed 5% provide reasonable efficiency of operation. *

This means that the voltage drop should be fairly equally divided between the feeder and the branch-circuit conductors. If the branch-circuit voltage drop is 3% of the rated voltage, then the voltage drop in the feeder should not exceed more than 2%. A voltage drop greater than 5% of line voltage can result in a variation greater than 10% from designed operating voltage, as in the case of a 240-volt motor connected to a 208 branch-circuit having an excessive voltage drop.

Available heat from an electric heating appliance is proportional to the square of the applied voltage. If a heating unit is rated at 240 volts and the terminal voltage is 5% less, or 228 volts, the heat generated at the appliance is reduced by 10%. At a 10% reduction, or 216 volts, the heat generated at the appliance is reduced by nearly 20%.

It is important, therefore, that proper precautions be taken to avoid excessive voltage drop in all types of circuits, whether for lighting, power, or heating. The basic formula for determining voltage drop in a 2-wire, direct current circuit, a 2-wire, single-phase, or a 3-wire, single-phase alternating current circuit with a balanced load at 100% power factor (where reactance is negligible) is:

$$\text{Voltage Drop} = \frac{2 \times R \times L \times I}{\text{Circular Mils}}$$

where: R = Resistivity of conductor material (12 Ω per circular mil foot for copper, 19 Ω per circular mil foot for luminarie); L = one way length of circuit (feet); I = current in one conductor (amperes).

For 3-phase, 4-wire, alternating current circuits, the voltage drop between 1-phase wire and the neutral is ½ of the value derived from this formula. For 3-phase circuits at 100% power factor, the voltage drop between any 2-phase wires is 0.866 × the voltage drop calculated by the formula.

In circuits where each phase and neutral conductor consists of two or more wires, the voltage drop is determined by using the current in only one wire of a phase conductor and the neutral. For loads of other than 100% (unity) power factor, the formula above does not apply. For example, where motors are a major part of a circuit load, the power factor may be considerably less than one. Determining voltage drop in circuits having a load at less than unity power factor involves the use of a more complicated formula.

For 3-phase circuits, the voltage drop may be computed by multiplying the single-phase voltage drop by .86. Should the voltage drop so determined be greater than the suggested desirable percentage of the circuit voltage, a larger size conductor should be used, the circuit length shortened, or the circuit load reduced until the voltage drop is within the required or desired limits.

QUESTION REVIEW

Each lesson is purposely designed to require the student to apply the entire *NEC®* text and not specific chapters or articles. It has been found that when studying for a timed, open-book examination, the student must gain proficiency in the Table of Contents, the Index, and the ability to move quickly from cover to cover to find the correct answer to each question in a timely fashion.

1. A 240/480 3-phase panelboard supplies only one 15,000 VA, 480-volt, 3-phase balanced resistive load. Each ungrounded conductor in the subfeeder to this power panel has a total net computed load of _____ amperes.

   Answer: _____

   _____

   _____

   _____

   Reference: _____

2. Fluorescent luminaries, each containing two ballasts rated 0.8 A each at 120 volts, are to be installed for general lighting in a store. The overcurrent protection devices are not listed for continuous operation at 100% of its rating. What is the maximum number of these luminaries that may be permanently wired to a 20-A, 120-volt branch-circuit?

   Answer: _____

   _____

   _____

   _____

   Reference: _____

The following drawing and the drawing at the top of the next page apply to Questions 3 through 22.

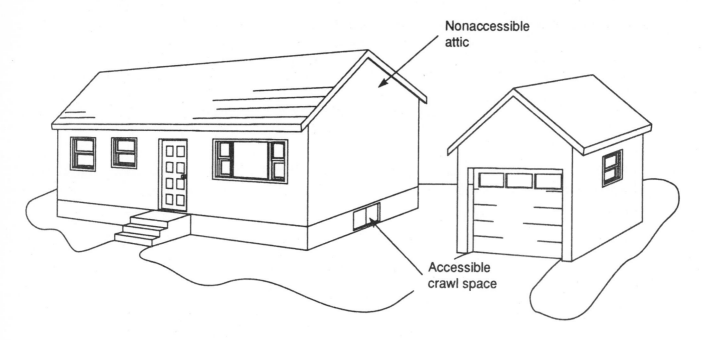

Example for Questions 3 through 22: You are going to build a single-family dwelling 28 ft by 40 ft (8.5 m by 12 m). There will be three bedrooms and one bathroom with central natural gas furnace located under the floor in the crawl space. There will be no air conditioning. There will be an electric water heater (40-gallon, quick recovery) with two 4500-W elements, a gas cooking range, and electric dryer. There will be a detached single-car garage. The electrical installation will be made in accordance with the 2011 *NEC*®. The attic in this dwelling is not suitable for the location of equipment or for storage.

3. How many square feet (square meters) will this house contain?

   Answer: _____

   _____

   _____

   _____

   Reference: _____

4. How many 15-A general lighting and receptacle branch circuits are required?

   Answer: _____

   _____

   _____

   _____

   Reference: _____

6. How many small appliance branch circuits are required?

Answer: _____

_____

_____

Reference: _____

7. Is a special circuit required for the laundry?

Answer: _____

_____

_____

Reference: _____

8. What is the maximum standard size overcurrent device permitted to protect the electric water heater?

Answer: _____

_____

_____

Reference: _____

9. How many branch circuits are required to supply the detached garage?

Answer: _____

_____

_____

Reference: _____

10. How many outdoor receptacles are required?

Answer: _____

_____

_____

Reference: _____

11. What is the minimum number of ground-fault circuit interrupter (GFCI) devices required?

    Answer: _____

    _____

    _____

    _____

    Reference: _____

12. Are lighting outlets required at each outside door?

    Answer: _____

    _____

    _____

    _____

    Reference: _____

13. Are lighting outlets required in the attic not used for storage?

    Answer: _____

    _____

    _____

    _____

    Reference: _____

14. Is a receptacle required under the floor containing equipment requiring servicing?

    Answer: _____

    _____

    _____

    _____

    Reference: _____

15. If the answer to Question 7 is yes, what size branch circuit would be required?

    Answer: _____

    _____

    _____

    _____

    Reference: _____

16. Would it be permissible to connect the furnace to a general lighting and receptacle branch circuit, provided it does not overload the circuit?

    Answer: _____

    _____

    _____

    _____

    Reference: _____

17. Are lighting outlets required in each closet?

    Answer: _____

    _____

    _____

    _____

    Reference: _____

18. Is it permissible to connect the electric ignition system for gas ranges to one of the required small-appliance branch circuits?

    Answer: _____

    _____

    _____

    _____

    Reference: _____

19. If an outdoor receptacle is located 6 feet (1.8 m) above the ground level, is it required to be GFCI protected?

    Answer: _____

    _____

    _____

    _____

    Reference: _____

20. Could the outdoor receptacle described in Question 19 serve as one of the required outdoor receptacle?

    Answer: _____

    _____

    _____

    _____

    Reference: _____

21. Which areas in the drawing would not require the branch circuits supplying single-phase, 15- and 20-A outlets to be protected by arc-fault circuit protection?

Answer: _____

_____

_____

_____

Reference: _____

22. What is the minimum size service required for this dwelling? (Calculate answers where required.)

   (a) General purpose lighting and receptacle load
   (b) Small appliance branch-circuit load
   (c) Water heater
   (d) Laundry
   (e) Dryer
   (f) Detached garage
   (g) Furnace
   (h) Outdoor receptacle
   (i) Air conditioning
   (j) Range
   (k) Bathroom

   (a) _____

   (b) _____

   (c) _____

   (d) _____

   (e) _____

   (f) _____

   (g) _____

   (h) _____

   (i) _____

   (j) _____

   (k) _____

23. A branch circuit supplies a single, continuous-duty pump motor for a residential water supply. The motor-circuit conductors must have an ampacity of at least _____ percent of the motor full-load current rating.

Answer: _____

_____

_____

_____

Reference: _____

24. The minimum lighting load required for the general lighting (only) of a church building having outside dimensions of 100 ft × 200 ft (30.5 m × 61 m) is _____ volt-amps.

    Answer: _____

    _____

    _____

    _____

    Reference: _____

25. For the kitchen small-appliance load, a dwelling unit requires at least _____ circuit(s).

    Answer: _____

    _____

    _____

    _____

    Reference: _____

26. To lower the voltage drop in a wire, reduce the conductors' length or increase their _____.

    Answer: _____

    _____

    _____

    _____

    Reference: _____

CHAPTER 7

OBJECTIVES

Studying this chapter along with the 2011 *NEC®* gives the reader a basic introduction to the service requirements in the *Code* that apply to all installations for above and below 600 volts.

After studying this chapter, you should know:

- The definitions and the application of terms that are applicable and are unique to the 2011 *NEC®*

- How to find the many *Code* requirements in which services are addressed as they apply to the *NEC®*

- Service circuit sizing

- Service conductor sizing

- Service overcurrent sizing

- Service raceway sizing

- How to find the *Code* requirements in which services above 600 volts are addressed as they apply to the *NEC®*

SERVICES 600 VOLTS OR LESS

NEC® Article 230 covers the requirements for services, service conductors, and equipment for the control and protection of services and their installation requirements. Parts I through VII cover services, 600 volts, nominal, or less; Part VIII has additional requirements for services exceeding 600 volts, nominal. It should be remembered that all of Article 230 applies to these services, and Part VIII contains provisions that supplement or modify the rest of Article 230. Part VIII does not apply to the equipment on the supply side of the service point. Clearances for conductors over 600 volts are found in ANSI C2, 2007, the *National Electrical Safety Code. NEC®* 230.2 generally limits a building or structure to one service. There are several special conditions that permit more than one service per building or structure. It is necessary to comply with these special conditions any time two or more services are being installed on a building or structure. Confusing to many is the permissiveness for more than one set of service conductors permitted by the Exceptions of 230.40. These exceptions are frequently used on strip shopping centers, condominium complexes, and apartment buildings. The rule satisfies the needs for individual supply, control, and metering of occupancies in these multioccupancy buildings.

However, it does not relieve the requirement of 230.2 for one service for each building or structure (see Figure 7–1 and Figure 7–2). The service conductors, as defined in Article 100, are required to go directly to the disconnecting means of that building or structure because the disconnecting means is required to be installed at a readily accessible location either outside of the building or structure or inside nearest the point of the entrance of the service conductors. *NEC®* 230.6 states that conductors are considered outside of a building or other structure where installed under not less than 2 in (50 mm) of concrete beneath a building or structure, where installed within a building or structure in a raceway that is encased in concrete or brick not less than 2 in (50 mm) thick, or where installed in a vault that complies with the requirements of *NEC®* Article 450, Part III. A typical service supplying a building or structure contains several carefully defined parts. These parts are defined in Article 100 as service, service cable, service conductors, service drop, service-entrance conductors, overhead service, underground service conductors, service lateral, service equipment, and service point. Each service will contain several of these parts—however, not all these parts. It is important that the components being installed in accordance with the *NEC®* be correctly

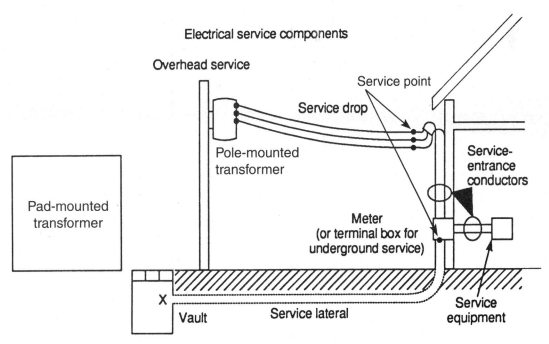

Figure 7–1 *NEC®* 230.2 permits only one service to each building. There are several conditions to this rule where more than one service is permitted. The service may be supplied with aerial spans overhead or by underground "service laterals" from the "service point." (See Article 100—Definitions.) *Note:* A vault is not required for underground runs but may be installed where necessary.

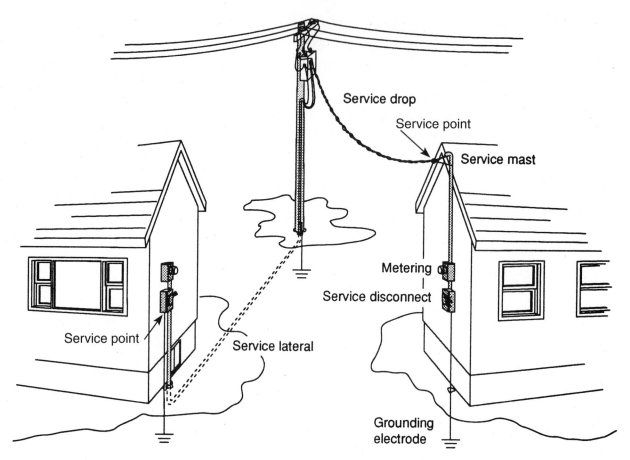

Figure 7–2 Illustration of two common types of residential services: overhead service drop and underground service lateral.

defined before sizing, calculating, or installing. For instance, a typical residential service can be supplied by the serving utility either overhead or underground. When supplied overhead, it shall be supplied as a service drop. When supplied underground, it shall be supplied as a service lateral. These terms should be studied carefully in *NEC®* Article 100 so that calculations and design can be made correctly. The diagram in 230.1 should help the reader identify and use these terms correctly. (See Figure 7–3.)

Part II of Article 230 covers the overhead service drop conductors, size and rating, and clearances required for making this installation. These conductors are generally installed by the serving utility; however, it is the responsibility of the installer to locate the service mast so that these clearances will be met. Careful and close coordination between the serving utility, the customer to be served, and the installer is necessary to ensure compliance with this section of the *NEC®*. (Refer to Figure 6–5 for clearance examples.)

Part III of Article 230 covers underground service-laterals (see Figure 7–4). Care should be taken so that the service lateral is installed correctly and so that where it emerges from the ground is amply protected by rigid or IMC steel conduit or Schedule 80 PVC to provide these conductors with physical protection against damage. Article 230, Part IV, covers the requirements for the service entrance conductors. These service entrance conductors, as defined in *NEC®* Article 100, must be installed in one of the wiring methods listed in 230.43 and sized and rated in accordance with 230.42. Specific requirements for these service entrance conductors are covered in Article 230 Part IV, and must be complied with regardless of how they are routed. The general requirements for service equipment can be found in Article 230, Part V, and the disconnecting means is found in Part VI.

Example: Where a trade size 3 (78) steel intermediate metal conduit (IMC) contains 120/208 3-phase service conductors consisting of eight size 3/0 AWG

Article 230 — Services

General . . . . **Part I**
Overhead service drop conductors . . . . **Part II**
Underground service-lateral conductors . . . . **Part III**
Service-entrance conductors . . . . **Part IV**
Service equipment – General . . . . **Part V**
Service equipment – Disconnecting means . . . . **Part VI**
Service equipment – Overcurrent protection . . . . **Part VII**
Services over 600 volts, nominal . . . . **Part VIII**

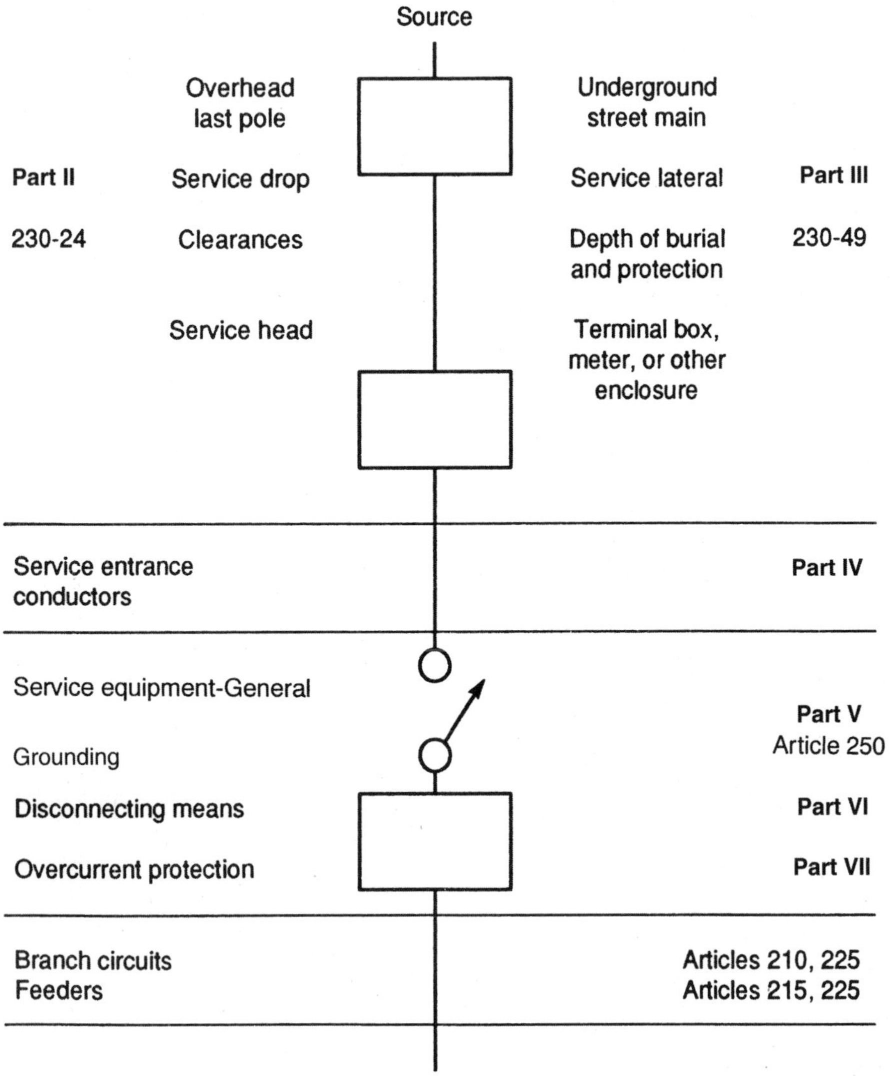

Figure 7–3 Article 230 is divided into several parts. Each covers a specific part of service installation. (Reprinted with permission from NFPA 70-2011, the *National Electrical Code®*, © 2011, National Fire Protection Association, Quincy, MA 02269. This reprinted material is not the complete and official position of the National Fire Protection Association on the referenced subject, which is represented only by the standard in its entirety.)

THWN copper conductors, each phase consists of two conductors per phase, and the full-size neutral (grounded) conductor consists of two conductors terminating in a 400-A service-disconnecting means consisting of an inverse time circuit breaker. The data processing, fluorescent lighting, and air-conditioning load is evenly divided. Does this installation meet *Code*?

Electrical service components

Overhead service*

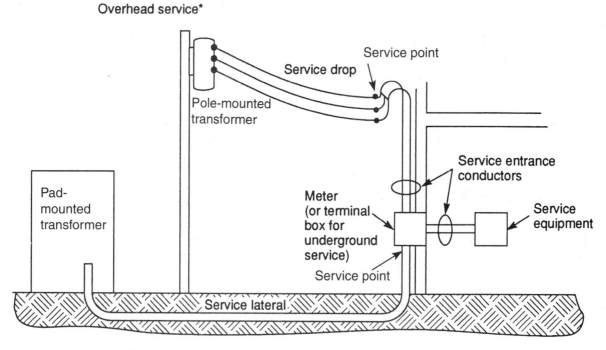

Figure 7–4 *NEC*® 230.2 permits only one service per building. It is permissible to supply overhead or underground.

Step 1. Chapter 9, Table 1, permits a 40% fill of this conduit. 3/0 AWG THWN conductors; Annex C, Table C4A, would allow eight 3/0 AWG THWN conductors in a trade size 2½ (63) conduit. Therefore, the trade size 3 (78) IMC conduit is acceptable.

> *Note:* Due to the change in the tables in Chapter 9 and the addition of Annex C in the *NEC*®, each type of raceway must be calculated individually. For example, in this problem, if PVC conduit Schedule 40 were used, a trade size 3 (78) would be required. If more than one size conductor was contained within the raceway, calculations would be necessary using Chapter 9, Tables 1, 4, and 5, for the number of conductors, the specific type and size, and the specific raceway type.

Step 2. The ampacity of the 3/0 AWG THWN conductors is 200 amperes each based on Table 310.15(B)(16) (summary 310.16). But Table 310.15(B)(3)(a) requires an ampacity

adjustment to 70% where eight conductors are in one raceway. Therefore, 200 A × 2 = 400 × 70% = 280-A ampacity. *NEC*® 230.90(A), Exception 2 references 240.4(C) and 240.6, which permits overload protection of the conductors at the next higher standard size overcurrent device in accordance with 240.6. Thus, the largest size overcurrent device permitted would be a 300-A device.

Step 3. Assuming the calculated load required the 400-A device, you would be required to either increase the conductor size, which would indubitably require a larger conduit or two conduits in parallel with conductors sized to carry the calculated load after the ampacity adjustment factors have been applied.

Step 4. 400-ampere calculated load would require a trade size 3½ (91) IMC and a trade size 4 (103) PVC Schedule 40 conduit to contain the required eight 350-kcmil type THWN conductors.

Answer: Table 310.15(B)(16) and Table 310.15(B)(3)(a): 310 × 2 = 620 × 70% = 434.

Each service disconnecting means is required to be suitable for the prevailing equipment conditions and when installed in hazardous locations must comply with Chapter 5 of the *NEC*®. Each service disconnecting means permitted by 230.2, or each set of service entrance conductors permitted by 230.40 Exception 1. (see Figure 7–5) shall consist of not more than six switches or six circuit breakers mounted in a single enclosure, in a group of separate enclosures, or on a switchboard. (See Figure 7–6.) There shall be no more than six disconnects per service grouped in any one location. These requirements are critical and must be followed. The disconnecting means must be grouped; each disconnect must be marked to indicate the load served. An exception permits one of the six disconnecting means permitted where used for a water pump intended to provide fire protection to be located remotely from the other disconnecting means. In multiple-occupancy buildings, each occupant shall have access unless local building management, which supplies continuous supervision, is present. In such a case, the service disconnecting means shall be permitted to be accessible to the authorized management personnel only. Each service disconnecting means shall simultaneously disconnect the ungrounded service conductors from the premise wiring. Where the service disconnecting means does not disconnect the grounded conductor from the premise wiring, other means shall be provided for this purpose in the service equipment. A terminal or bus to which all grounded conductors can be attached by means of pressure connections shall be permitted for this purpose.

Part VII covers the overcurrent protection and the location of the overcurrent protection requirements for services. 230.95 states that all solidly grounded wye services of more than 150 volts to ground but not exceeding 600 volts phase-to-phase rated 1000 A or more must be provided with ground-fault protection of equipment. The exception to this states that ground-fault protection does not apply to the service disconnecting means of a continuous industrial process where a non-orderly shutdown will introduce additional or increased hazard. Specific settings and testing procedures for this ground-fault protection of equipment are covered in 230.95. Additional ground-fault protection of feeders where ground-fault protection has not been provided on the service is covered in 215.10 and 240.13.

Example: Where a rigid metal conduit contains eight (8) 3/0 AWG THWN insulated copper conductors with two of the 3/0 AWG conductors being in parallel for each phase and two 3/0 AWG conductors being in parallel for the neutral,(a) would this installation be permitted to be terminated into a 400-A main service breaker where the system voltage is 120/208? (The expected loads are evenly divided between fluorescent lighting, data processing equipment, and air-conditioning equipment.) (b) If not, what would be the maximum ampere rating for this service?

Answer: Eight (8) 3/0 AWG THWN conductors in a single conduit, Annex C, Table C8, would require a trade size 3 (78) rigid metal conduit or tubing.

Example: A 277/480-volt, 3-phase service to a building is rated at 1600 A. The required main disconnecting means consists of two 800-A fused disconnects. Is this service required to have ground-fault protection of equipment (GFPE)?

Index "Ground-Fault Protection," *Ground-fault protection for service disconnects, 230.95.*

> *Note:* To answer this question, some basic electrical knowledge is required. First, the voltage tells the story. 277/480V indicates it is first a solidly grounded system. It also indicates it is a wye system because 480V equals 1.73 of 277V. This information can be obtained from additional reference books listed as study material.

Answer: No. Reference: 230.95 states that solidly grounded wye services of more than 150 volts to ground not exceeding 600 volts phase-to-phase and over 1000 A requires ground-fault protection.

Table 310.15(B)(16)—3/0 AWG THWN has an ampacity of 200 A (from 75°C column). Where the number of current-carrying conductors exceeds three, the allowable ampacity shall be reduced as shown in the table with seven to nine conductors, an ampacity adjustment of 70 % shall be applied. See Table 310.15(B)(3)(a).

Two conductors per phase:

$$2 \times 200 \times 70\% = 280 \text{ A}$$

The next higher overcurrent device per 240.6(A) would be 300 A The service would be a 300-A service.

The 400-A overcurrent device would require 2-350 kcmil per phase.

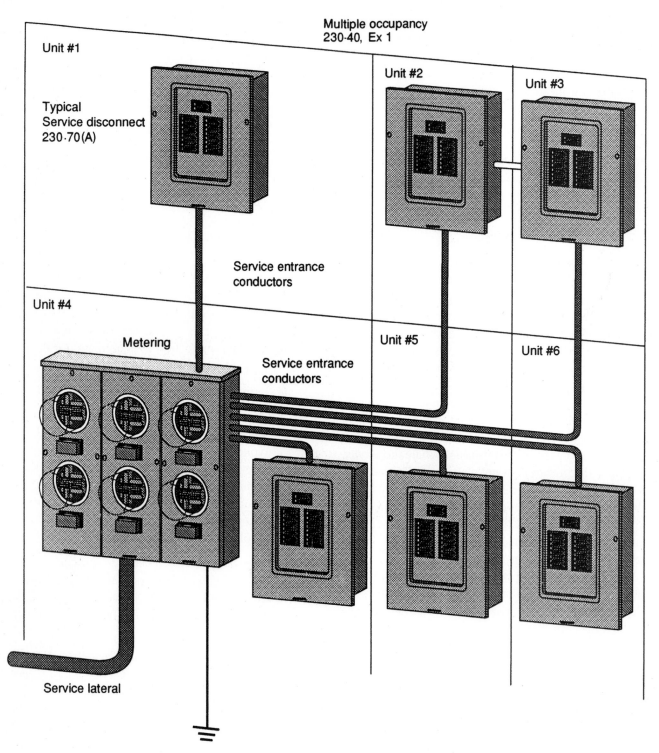

Figure 7–5 Typical riser sketch of a multiple occupancy building. *NEC*® 230.40, Exception 1, permits the main disconnecting means for that occupancy to be located on the load end of the feeder within the occupancy.

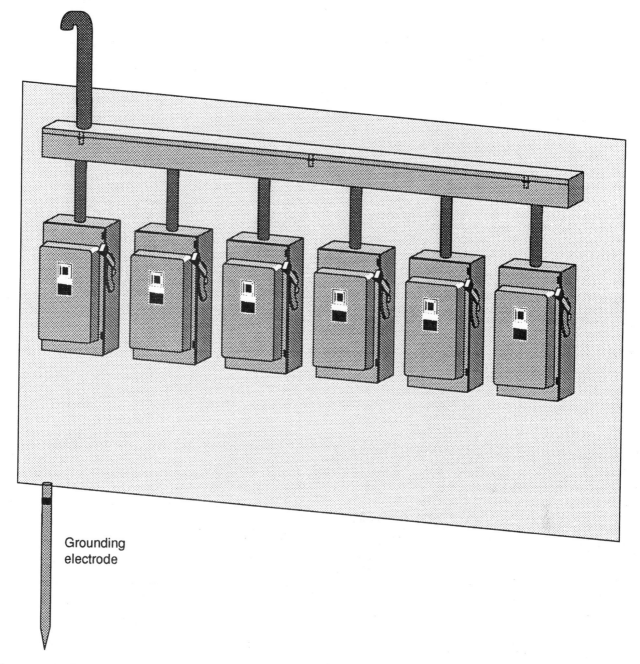

Grounding
electrode

Figure 7–6 Typical sketch of six individual disconnecting means on one service as permitted by *NEC®* 230.71(A).

SERVICES OVER 600 VOLTS

Services over 600 volts nominal introduce an additional level of hazard. Article 230, Part VIII, covers the requirements for services exceeding 600 volts nominal, which modify or amend the rest of Article 230. All of Article 230 is applicable and, in addition, Part VIII must be followed. Clearance requirements for over-600-volts services can be governed by ANSI C2, 2007, the *National Electrical Safety Code*, as noted in the informational note in 230.200. The service entrance conductors to buildings or enclosures shall not be smaller than 6 AWG unless in cable, and in cable not smaller than 8 AWG, and installed by the specific methods listed in 230.202(B). It should be noted that Article 300, Part II and Article 110, Part II, will also need to be regularly referenced in making installations over 600 volts. The requirements for support guarding and draining cables are covered in this section. Warning signs with the words "Danger-High Voltage-Keep Out." shall be posted where unauthorized people come in contact with energized parts 110,31(c). The disconnecting means must comply with 230.70 or 230.208(B) and shall simultaneously disconnect all underground conductors and shall have a fault closing rating of not less than the maximum short circuit available in supply terminals, except where used switches or separate mining fuses are installed, and the fuse characteristic shall be permitted to contribute to the fault closing rating of the disconnecting means. These requirements for over-600-volts services create the need for your studies to take you into new areas of the *Code*, such as Part II of Article 110, which covers the general requirements for over 600 volts nominal; Part II of Article 100, which gives you the unique definitions for those circuits; Part IX of Article 240; Part XI of Article 250 for grounding of system; Part II of Article 300; and the requirements of Article 490. You must be aware that these additional requirements amend or augment the primary requirements in each article. In most instances, those requirements also apply to over-600-volts nominal services. Therefore, as you study this section, familiarize yourself with these new parts as they apply to your studies.

QUESTION REVIEW

Each lesson is purposely designed to require the student to apply the entire *NEC*® text and not specific chapters or articles. It has been found that when studying for a timed, open-book examination, the student must gain proficiency in the Table of Contents, the Index, and the ability to move quickly from cover to cover to find the correct answer to each question in a timely fashion.

1. Where a circuit breaker is provided as the short-circuit protective device for service entrance conductors exceeding 600 volts, the breaker shall have a trip setting of not more than _____ times the ampacity of the conductors.

   Answer: _____

   _____

   _____

   _____

   Reference: _____

2. An electrical service transmission line drops the voltage from 255 volts to 240 volts. The service transmission line efficiency is _____ %.

   Answer: _____

   _____

   _____

   _____

   Reference: _____

3. Where a group of buildings are served by electrical feeders from a single service drop or electrical service point in another building, is a disconnecting means required at each of the other buildings where the feeders terminate or can the feeders terminate in main lug-only panels?

   Answer: _____

   _____

   _____

   _____

   Reference: _____

4. Is there a limit to the number of overcurrent devices that can be used on the secondary side of a transformer if the total of the device ratings does not exceed the allowed value for a single secondary overcurrent device?

   Answer: _____

   _____

   _____

   _____

   Reference: _____

5. Does it make a difference which opening in the service head the service entrance conductors are brought out of? If so, why?

Answer: _____

_____

_____

_____

Reference: _____

6. If the clearances required by *NEC*® 110.26 can be maintained in front of a panelboard or a service disconnect to be installed in a bathroom in a dwelling, would this installation be permitted by the *NEC*®?

Answer: _____

_____

_____

_____

Reference: _____

7. When service entrance cable is used within a dwelling for branch-circuit wiring, is the Type SE cable considered the same as Type NM non–metallic-sheathed cable?

Answer: _____

_____

_____

_____

Reference: _____

8. Can a 120/240-volt, single-phase, 3-wire overhead service be located closer than 3 feet (900 mm) to a window of a dwelling?

Answer: _____

_____

_____

_____

Reference: _____

9. The *Code* generally prohibits splicing a grounding-electrode conductor.
   (a) Is this conductor permitted to be tapped?
   (b) If so, how are these taps sized?

Answer: _____

_____

_____

_____

Reference: _____

10. Circuit breakers are required to open all ungrounded conductors of a multiwire circuit simultaneously. Obviously, fuses cannot be required to open all ungrounded conductors of a circuit simultaneously. Why the difference in the requirements?

Answer: _____

_____

_____

_____

Reference: _____

11. What is a separately derived system? Where is it mentioned in the *Code*?

Answer: _____

_____

_____

_____

Reference: _____

12. Where secondary protection is required many overcurrent devices are permitted to be used on the secondary side of a transformer if the total of the device ratings does not exceed the allowed value for a single overcurrent device?

Answer: _____

_____

_____

_____

Reference: _____

13. Is it necessary to identify the higher voltage-to-ground phase (hi-leg) at the disconnect of a 3-phase motor where the service available is 120/240-volt, 3-phase, 4-wire?

Answer: _____

_____

_____

_____

Reference: _____

14. A service panel is securely bolted to the metal frame of a building and the panelboard's main bonding jumper is properly installed. Is this sufficient to ground the panelboard to the metal building grounding electrode as required in *NEC*® 250.64?

Answer: _____

_____

_____

_____

Reference: _____

15. What is the maximum permitted contraction or expansion for a run of Schedule 40 PVC conduit before the installation of an expansion coupling is required?

    Answer: _____

    _____

    _____

    _____

    Reference: _____

16. How can one distinguish between the mandatory rules of the *NEC*® and the explanatory rules of the *NEC*®?

    Answer: _____

    _____

    _____

    _____

    Reference: _____

17. On a recent installation, the plans called for the building steel to be used as a grounding electrode. Where is it permitted to be used in the *NEC*®?

    Answer: _____

    _____

    _____

    _____

    Reference: _____

18. Which article in the *NEC*® covers type non–metallic-sheathed cable?

    Answer: _____

    _____

    _____

    _____

    Reference: _____

19. What is the maximum size flexible metallic tubing permitted?

    Answer: _____

    _____

    _____

    _____

    Reference: _____

20. Which article in the *NEC®* contains the requirements for fusible safety switches and other type snap switches?

    Answer: _____

    _____

    _____

    _____

    Reference: _____

21. In making an installation in a residential garage, you have determined that physical protection is needed for the non–metallic-sheathed cable and that it must be installed in metal conduit or tubing. How do you size the conduit for this non–metallic-sheathed cable?

    Answer: _____

    _____

    _____

    _____

    Reference: _____

22. In making an installation for a large parking garage adjacent to a hotel building, which article in the *NEC®* covers these wiring methods?

    Answer: _____

    _____

    _____

    _____

    Reference: _____

23. You have encountered a nonmetallic wireway to be installed in a wet location. Which article in the *NEC®* covers this nonmetallic wireway?

    Answer: _____

    _____

    _____

    _____

    Reference: _____

24. In attempting to calculate the load on a small office building, the receptacles are to be calculated at how many VA each?

Answer: _____

_____

_____

_____

Reference: _____

25. In wiring a new residence, the most convenient location for the panelboard is in a bathroom wall. Is this an acceptable location?

Answer: _____

_____

_____

_____

Reference: _____

CHAPTER 8

OBJECTIVES

Studying this chapter along with the 2011 *NEC®* gives the reader a basic introduction to the overcurrent requirements in the *Code* that apply to all installations for above and below 600 volts.

After studying this chapter, you should know:

- The definitions and the application of terms that are applicable and are unique to the 2011 *NEC®*

- The many *Code* articles in which overcurrent protection is addressed as it applies to the *NEC®*

- Overcurrent service sizing

- Fuse applications and types

- Circuit breaker types and applications

- Short circuit/fault current calculations

OVERCURRENT PROTECTION

NEC® Article 240 provides the general requirements for overcurrent protection and overcurrent protective devices not more than 600 volts, nominal in part I-VIII. Part IX of *NEC®* Article 240 covers the overcurrent protection for over 600 volts, nominal. Overcurrent protection for conductors and equipment is provided to open the circuit if the current reaches a value that will cause an excessive or dangerous temperature in conductors or conductor insulation. *NEC®* 110.9 and 110.10 cover the requirements for the interrupting capacity and protection against fault currents (see Figure 8–1). Article 240 covers the general requirements. The specific overcurrent requirements for equipment can be found in each individual article as it pertains to that specific equipment or circuitry; for example, the overcurrent protection for air-conditioning and refrigeration equipment is covered in Article 440 and appliances in Article 422. A listing of the specific overcurrent requirements can be found in 240.3. *NEC®* 240.4 outlines the protection of conductors. This section provides the rules for the many different applications throughout the *Code. NEC®* 240.5 covers the requirements for protecting flexible cords and fixture wires, and standard overcurrent devices are listed in *NEC®* 240.6 for fuses and fixed-trip circuit breakers. See Figure 8–2 and Figure 8–3 for examples. Adjustable-trip circuit breakers are described and covered in 240.6(B). Part II of Article 240 specifies the location that the overcurrent device must be placed in the circuit. Part III specifies the enclosure required for enclosing the overcurrent devices. Part IV covers the disconnecting requirements for disconnecting

Figure 8–2 Types IFL, IKL, and ILL "1 LIMITER" molded case current-limiting circuit breakers. (*Courtesy of Square D Company*)

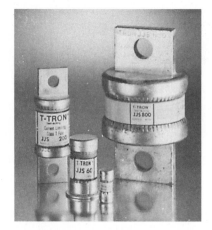

Figure 8–3 Class T current-limiting, fast-acting fuse; 200,000-ampere interrupting rating, 300 and 600 volts. Has little time delay. Generally used for protection of circuit-breaker panels and for circuits that do not have high-inrush loads, such as motors. When used on high-inrush loads, generally size at 300% so as to be able to override the momentary inrush current. These fuses have different dimensions compared with ordinary fuses. They will not fit into switches made for other classes of fuses nor will other classes of fuses fit into a Class T disconnect switch. (*Courtesy of Cooper Bussmann*)

Figure 8–1 Panelboard. (*Courtesy of Square D Company*)

the overcurrent devices so that they are accessible to the maintenance personnel performing work on those devices. Parts V and VI cover fuses, fuseholders, and adapters for both plug-and cartridge-type fuses; circuit breakers are covered in Part VII. Article 240, Part VIII, covers supervised industrial installations. Part IX covers the overcurrent protection for over 600 volts, nominal, and states that feeders shall have short-circuit protection devices in each ungrounded conductor or comply with 230.208(A) and (B). The protective device(s) shall be capable of detecting and interrupting all values of current that can occur at their location in excess of their trip setting or melting point. In no case shall the fuse rating exceed three times the long-time trip element setting of the breaker, or six times the ampacity of the conductor. Branch-circuit requirements for over-600 volts circuits are covered in 240.100.

CAPABLE OF CARRYING THE AVAILABLE FAULT CURRENT
Short-Circuit Calculations

Throughout the *NEC®* as well as in industry standards, you will find references to fault currents, short-circuit currents, ground-fault currents, interrupting ratings, impedances of transformers, and the like.

NEC® 110.9 states that *equipment intended to interrupt current at fault levels shall have an interrupting rating nor less than the nominal circuit voltage and the current that is available at the line terminals of the equipment.**

NEC® 110.10 states that *the overcurrent protective devices, the total impedance, the equipment short-circuit ratings, and other characteristics of the circuit to be protected shall be selected and coordinated to permit the circuit-protective devices used to clear a fault to do so without extensive damage to the electrical equipments of the circuit.**

These tough code requirements mean that we need to know the available fault currents at the main service equipment as well as at all points in the system. Too often, engineers, electricians, and contractors look for adequate interrupting ratings of the fuses or breakers in the main service equipment only. This is wrong. The *Code* tells us to observe available fault currents throughout the electrical system. This means panelboards, motor controllers, motor control centers, distribution centers, and so on.

*Reprinted with permission from NFPA 70-2011.

Determining Short-Circuit Current at Various Distances from Transformers, Switchboards, Panelboards, and Load Centers Using the Point-to-Point Method

A relatively simple method of determining the available short-circuit current (also called fault current) at various distances from a given location is the "**point-to-point method**." Reasonable accuracy is obtained when this method is used with 3-phase and single-phase systems.

The following procedure demonstrates the use of the "**point-to-point method**."

Step 1. Determine the full-load rating of the transformer—in amperes—from the transformer nameplate, tables, or the following formulas:

a) For 3-phase transformers:

$$I_{FLA} = \frac{kVA \times 1000}{E_{L-L} \times 1.73}$$

where E_{L-L} = Line-to-line voltage

b) For single-phase transformers:

$$I_{FLA} = \frac{kVA \times 1000}{E_{L-L}}$$

Step 2. Find the percent impedance (Z) on the nameplate of the transformer.

Step 3. Find the transformer multiplier "M$_1$":

$$M1 = \frac{100}{transformer\ \%\ impedance(Z) \times 0.9}$$

(*Note:* Because the marked transformer impedance can vary ±10% per the UL standard, the 0.9 factor takes this into consideration to show "worst case" conditions.)

Step 4. Determine the transformer let-through short-circuit current at the secondary terminals of transformer. Use tables or the following formula:

a) For 3-phase transformers (L-L-L):

I_{SCA} = transformer$_{FLA}$ × multiplier "M$_1$"

b) For 1-phase transformers (L-L):

I_{SCA} = transformer$_{FLA}$ × multiplier "M$_1$"

c) For 1-phase transformers (L-N):

I_{SCA} = transformer$_{FLA}$ × multiplier "M$_1$" × 1.5

(*Note:* The 1.5 factor is explained in Step 5 in the paragraph marked "X.")

Step 5. Determine the "f" factor:

a) For 3-phase faults:

$$f = \frac{1.73 \times L \times I}{N \times C \times E_{L-L}}$$

b) For 1-phase, line-to-line (L-L) faults on 1-phase, center-tapped transformers:

$$f = \frac{2 \times L \times I}{N \times C \times E_{L-L}}$$

c) For 1-phase, line-to-neutral (L-N) faults on 1-phase, center-tapped transformers:

$$f = \frac{2 \times L \times I}{N \times C \times E_{L-N}}$$

where

$L =$ the length of the circuit to the fault, in feet

$I =$ the available fault-current—in amperes—at the beginning of the circuit.

$C =$ the constant derived from the "C" tables (see tables on pages 147–148) for the specific type of conductors and wiring method.

$E =$ the voltage, line-to-line or line-to-neutral. See Step 4a, b, and c to decide which voltage to use.

$N =$ the number of conductors in parallel.

The L-N fault current is higher than the L-L fault current at the secondary terminals of a single-phase, center-tapped transformer. At some distance from the terminals, depending on the wire size, the L-N fault current is lower than the L-L fault current. This can vary from 1.33 to 1.67 times. These figures are based on the change in the turns ratio between primary and secondary, infinite-source impedance, a distance of zero feet from the terminals of the transformer, and 1.2% resistance (X) and 1.5% resistance (R) for the L-N versus L-L resistance and reactance values.

For simplicity, in Step 4c, we used an approximate multiplier of 1.5. First, do the L-N calculation at the transformer secondary terminals, Step 4c, then proceed with the point-to-point method.

Step 6. After finding the "f" factor, refer to Chart M (see page 148) for the appropriate value of the multiplier "M_2" for the specific "f" value. Or calculate as follows:

$$M_2 = \frac{1}{1+f}$$

Step 7. Multiply the available fault current at the beginning of the circuit by the multiplier "M_2" to determine the available symmetrical fault current at the fault.

I_{SCA} at fault $= I_{SCA}$ at beginning of circuit $\times$ "M_2"

Example: Provide the fuse ratings for a 10-horsepower, 230-volt, 3-phase motor with a code letter G and a service factor of 1.15. Using *Table 430.250*, the full load current is 28 A. [See *430.6(A)(1)*.]

(A) Using *Table 430.52*, the maximum rating of a time-delay fuse is 175% of the motor full-load current. The fuse size, using time-delay fuses for the branch-circuit, short-circuit, and ground-fault protection, is calculated as follows: 28 × 1.75 = 49 A. *NEC® 430.52, Exception No. 1*, permits that where the values for branch-circuit, short-circuit, and ground-fault protective devices determined by *Table 430.52* do not correspond to the standard sizes of fuses, a higher size that does not exceed the next higher standard ampere rating shall be permitted. The next higher fuse rating as shown in *240.6* is 50 A. *NEC® 430.52(C)(1), Exception No. 2(b)*, permits that where this rating is not sufficient for the starting current of the motor, the rating of the time-delay fuse shall be permitted to be increased but shall in no case exceed 225% of the full-load current.

(B) Motor overload protection based on the motor service factor of 1.15% is calculated in accordance with *430.32* as follows: 28 × 1.25 = 35 A.

(C) Using non–time-delay fuses, sizes for branch-circuit, short-circuit, and ground-fault protective devices are calculated as follows: 28 × 300% = 84 A The next higher rating as permitted by *430.52(C)(1), Exception No. 2(a)*, is 90 amperes. Where this rating is not sufficient for the starting current of the motor, the rating of the non–time-delay fuse shall be permitted to be increased but shall in no case exceed 400% of the full-load current.

"C" VALUES—Copper Conductors

AWG or kcmil	Three Single Conductors						Three Conductor Cable					
	Steel Conduit			Nonmagnetic Conduit			Steel Conduit			Nonmagnetic Conduit		
	600 V	5kV	15kV	600 V	5kV	15kV	600 V	5kV	15kV	600 V	5kV	15kV
14	389	—	—	389	—	—	389	—	—	389	—	—
12	617	—	—	617	—	—	617	—	—	617	—	—
10	981	—	—	981	—	—	981	—	—	981	—	—
8	1557	1551	1557	1556	1555	1558	1559	1557	1559	1559	1558	1559
6	2425	2406	2389	2430	2417	2406	2431	2424	2414	2433	2428	2420
4	3806	3750	3695	3825	3789	3752	3830	3811	3778	3837	3823	3798
3	4760	4760	4760	4802	4802	4802	4760	4790	4760	4802	4802	4802
2	5906	5736	5574	6044	5926	5809	5989	5929	5827	6087	6022	5957
1	7292	7029	6758	7493	7306	7108	7454	7364	7188	7579	7507	7364
1/0	8924	8543	7973	9317	9033	8590	9209	9086	8707	9472	9372	9052
2/0	10755	10061	9389	11423	10877	10318	11244	11045	10500	11703	11528	11052
3/0	12843	11804	11021	13923	13048	12360	13656	13333	12613	14410	14118	13461
4/0	15082	13605	12542	16673	15351	14347	16391	15890	14813	17482	17019	16012
250	16483	14924	13643	18593	17120	15865	18310	17850	16465	19779	19352	18001
300	18176	16292	14768	20867	18975	17408	20617	20051	18318	22524	21938	20163
350	19703	17385	15678	22736	20526	18672	22646	21914	19821	24904	24126	21982
400	20565	18235	16365	24296	21786	19731	24253	23371	21042	26915	26044	23517
500	22185	19172	17492	26706	23277	21329	26980	25449	23125	30028	28712	25916
600	22965	20567	17962	28033	25203	22097	28752	27974	24896	32236	31258	27766
750	24136	21386	18888	28303	25430	22690	31050	30024	26932	32404	31338	28303
1000	25278	22539	19923	31490	28083	24887	33864	32688	29320	37197	35748	31959

"C" VALUES—Aluminum Conductors

AWG or kcmil	Three Single Conductors						Three Conductor Cable					
	Steel Conduit			Nonmagnetic Conduit			Steel Conduit			Nonmagnetic Conduit		
	600 V	5kV	15kV	600 V	5kV	15kV	600 V	5kV	15kV	600 V	5kV	15kV
14	236	—	—	236	—	—	236	—	—	236	—	—
12	375	—	—	375	—	—	375	—	—	375	—	—
10	598	—	—	598	—	—	598	—	—	598	—	—
8	951	950	951	951	950	951	951	951	951	951	951	951
6	1480	1476	1472	1481	1478	1476	1481	1480	1478	1482	1481	1479
4	2345	2332	2319	2350	2341	2333	2351	2347	2339	2353	2349	2344
3	2948	2948	2948	2958	2958	2958	2948	2956	2948	2958	2958	2958
2	3713	3669	3626	3729	3701	3767	3733	3719	3693	3739	3724	3709
1	4645	4574	4497	4678	4631	4580	4686	4663	4617	4699	4681	4646
1/0	5777	5669	5493	5838	5766	5645	5852	5820	5717	5875	5851	5771
2/0	7186	6968	6733	7301	7152	6986	7327	7270	7109	7372	7328	7201
3/0	8826	8466	8163	9110	8851	8627	9077	8980	8750	9242	9164	8977
4/0	10740	10167	9700	11174	10749	10386	11184	11021	10642	11408	11277	10968
250	12122	11460	10848	12862	12343	11847	12796	12636	12115	13236	13105	12661
300	13909	13009	12192	14922	14182	13491	14916	14698	13973	15494	15299	14658
400	16670	15355	14188	18505	17321	16233	18461	18063	16921	19587	19243	18154
500	18755	16827	15657	21390	19503	18314	21394	20606	19314	22987	22381	20978
600	20093	18427	16484	23451	21718	19635	23633	23195	21348	25750	25243	23294
750	21766	19685	17686	25976	23701	20934	26431	25789	23750	29036	28262	25976
1000	23477	21235	19005	28778	26109	23482	29864	29049	26608	32938	31919	29135

"C" VALUES—Busways

Ampacity	Plug-In Busway		Feeder Busway		High Imped. Busway
	Copper	Aluminum	Copper	Aluminum	Copper
225	28700	23000	18700	12000	—
400	38900	34700	23900	21300	—
600	41000	38300	36500	31300	—
800	46100	57500	49300	44100	—
1000	69400	89300	62900	56200	15600
1200	94300	97100	76900	69900	16100
1350	119000	104200	90100	84000	17500
1600	129900	120500	101000	90900	19200
2000	142900	135100	134200	125000	20400
2500	143800	156300	180500	166700	21700
3000	144900	175400	204100	188700	23800
4000	—	—	277800	256400	—

CHART M (multiplier)

f	M	f	M
0.01	0.99	1.20	0.45
0.02	0.98	1.50	0.40
0.03	0.97	2.00	0.33
0.04	0.96	3.00	0.25
0.05	0.95	4.00	0.20
0.06	0.94	5.00	0.17
0.07	0.93	6.00	0.14
0.08	0.93	7.00	0.13
0.09	0.92	8.00	0.11
0.10	0.91	9.00	0.10
0.15	0.87	10.00	0.09
0.20	0.83	15.00	0.06
0.30	0.77	20.00	0.05
0.40	0.71	30.00	0.03
0.50	0.67	40.00	0.02
0.60	0.63	50.00	0.02
0.70	0.59	60.00	0.02
0.80	0.55	70.00	0.01
0.90	0.53	80.00	0.01
1.00	0.50	90.00	0.01
		100.00	0.01

$$M = \frac{1}{1+f}$$

Example: To calculate the available short-circuit current at a panelboard located 20 feet (6 m) away from a transformer when the transformers are 500 kcmil copper in steel conduit. The "C" value for the conductors is 22,185. The transformer is marked 300 kVA, 208/120-volt, 3-phase, 4-wire.

The transformer impedance is 2%. Consider the source to have an infinite amount of fault current available (often referred to as "infinite primary").

(1) To find the transformer's full-load secondary current:

$$I_{fla} = \frac{kVA \times 1000}{E \times 1.73} = \frac{300 \times 1000}{208 \times 1.73} = 834 \text{ A}$$

(2) To find the transformer SCA:
(multiplier = 100/2 = 50)

Transformer SCA =
$I_{FLA} \times$ multiplier $= 834 \times 50 = 41{,}700$ A

(3) To find the (F) factor:

$$F = \frac{1.73 \times L \times I}{C \times E_{L-L}}$$

$$= \frac{1.73 \times 20 \times 41{,}700}{22{,}185 \times 208} = 0.3127$$

(4) To find the (M) multiplier:

$$M = \frac{1}{1+F} = \frac{1}{1+0.3127} = 0.76$$

(5) To find the short-circuit current of the panelboard:

$$I_{SCA} = \text{transformer}_{SCA} \times (M) = 41{,}700 \times 0.76$$
$$= 31{,}692 \text{ A}$$

The fault current available at the panelboard where it is located 20 feet (6 m) away from the transformer is equal to 31,692 A.

QUESTION REVIEW

Each lesson is purposely designed to require the student to apply the entire *NEC®* text and not specific chapters or articles. It has been found that when studying for a timed, open-book examination, the student must gain proficiency in the Table of Contents, the Index, and the ability to move quickly from cover to cover to find the correct answer to each question in a timely fashion.

1. Does the *Code* permit the use of two single-pole circuit breakers in a panelboard to serve a line-to-line connected load such as a household electrical range rated at 120/240 volts or a hot water heater rated at 240 volts?

Answer: _____

_____

_____

_____

Reference: _____

2. Where outdoor conductors are tapped and these tapped conductors terminate into a single overcurrent device designed to limit the load so as not to exceed the ampacity of the tap conductors, how long can these tap conductors be?

Answer: _____

_____

_____

_____

Reference: _____

3. Is it permissible to use 300-volt cartridge-type fuses and fuseholders to protect a multiwire, 4-wire circuit such as an emergency circuit in a food store where the electrical service is 277/480-volt, 3-phase, 4-wire?

Answer: _____

_____

_____

_____

Reference: _____

4. Service entrance conductors are required to be protected at their rated ampacity with exception—the exception being the next standard size overcurrent device when the service conductors are protected by a single device. When are multiple overcurrent devices supplied by a service allowed to exceed the ampacity rating of the service entrance conductors?

Answer: _____

_____

_____

_____

Reference: _____

5. Fixed electric space heating loads shall be computed at _____ percent of the total connected load, generally.

   Answer: _____

   _____

   _____

   _____

   Reference: _____

6. A motor is to be installed in a Class I, Division 2 location. It is to be connected with 3 feet (900 mm) of trade size ¾ (21) liquidtight flexible metal conduit. Is it permissible to use the liquidtight flexible metal conduit as the grounding path for this installation?

   Answer: _____

   _____

   _____

   _____

   Reference: _____

7. Are feeder conductors required to be rated for 125% of the continuous load plus the noncontinuous load?

   Answer: _____

   _____

   _____

   _____

   Reference: _____

8. How many duplex receptacles can be installed on a 20-A branch circuit in a dwelling?

   Answer: _____

   _____

   _____

   _____

   Reference: _____

9. Table 310.15(B)(16) lists the ampacity of 12 AWG THWN conductors as 25 A. Can 16 general-use receptacles in an office building be connected to a 20-A circuit?

   Answer: _____

   _____

   _____

   _____

   Reference: _____

10. Should you use Article 310 or Annex B to calculate underground conductor ampacities?

    Answer: _____

    _____

    _____

    _____

    Reference: _____

11. When an electric heat pump is installed with backup resistance heat in a building, (a) could one of the loads ever be considered dissimilar for the purposes of calculating the service loads? (b) Would the heat pump load have to be calculated at 125% for the service sizing?

    Answer: _____

    _____

    _____

    _____

    Reference: _____

12. Are all 20-A residential underground circuits required to have GFCI protection?

    Answer: _____

    _____

    _____

    _____

    Reference: _____

13. Can two circuit breakers with handle ties be used in place of a two-pole breaker to supply a 240-volt electric baseboard heater?

    Answer: _____

    _____

    _____

    _____

    Reference: _____

14. Are water heaters considered to be a continuous load so that a 125% load is required for calculating feeder and service on an installation?

    Answer: _____

    _____

    _____

    _____

    Reference: _____

15. Do the secondary conductors from the secondary of a transformer always require secondary overcurrent protection? How about a 10-foot (3 m) secondary conductor? A 25-foot (7.5 m) secondary conductor?

    Answer: _____

    _____

    _____

    _____

    Reference: _____

16. Is it permitted to plug a microwave oven with a nameplate rating of 13 amperes into a 15-ampere receptacle protected on a 20-ampere circuit?

    Answer: _____

    _____

    _____

    _____

    Reference: _____

17. Can a 1400 VA load—a cord-and-plug-connected load—be supplied by a 240-volt circuit in a residence?

    Answer: _____

    _____

    _____

    _____

    Reference: _____

18. Three receptacles on a single yoke or strap are to be installed in a commercial or industrial facility. Are these outlets to be calculated at 180 VA or 540 VA?

    Answer: _____

    _____

    _____

    _____

    Reference: _____

Use the following diagram for Questions 19–21.

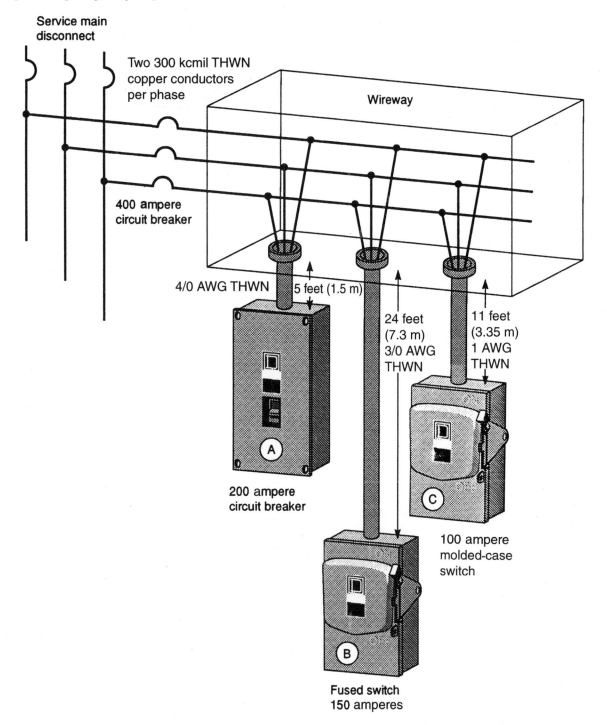

Service main disconnect

Two 300 kcmil THWN copper conductors per phase

Wireway

400 ampere circuit breaker

4/0 AWG THWN — 5 feet (1.5 m)

24 feet (7.3 m) 3/0 AWG THWN

11 feet (3.35 m) 1 AWG THWN

A

200 ampere circuit breaker

C

100 ampere molded-case switch

B

Fused switch 150 amperes

19. Does the tap supplying the circuit breaker marked Ⓐ comply with the *NEC*®? If so, which section?

Answer: _____

_____

_____

_____

Reference: _____

20. Does the tap supplying the switch marked ® comply with the *NEC*®? If so, which section?

    Answer: _____

    _____

    _____

    _____

    Reference: _____

21. Does the tap supplying the molded-case switch marked © comply with the *NEC*®?

    Answer: _____

    _____

    _____

    _____

    Reference: _____

22. The *NEC*® requires overcurrent protection where a busway is reduced in size for the last 50 feet (15.24 m) or more in other than industrial installations. True or False?

    Answer: _____

    _____

    _____

    _____

    Reference: _____

23. A 10-foot (3 m) tap conductor must be:
    (a) at least ⅟₁₀ the size of the conductor to which it is tapped.
    (b) at least ⅟₁₀ the ampere rating of the overcurrent device that protects the conductors from which the tap is made.

    Answer: _____

    _____

    _____

    _____

    Reference: _____

24. In an extremely long feeder circuit, the circuit conductors have been increased in size to compensate for voltage drop. Is the equipment-grounding conductor also required to be increased?

    Answer: _____

    _____

    _____

    _____

    Reference: _____

25. The nameplate on an air-conditioning unit is marked "maximum size fuse 40-amperes." Is it permissible to install a 40-ampere circuit breaker or must a 40-ampere fuse protection be installed?

    Answer: _____

    _____

    _____

    _____

    Reference: _____

CHAPTER 9

OBJECTIVES

Studying this chapter along with the 2011 *NEC*® gives the reader a basic introduction to the grounding and bonding requirements in the Code that apply to all installations, above and below 600 volts.

After studying this chapter, you should know:

■ A brief history on grounding and bonding, and why the system used in the United States was chosen over other systems

■ The definitions and the application of terms that are applicable and that are unique to the 2011 *NEC*®

■ The many *Code* requirements in which grounding and bonding are addressed as it applies to the *NEC*®

■ The differences between grounding and bonding

■ The differences between the grounded conductor and the neutral conductor

■ The importance of equipment grounding and the proper sizing of the equipment grounding conductor

■ The charts and software available for proper sizing of the equipment grounding conductor

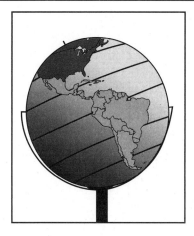

Ground. *The earth*

Bonded (Bonding). *Connected to establish electrical continuity and conductivity*

Grounded (Grounding). *Connected to ground or to a conductive body that extends the ground connection*

Equipment Grounding Conductor (EGC). *See 250.118 for a list of acceptable EGCs. The conductive path installed to connect normally non–current-carrying metal parts of equipment to the system grounded conductor or to the grounding electrode conductor.*

Grounding Electrode. *A conducting object through which a direct connection to earth is established.*

Grounding Electrode Conductor. *A conductor used to connect the system grounding conductor or the equipment to a grounding electrode or to a point on the grounding electrode system.*

Neutral Conductor. *The conductor connected to the neutral point of a system that is intended to carry current under normal conditions.*

Neutral Point. *The common point on a wye connection in a polyphase system, the midpoint on a single-phase, 3-wire system, the midpoint of a single-phase portion of a 3-phase delta system, or the midpoint of a 3-wire, direct current system.*

Informational Note: At the neutral point of the system, the vectorial sum of the nominal voltages from all other phases within the system that utilize the neutral, with respect to the neutral point, is zero potential.

GROUNDING ELECTRICAL SYSTEMS

Important reasons for achieving a properly grounded installation are the safety and property protection afforded by a grounded system.

The key to a reliable electrical system is a well-designed grounded electrical system with all materials properly selected for the application and installed in accordance with the manufacturer's recommended installation guidelines, product "listing," and the National Electrical Code.

Note: There are exceptions for unique installations in which an ungrounded system with reliable ground-fault detectors and assured maintenance are acceptable.

It is important to note that although the ground element of all electrical systems is one of the most important elements, it still remains one of the most misunderstood and most errant parts of any electrical system. Many engineers, teachers, students, and skilled workers regularly design, teach, and install grounding systems incorrectly, which often negates or exacerbates the systems' operating characteristics. (See Figure 9–1.)

History of Grounding

Since before the turn of the century when electricity began being used for commercial purposes, the topic of grounding and the mysteries surrounding this subject have occupied the minds of those considered to be experts in the electrical field. The mysteries of grounding become a major topic at nearly every meeting of any kind involving the electrical community, regardless of whether it is on the agenda. The discussion varies from the mythology to many real-life problem scenarios encountered by those involved in the use of day-to-day electricity.

It was decided many years ago that a grounded system provides the best system, both economically and from a safety point of view. A grounded system provides for a safer system and also stabilizes the voltage levels so as to protect the equipment. Even the earliest leaders in this industry seemed to generally agree that with an AC system, all low-voltage secondaries should be grounded; this is necessary to reduce the many risks of fatal shock, damage to equipment, and to reduce electrical fire hazards. Many heated discussions took place in the first 20 years of the twentieth century regarding this subject. Although the experts did not agree initially, as time went by, it was finally resolved that the grounded system provided the safest economical system for general use. Early discussions on electrical circuit grounding brought about such comments as those from Killingsworth Hedges, an English scientist, who stated, "One precaution is to earth the secondary circuit, another is to connect one or both leads to a safety appliance which would automatically divert any

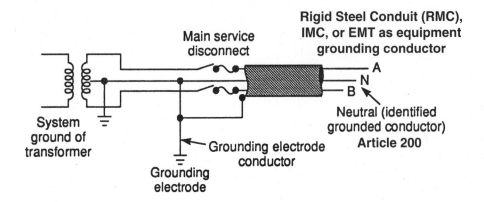

Requirements for grounding electrode conductor, grounded system

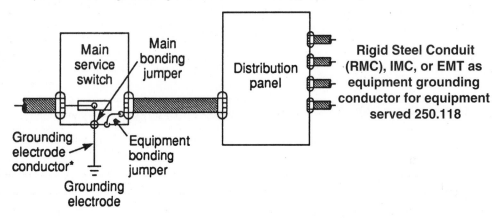

*Grounds both the identified grounded conductor and the equipment grounding conductor. It is connected to the grounded metallic cold water piping on premises, or other available grounding electrode, such as metal building frame if it qualifies per 250.52(A)(2), or a made electrode.

Figure 9–1 Grounding principles.

excess current to earth and at the same time shut off the supply." Another comment from Professor Elihu Thomson suggested grounding the transformer secondary winding or surrounding it with a grounded sheath; as alternatives, he suggested cutting off the secondary by automatic means or grounding it automatically by fuse cutouts when excessive voltage appeared.

AIEE, the predecessor to IEEE, resolved early on that the recommendation of grounding low-tension secondary AC systems was strongly recommended in all cases where a reliable grounding connection could be secured and in 3-wire systems with a grounded neutral and that solid connections without fuse be permitted on the neutral wire. They further stated, "The grounding of the secondary AC systems inside of buildings to water pipes provided such connection

is made nearest the point to cellar wall on water pipes outside of meter is not only safe, but places no burden on the water pipes."

In 1911, the National Fire Protection Association met and resolved that grounding of transformer secondary circuits was highly recommended. In the 1913 *NEC®*, a mandatory grounding rule, *Rule 15*, was adopted, which stated:

Grounding Low Potential Circuits-Alternating current secondary systems, transformer secondary of distribution systems except when supplied from private, industrial power or lighting plants, when the primary voltage does not exceed 550 volts, must be grounded provided the maximum difference of potential between the grounded point of any other point in the circuit does not exceed 150 volts and may be grounded when the maximum difference of potential between the grounded point and any other point in the circuit exceeds 150 volts. In either case, the following rules must be complied with: (1) The grounding must be made at the neutral point or wire whenever the neutral point or wire is accessible; (2) When no neutral point or wire is accessible, one side of the secondary circuit must be grounded; (3) The ground connection must be at the transformer or on the individual service as provided in Parts III–VII and when transformers feed systems with a neutral wire, the neutral must also be grounded at least every 500 feet (152.4 m).

This rule has not been changed in substance since that time. The main difference is that in the 1923 and 1940 *NECs®*, grounding for voltages above 150 to 300 volts was recommended instead of merely permitted.

Since the 1913 resolution requiring mandatory grounding, experts have had many discussions on the merits and methods of grounding. There is no reason to believe that many more discussions on this great subject will not occur. However, it continues to contain mandatory rules for safe grounding practices. These rules are essentially the same as the original rules that went into the 1913 *NEC®*. Although experts have continued to discuss and debate on the methods and merits, the science has not changed, and the general agreement is consistent with those of the early fathers of the electrical community:

Grounding provides the safest, most economical electrical system for general use known to humankind.

Most Use the Solidly Grounded System

It is enlightening to note that the major countries in North America—Canada, the United States, and Mexico—all supply predominantly grounded systems to residential and commercial facilities; therefore, most electrical equipment manufactured in these three countries can readily be transported between the countries without alteration of the equipment. However, it is equally important to note that each country—Canada, the United States, and Mexico—has quality

installation problems. In many areas across the North American continent, good grounding and bonding practices are not followed, and installations are not made in accordance with all good engineering guidelines; nor is the *NEC®* closely followed where adopted. Even though it is considered a minimum standard for an electrical installation, many installers are negligent in making the grounding and bonding installations in accordance with those minimum standards or better. (See Figure 9–1 and Figure 9–2.)

**Types of AC Utilization Systems
600 Volts or Less**

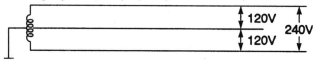

a. Single-phase, 3-wire, 120, 240 volts

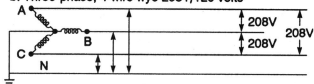

b. Three-phase, 4-wire wye 208Y/120 volts

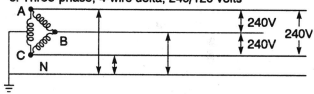

c. Three-phase, 4-wire delta, 240/120 volts

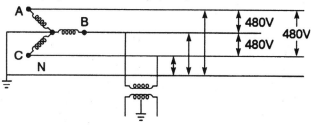

d. Three-phase, 4-wire wye, 480/277 volts

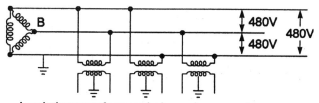

e. Three-phase, 3-wire, 120 delta, 240 or 480 volts

Insulating transformers balanced across phases; secondaries must be grounded

Figure 9–2 Examples of grounded systems.

General Requirements

The requirements for systems that must be grounded are covered in *NEC*® 250.20 and 250.21. (See Figure 9–2.) 250.22 lists systems that are not permitted to be grounded.

250.20 Alternating-Current Systems to Be Grounded

Alternating-current systems shall be grounded as provided for in **250.20(A)**, **(B)**, **(C)**, or **(D)**. Other systems shall be permitted to be grounded. If such systems are grounded, they shall comply with the applicable provisions of this article.

Informational Note: An example of a system permitted to be grounded is a corner-grounded delta transformer connection. See **250.26**(4) for conductor to be grounded.

(A) Alternating-Current Systems of Less Than 50 Volts. Alternating-current systems of less than 50 volts shall be grounded under any of the following conditions:

(1) Where supplied by transformers, if the transformer supply system exceeds 150 volts to ground

(2) Where supplied by transformers, if the transformer supply system is ungrounded

(3) Where installed outside as overhead conductors

(B) Alternating-Current Systems of 50 Volts to 1000 Volts. Alternating-current systems of 50 volts to less than 1000 volts that supply premises wiring and premises wiring systems shall be grounded under any of the following conditions:

(1) Where the system can be grounded so that the maximum voltage to ground on the ungrounded conductors does not exceed 150 volts

(2) Where the system is 3-phase, 4-wire, wye connected in which the neutral conductor is used as a circuit conductor

(3) Where the system is 3-phase, 4-wire, delta connected in which the midpoint of one phase winding is used as a circuit conductor

(C) Alternating-Current Systems of 1 kV and Over. Alternating-current systems supplying mobile or portable equipment shall be grounded as specified in 250.188. Where supplying other than mobile or portable equipment, such systems shall be permitted to be grounded.

(D) Impedance Grounded Neutral Systems. Impedance grounded neutral systems shall be grounded in accordance with **250.36** or **250.186**.

Ungrounded Systems Permitted

There are delta systems that are permitted by the *NEC*® to be ungrounded. These are generally systems used for industrial or agriculture applications, and where the ungrounded system is installed, usually an additional degree of service continuity is needed. It is important, however, to note that a lesser degree of safety is ensured with this system, and an occurrence of a ground fault on a system will not cause an overcurrent protective device to open. It does, however, ground the system. If the second fault is in another feeder or circuit on the same system, one or more of the overcurrent devices will open. For safety reasons as well as for system continuity, it is necessary that maintenance personnel locate and correct these ground faults as soon as practicable. Systems permitted by the *NEC*® to be installed as ungrounded systems are delta systems, such as 480-volt, 3-phase, 3-wire; 2300-volt, 3-phase, 3-wire; 4600-volt, 3-phase, 3 wire; and 13.8 kV, 3-phase, 3-wire systems. Ground detector lights that indicate that a ground fault has occurred on an ungrounded system are required. The light will burn until the phase goes to ground, thus identifying the faulted phase. This indicator light is to alert maintenance personnel so that the ground fault can be corrected, preferably immediately or during the hours while production is not in operation, and the plant can continue operations with the single fault (one phase grounded), thus preventing a costly shutdown. State-of-the-art equipment is available, which will help identify the exact location of the fault.

250.21 Alternating-Current Systems of 50 Volts to Less Than 1000 Volts Not Required to Be Grounded

(A) General. The following AC systems of 50 volts to less than 1000 volts shall be permitted to be grounded but shall not be required to be grounded:

(1) Electrical systems used exclusively to supply industrial electric furnaces for melting, refining, tempering, and the like

(2) Separately derived systems used exclusively for rectifiers that supply only adjustable-speed industrial drives

(3) Separately derived systems supplied by transformers that have a primary voltage rating less than 1000 volts, provided that all the following conditions are met:
   a. The system is used exclusively for control circuits.
   b. The conditions of maintenance and supervision ensure that only qualified persons service the installation.
   c. Continuity of control power is required.

(4) Other systems that are not required to be grounded in accordance with the requirements of **250.20(B)**

(B) Ground Detectors. Ground detectors shall be installed in accordance with 250.21(B)(1) and (B)(2).

(1) Ungrounded alternating current systems as permitted in **250.21(A)(1)** through (A)(4) operating at not less than 120 volts and not exceeding 1000 volts shall have ground detectors installed on the system.

(2) The ground detection sensing equipment shall be connected as close as practicable to where the system receives its supply.

(C) Marking. Ungrounded systems shall be legibly marked "Ungrounded System" at the source or first disconnecting means of the system. The marking shall be of sufficient durability to withstand the environment involved.

250.22 Circuits Not to Be Grounded

The following circuits shall not be grounded:

(1) Circuits for electric cranes operating over combustible fibers in Class III locations, as provided in **503.155**
(2) Circuits in health care facilities as provided in **517.61** and **517.160**
(3) Circuits for equipment within electrolytic cell working zone as provided in Article **668**
(4) Secondary circuits of lighting systems as provided in **411.5(A)**
(5) Secondary circuits of lighting systems as provided in **680.23(A)(2)**.

Corner Grounded Delta

Many power companies in the United States occasionally furnish a 480-volt corner grounded delta service for secondary use in agriculture and oil field installations. Many questions arise as to the use of this system, which employs the 3-wire delta secondary with one phase intentionally grounded. (See Figure 9–3.) It is not the most common service used today; however, it has provided a grounded system in locations formerly supplied by a 3-wire ungrounded delta system. It should be emphasized, however, that all metal parts that are not intended to carry current must be permanently bonded to an unfused, unswitched conductor, which extends back to the transformer secondary ground. It must be run with the circuit conductors and sized to the overcurrent protection protecting the circuit. It is not acceptable to bond the equipment to a driven grounding rod at the equipment location. This means that the user's wiring, beyond the service main disconnecting means, must consist of four wires, the three current-carrying phase conductors, and a fourth wire called an equipment-grounding conductor (EGC). The EGC is the conductor that is not intended to carry load current but instead provides an interconnection between the transformer bank ground and all grounds on the user's equipment. All grounding of the electrical circuit conductors, surge arrestors, and conductive non–current-carrying materials and equipment are required to be installed and arranged in a manner that will prevent objectionable flow of current over the grounding conductor or grounding paths under normal operating conditions. If these objectionable currents are present, one or more alterations must be made. One must discontinue one or more such grounding connections, change the location of the grounding connections, interrupt the continuity of one or more of the conducting paths interconnecting the grounding

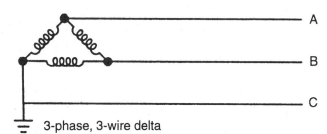

Figure 9–3 3-phase, 3-wire delta corner grounded transformer. (See *NEC*® 250.20 [Informational Note] and 250.26(4).)

connections, or take other suitable remedial action satisfactory to the authority having jurisdiction.

The advantages for this 480-volt corner-grounded system are that two element meters can be used and are less expensive than three element meters and require less maintenance. Triplex (three-conductor) wire is less expensive than quadplex (four-conductor) wire. The meter base is also less expensive. However, dangerous step-and-touch potentials can exist, such as nonautomatic isolation and location of phase-to-ground shorts and ground currents objectionable to residents living in the area. Ground potentials have been known to cause solid-state starters to start in the "off" position. Lightning arrestor applications may be unsatisfactory. There may be interference to communication equipment nearby. This system places a higher dielectric stress on the two ungrounded phases and higher magnitudes of transient overvoltages. In conclusion, for the corner-grounded 480-volt service, the most significant reason for use is that it is less expensive. However, it may not provide a safe working environment, and it is advisable to request service from a 4-wire system from the utility company if it is available. This will reduce normal current flows in the grounds due to operational problems. It is advisable to consider other types of 480-volt delta grounding or solidly grounded wye systems where available.

As you can see, the ungrounded system for some specialty systems is still permitted, and the corner-grounded delta provides an additional level of safety where the traditional grounded system is not available. But, generally, it is not nearly as desirable as a solidly grounded system and should only be selected under engineering guidance.

Grounded Conductor (Neutral)

The grounded conductor of the system is one of the most important elements. It is that conductor derived from the supplying power source, transformer, generator, or other derived source that is required by the NEC® to be grounded at an accessible location from the load end of the service drop or service lateral to and including the terminal bus at the service disconnecting means. These locations would include current transformer enclosures, meter enclosures, pull boxes, junction boxes, busways, auxiliary gutters, or in the panel box acceptable to the authority having jurisdiction. This grounded service conductor provides a vital good low impedance path for the fault current to

return to the service. It must be run to and be installed in each service disconnecting means regardless of whether it is needed for the service. If the grounded conductor is not run to the service as required, it is likely for ground fault to have a high impedance path back to the source. It is a violation of NEC® 250.4(A) and becomes virtually impossible to clear a ground fault, thus introducing a hazard in the system. However, if we comply with NEC® 250.24 for a single building installation and 250.32 for multiple buildings supplied by one service, we have a low-impedance ground fault path and have complied with 250.4 of the NEC®. The safety of the service is improved immeasurably because the grounded conductor is run to provide a ground-fault path. The size of the conductor depends on the size of the phase conductors and must never be smaller than the grounding-electrode conductor, the size of which is given in Table 250.66 of the NEC®. (See Figure 9–4.)

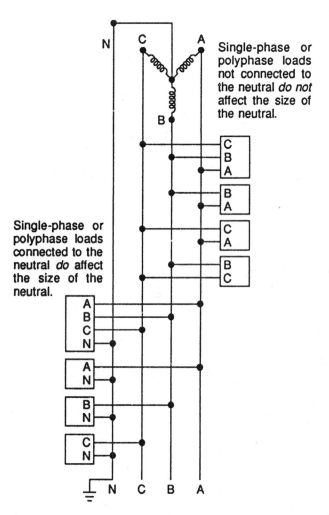

Figure 9–4 Size of a neutral conductor as in NEC® 220.61.

Grounding-Electrode System (Conductor)

Earth is considered to be at zero potential. The grounding electrode—see *NEC*® 250.50 and 250.52(A)(1) through (8)—consisting of all grounded components present and where one or more grounded electrodes are not present, a made electrode must be used; see *NEC*® 250.52(A)(4) through (8). Where metal water pipe is the only electrode available, it must also be supplemented as per *NEC*® 250.53(A)(2). The electrodes are bonded together to form one common electrode system, as stated in *NEC*® 250.58. The grounding electrode must be large enough to provide a low-impedance path to the electrode system for lightning strikes, voltage surges, or in the event of a large fault. (See Figure 9–5.)

It is important to note that under normal conditions, no current potential exists between any conductor or equipment where the system is properly designed, installed, and maintained. Only in abnormal conditions, such as lightning, line surges, and high-fault conditions, does current flow through this system to earth.

Although it is desirable to have a grounding-electrode system have an impedance as low as practicable, it is imperative that we design and install systems that provide a low-impedance path back to the source that will clear a fault when it occurs without resulting in loss of property or life. In a properly designed grounded system where a hazard can exist for only the time it

NEC® Article 250, Part V

Bonding jumper around water meter and other fittings in water piping likely to be disconnected, such as unions, where grounding connection is not on street side of water meter.

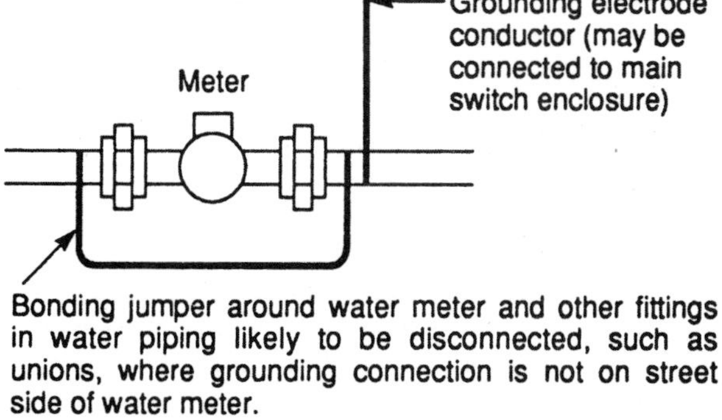

Figure 9–5 Bonding jumpers as in *NEC*® Article 250, Part V.

takes to clear the fault, it is extremely unlikely that any loss of life would occur and damage to property and equipment would be kept to a minimum.

Equipment-Grounding System (Conductor)

Equipment grounding is somewhat different than the system-grounded conductor. It relates to the methods for grounding and bonding nonelectrical conductive materials, which either enclose energized conductors or are adjacent thereto. These conductive materials are required to be interconnected and grounded and bonded for several reasons to achieve the end result of a safe installation. The requirement is to bond all metal enclosures and objects adjacent to conductive energized conductors; to provide current-carrying capabilities large enough to adequately accept any ground-fault current permitted by the overcurrent protective system without creating a fire or explosive hazard to life or property; to ensure freedom from dangerous electrical shock exposure to persons in the area; and to contribute to a superior performance of the system. The equipment ground plays no part in the lighting or power circuit; however, it plays a very important part in ensuring the safety and equipment protection needed for every good reliable system (see the definition of "**Grounding conductor, Equipment**"). Some examples of components of the equipment-grounding system are metallic conduit, motors, equipment enclosures, and equipment grounding conductor sized as per 250.122. *NEC*® 250.122 and Table 250.122 provide for a minimum size only. The size of the conductor may have to be increased due to the length of the circuit or the amount of fault current that is available and may be imposed upon it. This is clarified in the general requirements of *NEC*® Article 110, and the requirements of 110.10 which must be complied with.

> *Note:* An equipment grounding conductor is part of the equipment-grounding system as distinguished from a grounded conductor, which is part of the power distribution system. The requirements for this grounding system are covered in Article 250 of the *NEC*®. See 250.4(A).

110.10 Circuit Impedance, Short-Circuit Current Ratings, and Other Characteristics
The overcurrent protective devices, the total impedance, the equipment short-circuit current ratings,

(Continued)
---

and other characteristics of the circuit to be protected shall be selected and coordinated to permit the circuit protective devices used to clear a fault to do so without extensive damage to the electrical equipment of the circuit. This fault shall be assumed to be either between two or more of the circuit conductors or between any circuit conductor and the equipment grounding conductor(s) permitted in **250.118**. Listed equipment applied in accordance with their listing shall be considered to meet the requirements of this section.

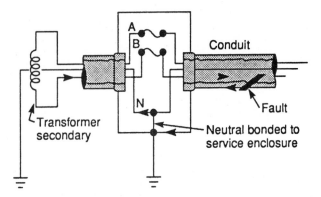

Figure 9–6 Bonding at service disconnect enclosure as required in *NEC®* 250.92.

NEC® 250.4 states that the path to ground from circuits, equipment, and metal enclosures for conductors shall:

- be permanent and continuous.

- have the capacity to safely conduct any fault current likely to be imposed on it.

- have sufficiently low impedance to limit the voltage to ground and to facilitate the operation of the circuit protective devices in the circuit.

NEC® 250.90, 250.96, and 250.97 require bonding to ensure the continuity and the capacity to safely carry the fault currents likely to be imposed. *NEC®* 250.118 lists the types of EGCs with exceptions. Table 250.122 lists the minimum allowable size of copper and aluminum conductors. There are three basic components to a complete equipment grounding system. Each is composed of one or more parts that have different but related functions. In order to ensure this low-impedance path back to the service, it is necessary to comply with 300.3(B) of the *NEC®*, which states, *All conductors of the same circuit and, where used, the grounded conductor and all equipment grounding conductors and bonding conductors shall be contained within the same raceway, auxiliary gutter, cable tray, cablebus assembly, trench, cable, or cord, unless otherwise permitted in accordance with 300.3(B)(1) through (B)(4).* This becomes of extreme importance in AC applications. It is the total impedance that controls the current division among parallel paths. In 60 Hz circuits rated under 100 amperes, the circuit reactance is an insignificant part of the circuit impedance. However, reactance may be a prominent element of circuit impedance in circuits rated above 100 amperes, and reactance is the dominant element of

impedance for circuits rated above 200 amperes. The reactance of an AC circuit is determined mainly by the spacing between outgoing and return conductors and is only slightly affected by conductor size. However, circuit resistance is affected by conductor size. The ratio of inductive reactance to the resistance and the relative effect of reactance on circuit impedance increases as the conductor size increases. Increased separation (spacing between grounding and phase conductors) increases not only the inductive reactance of the equipment grounding conductor but also the zero sequence inductive reactance of the phase conductors. Hence, on a grounded system, this conductor plus the main bonding jumper at the service and the system bonding jumper between the transformer case and the transformer winding are all available as a low impedance path for fault currents. There is no dependence on the electrode earth path unless the grounded conductor if it is inadvertently interrupted. (See Figure 9–6.) If the grounded conductor, the main bonding jumper, and EGCs are provided, properly connected, and of adequate capacity, only three conditions can prevent rapid operation of the overcurrent device: (1) a faulty or improperly rated overcurrent device; (2) a high-resistance fault, such as a fault through high-resistance contact with a grounded object; or (3) a high-resistance EGC connecting the faulty equipment to the service equipment enclosure. The third condition can be prevented by proper selection of the equipment grounding conductor and correct installation.

Tubular metal raceways, such as rigid metal conduit, intermediate metal conduit, or electrical metallic tubing, fits the preferred conductor geometry perfectly. The phase conductors and the neutral conductors are enclosed within the metal raceway. (See Figure 9–7.) In this way, effectiveness is markedly improved. The

*Reprinted with permission from NFPA 70-2011.

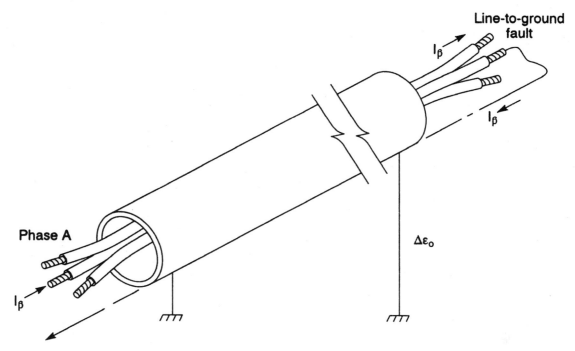

Figure 9–7 Tubular metal raceway used as an equipment grounding conductor. (See 250.4, 250.122(A), and Table 250.122.)

returning ground-fault current distributes itself about the entire enclosing metal shell in such a fashion as to result in a lower roundtrip voltage drop. The electrical behavior during the line-to-line fault is that of a coaxial line, except for the effects of resistivity in the metal shell. All electric and magnetic fields are minimal and contained inside the shell. The external space magnetic field becomes zero.

> **Warning:** It is extremely important that only one grounding system exist in an installation because multiple separated grounding systems create unwanted ground loops that may create errors in data transmission and malfunctions. Where a ground loop exists, there are known techniques that will solve the problem. Although the single-point ground is preferred, it may not always be feasible where long distances and very high frequencies are present.

Example: A dairy barn is fed underground from the farm house with a 225-ampere overcurrent device using 4/0 AWG copper phase conductors and a 1 AWG (neutral) grounded conductor. What size EGC is required for this installation?

Step 1. Index, Grounding, Separate buildings, 250.32; also EGC Sizes, Table 250.122.

Step 2. *NEC*® 547.5(F) requires an insulated or covered EGC where livestock is housed.

Step 3. Table 250.122; the minimum size permitted for a 225-ampere overcurrent device. The table lists only a 200 and a 300; therefore, the 300 would apply because the 200 would not be adequate.

Answer: 4 AWG copper

Computer Model Developed

It is with a great deal of satisfaction that I can report that the excellent work that has carried this industry for the past 40 years, developed and written by

R. H. "Dick" Kaufman (General Electric "GER 957A," *Some Fundamentals of Equipment Grounding Circuit Design,* IE 1058.33 November 1954, Applications and Industry, Vol. 73, Part II) and

Eustace C. Soares (Pringle Switch, *Grounding Electrical Distribution Systems for Safety,* 1966 through the 5th Edition *IAEI Soares Book on Grounding,* 1993, by J. Philip Simmons)

has been revisited.

In early 1993, at the request of J. Philip Simmons, author of the 4th and 5th editions of *IAEI Soares Book on Grounding*, the steel conduit and tubing producing and members of the National Electrical Manufacturers Association (NEMA) made funding available to the Georgia Institute of Technology's School of Electrical and Computer Engineering to develop a computer model. A. P. Sakis Meliopoulos, P. E., and Elias N. Glytsis, P. E., developed a model that was validated by field tests consisting of arc-voltage testing, and fault-current testing on thirteen 256-foot runs of electrical metallic tubing with both steel and die-cast set-screw and compression fittings, rigid steel conduit and intermediate steel conduit with conductors enclosed at Kearney Laboratories at McCook, Illinois. Mr. Meliopoulos stated in his final report that the techniques established by *Soares Book on Grounding* has served this industry well in providing primary design guidance. (See Figure 9–8, Figure 9–9, Figure 9–10, Figure 9–11, Figure 9–12, and Figure 9–13.)

Fortunately, computer and analytical means, which have been developed at a rapid pace, have now made it possible to develop and validate a model utilizing current data that can be used with confidence.

This new data will replace the only data available to date. Today's manufacturing processes have changed both the quality of material used to produce electrical materials and the consistency of those products. Although hand calculations were fairly accurate to evaluate the given criteria, this new model will offer a broad band of conditions and the user no longer has to extrapolate data. Mr. Meliopoulos and I presented a paper through IAS/IEEE in 1997 at the fall meeting, and the model is now available. The attachments to this report are excerpts from the final report. It is interesting to note that the tests and research of new materials and circuit-breaker test guidelines revealed that the criteria of 50 arc volt and 500% (5IP) of overcurrent device rating were overly conservative, and he recommended an arc voltage of 40 and a 4IP (4 times the rating of the overcurrent device; i.e., 30-ampere circuit-breaker, 4IP = 30 × 400% = 120) of the overcurrent device, and as stated, these are still conservative numbers.

"Modeling and Testing of Steel EMT, IMC and Rigid (GRC) Conduit." Prepared by A. P. Sakis Meliopoulos and Elias N. Glytsis, School of Electrical and Computer Engineering, Georgia Institute of Technology, Atlanta, GA 30332.

For copies of this report and the free software, contact: The Steel Tube Institute of North America, http://www.steelconduit.org.

Even when the raceway does not provide the sole equipment ground path but is supplemented by an EGC wire or a bus enclosed within, the raceway provides a parallel path and also protects the wire and bus from physical damage. Where connected as a parallel path, the raceway not only lowers the total impedance of the grounding path but carries 95% or more of the current. Furthermore, the raceway provides additional protection as a grounding shield; for instance, in the unlikely event that a phase conductor becomes damaged within a raceway and the conductor is exposed and comes in contact with a grounded raceway, the resulting fault current would be carried to the grounded service equipment enclosure through the main bonding jumper to the system and the overcurrent device, thus clearing the fault.

By contrast, a damaged phase conductor run as open wiring or in a nonmetallic raceway could remain exposed to accidental contact indefinitely. A fault within a nonmetallic raceway or open runs of conductors may or may not provide a low impedance return path to trip the overcurrent device protecting the circuit.

An EGC and a grounded circuit conductor are not the same electrically. It is unsafe to connect the grounded circuit conductor to the raceway or to any other non–current-carrying equipment at any point on the load side of the service equipment. The connections permitted to be made on the line and load sides of the service disconnecting equipment are specified in 250.142 of the *NEC®*, Parts I and II, which clarifies the preferred design method for location of these connections.

In 1998, the electronic and circuit-breaker industry introduced a new level of safety to the *NEC®*: the arc-fault circuit-interrupter.

The solution was to add arc-fault circuit interrupter (AFCI) protection. The AFCI device has electronic circuitry capable of recognizing the unique current and/or voltage signatures of an arcing fault. When the arc fault is detected, it acts similarly to the GFCI device and interrupts the circuit. This technology integrates a conventional thermal circuit breaker with an AFCI.

The *NEC®* has defined this device in article 100. This requirement only applies to dwellings. Current requirements are that all 15- and 20-ampere, 120 volt receptacles located in a dwelling unit family rooms, dining rooms, living rooms, parlors, libraries, dens, bedrooms, sunrooms, recreation rooms, closets, hallways, or similar rooms or areas are required to be

Examples of Maximum Length Equipment-Grounding Conductor (Steel EMT, IMC, GRC, and Copper or Aluminum Wire) Calculated as a Safe Return Fault Path to Overcurrent Device Based on 1997 Georgia Tech Software Version (GEMI 2.4,1998) With an Arc Voltage of 40 and 4 IP at 25°C Ambient Based on a Circuit Voltage of 120 Volts to Ground THHN/THWN Insulation

Overcurrent Device Rating Amperes (75°C)	400% (4IP) Overcurrent Device Rating Amperes	Circuit Conductor Size AWG/kcmil Copper or Aluminum	Steel EMT, IMC, GRC Trade Size	(1) Equipment-Grounding Conductor Size AWG Copper or Aluminum	Length of EMT Run Calculated Maximum (In Feet)	Length of IMC Run Calculated Maximum (In Feet)	Length of GRC Run Calculated Maximum (In Feet)	Copper Grounding Conductor w/o Steel Conduit Maximum Run (In Feet)	Aluminum Grounding Conductor w/o Steel Conduit Maximum Run (In Feet)
20	80	12	1/2 (16)	—	395	398	384	—	—
20	80	12	—	12	—	—	—	300	—
20	80	10 AL	—	10 AL	—	—	—	—	293
30	120	10	1/2 (16)	—	358	383	364	—	—
30	120	10	3/4 (21)	—	404	399	386	—	—
30	120	10	—	10	—	—	—	319	—
30	120	8 AL	—	8 AL	—	—	—	—	310
40	160	8	3/4 (21)	—	407	414	395	—	—
40	160	8	1 (27)	—	447	431	418	—	—
40	160	8	—	10	—	—	—	294	—
40	160	8 AL	—	8 AL	—	—	—	—	232
60	240	6	3/4 (21)	—	350	383	363	—	—
60	240	6	1 (27)	—	404	400	382	—	—
60	240	6	—	10	—	—	—	228	—
60	240	4 AL	—	8 AL	—	—	—	—	221
100	400	3	1 1/4 (35)	—	402	397 (4)	373	—	—
100	400	3	—	8	—	—	—	229	—
100	400	1 AL	—	6 AL	—	—	—	—	222
200	800	3/0	2 (53)	—	390	389	363	—	—
200	800	3/0	—	6	—	—	—	201	—
200	800	250 AL	—	4 AL	—	—	—	—	195

(1) Per 1999 NEC® Table 251-18

Note: Software is not limited to above examples.

Figure 9–8 *(Table derived from software [SCA] and testing developed at the Georgia Institute of Technology and sponsored by the producers of Steel EMT, IMC, and rigid steel conduit.)*

Examples of Maximum Length Equipment-Grounding Conductor (Steel EMT, IMC, GRC, and Copper or Aluminum Wire) Calculated as a Safe Return Fault Path to Overcurrent Device Based on 1997 Georgia Tech Software Version (GEMI 2.4, 1998) With an Arc Voltage of 40 and 4 IP at 25°C Ambient Based on a Circuit Voltage of 277 Volts to Ground THHN/THWN Insulation

Overcurrent Device Rating Amperes (75°C)	400% (4IP) Overcurrent Device Rating Amperes	Circuit Conductor Size AWG/kcmil Copper or Aluminum	Steel EMT, IMC, GRC Trade Size	(1) Equipment-Grounding Conductor Size AWG Copper or Aluminum	Length of EMT Run Calculated Maximum (In Feet)	Length of IMC Run Calculated Maximum (In Feet)	Length of GRC Run Calculated Maximum (In Feet)	Copper Grounding Conductor w/o Steel Conduit Maximum Run (In Feet)	Aluminum Grounding Conductor w/o Steel Conduit Maximum Run (In Feet)
20	80	12	1/2 (16)	—	1170	1179	1140	—	—
20	80	12	—	12	—	—	—	890	—
20	80	10 AL	—	10 AL	—	—	—	—	870
30	120	10	1/2 (16)	—	—	1135 (4)	—	—	—
30	120	10	3/4 (21)	—	1199	1182	1143	946	—
30	120	10	—	10	—	—	—	—	—
30	120	8 AL	—	8 AL	—	—	—	—	920
40	160	8	3/4 (21)	—	1208	1228	1170	—	—
40	160	8	1 (27)	—	1326	1276	1239	—	—
40	160	8	—	10	—	—	—	871	—
40	160	8 AL	—	8 AL	—	—	—	—	690
60	240	6	3/4 (21)	—	1039	1134	1075	—	—
60	240	6	1 (27)	—	1197	1186	1131	—	—
60	240	6	—	10	—	—	—	676	—
60	240	4 AL	—	8 AL	—	—	—	—	657
100	400	3	1 1/4 (35)	—	1192	1176 (4)	1107	—	—
100	400	3	—	8	—	—	—	680	—
100	400	1 AL	—	6 AL	—	—	—	—	659
200	800	3/0	2 (53)	—	1157	1155	1077	—	—
200	800	3/0	—	6	—	—	—	598	—
200	800	250 AL	—	4 AL	—	—	—	—	578

Note: Software is not limited to above examples.

(1) Per 1999 *NEC*®Table 251-18

Figure 9–9 *(Table derived from software [SCA] and testing developed at the Georgia Institute of Technology and sponsored by the producers of Steel EMT, IMC, and rigid steel conduit.)*

Maximum Length of Electrical Metallic Tubing That May Safely Be Used as an Equipment-Grounding Circuit Conductor.
Based on a Ground-Fault Current of 400% of the Overcurrent Device Rating.
Circuit 120 Volts to Ground; 40 Volts Drop at the Point of Fault.
Ambient Temperature 25°C

EMT Trade Size	Conductors AWG No.	Overcurrent Device Rating Amperes 75°C*	Fault Clearing Current 400% O.C. Device Rating Amperes	Maximum Length of EMT Run (In Feet)
1/2 (16)	3-12	20	80	395
	4-10	30	120	358
3/4 (21)	4-10	30	120	404
	4-8	50	200	332
1 (27)	4-8	50	200	370
	3-4	85	340	365
1 1/4 (35)	3-2	115	460	391
1 1/2 (41)	3-1	130	520	407
	3-2/0	175	700	364
2 (53)	3-3/0	200	800	390
	3-4/0	230	920	367
2 1/2 (63)	3-250 kcm	255	1020	406
3 (78)	3-350 kcm	310	1240	404
	3-500 kcm	380	1520	370
	3-600 kcm	420	1680	353
4 (103)	3-900 kcm	520	2080	353
	3-1000 kcm	545	2180	347

* 60°C for 20- and 30-ampere devices.

Based on 1994 Georgia Tech Model

Figure 9–10 *(Table derived from software [SCA] and testing developed at the Georgia Institute of Technology and sponsored by the producers of Steel EMT, IMC, and rigid steel conduit.)*

Maximum Length of Intermediate Metal Conduit That May Safely Be Used as an Equipment-Grounding Circuit Conductor. Based on a Ground-Fault Current of 400% of the Overcurrent Device Rating. Circuit 120 Volts to Ground; 40 Volts Drop at the Point of Fault. Ambient Temperature 25°C

IMC Trade Size	Conductors AWG No.	Overcurrent Device Rating Amperes 75°C*	Fault Clearing Current 400% O.C. Device Rating Amperes	Maximum Length of IMC Run (In Feet)
1/2 (16)	3-12	20	80	398
	4-10	30	120	383
3/4 (21)	4-10	30	120	399
	4-8	50	200	350
1 (27)	4-8	50	200	362
	3-4	85	340	382
1 1/4 (35)	3-2	115	460	392
1 1/2 (41)	3-1	130	520	402
	3-2/0	175	700	377
2 (53)	3-3/0	200	800	389
	3-4/0	230	920	375
2 1/2 (63)	3-250 kcm	255	1020	368
3 (78)	3-350 kcm	310	1240	367
	3-500 kcm	380	1520	338
	3-600 kcm	420	1680	325
4 (103)	3-900 kcm	520	2080	320
	3-1000 kcm	545	2180	314

* 60°C for 20- and 30-ampere devices.

Based on 1994 Georgia Tech Model

Figure 9–11 *(Table derived from software [SCA] and testing developed at the Georgia Institute of Technology and sponsored by the producers of Steel EMT, IMC, and rigid steel conduit.)*

**Maximum Length of Galvanized Rigid Conduit That May Safely
Be Used as an Equipment-Grounding Circuit Conductor.
Based on a Ground-Fault Current of 400% of the Overcurrent Device Rating.
Circuit 120 Volts to Ground; 40 Volts Drop at the Point of Fault.
Ambient Temperature 25°C**

GRC Trade Size	Conductors AWG No.	Overcurrent Device Rating Amperes 75°C*	Fault Clearing Current 400% O.C. Device Rating Amperes	Maximum Length of GRC Run (In Feet)
1/2 (16)	3-12	20	80	384
	4-10	30	120	364
3/4 (21)	4-10	30	120	386
	4-8	50	200	334
1 (27)	4-8	50	200	350
	3-4	85	340	357
1 1/4 (35)	3-2	115	460	365
1 1/2 (41)	3-1	130	520	377
	3-2/0	175	700	348
2 (53)	3-3/0	200	800	363
	3-4/0	230	920	347
2 1/2 (63)	3-250 kcm	255	1020	356
3 (78)	3-350 kcm	310	1240	355
	3-500 kcm	380	1520	327
	3-600 kcm	420	1680	314
4 (103)	3-900 kcm	520	2080	310
	3-1000 kcm	545	2180	304

* 60°C for 20- and 30-ampere devices.

Based on 1994 Georgia Tech Model

Figure 9–12 *(Table derived from software [SCA] and testing developed at the Georgia Institute of Technology and sponsored by the producers of Steel EMT, IMC, and rigid steel conduit.)*

Maximum Length of Equipment-Grounding Conductor That May Safely Be Used as an Equipment-Grounding Circuit Conductor. Based on a Ground-Fault Current of 400% of the Overcurrent Device Rating. Circuit 120 Volts to Ground; 40 Volts Drop at the Point of Fault. Ambient Temperature 25°C

Copper Equipment Grounding Conductor AWG Size***	Copper Circuit AWG Conductors	Maximum Length of Run (In Feet) Using Copper Equipment Ground Conductor	Aluminum Equipment Grounding Conductor AWG Size***	Aluminum Circuit AWG Conductors	Maximum Length of Run (In Feet) Using Aluminum Equipment Ground Conductor	Overcurrent Device Rating Amperes 75°C**	Fault Clearing Current 400% O.C. Device Rating Amperes
14	14	253	12	12	244	15	60
12	12	300	10	12	226	20	80
10	10	319	8	8	310	30	120
10	8	294	8	8	232	40	160
10	6	228	8	4	221	60	240
8	3	229	6	1	222	100	400
6	3/0	201	4	250 kcm	195	200	800
4	350 kcm	210	2	500 kcm	204	300	1200
3	600 kcm	195	1	900 kcm	192	400	1600
2	2-4/0	160	1/0	2-400 kcm	163	500	2000
1	2-300 kcm	160	2/0	2-500 kcm	161	600	2400
1/0	3-300 kcm	134	3/0	3-400 kcm	131	800	3200
2/0	4-250 kcm	114	4/0	4-400 kcm	115	1000	4000
3/0	4-300 kcm	106	250 kcm	4-500 kcm	107	1200	4800
4/0	4-600 kcm	93	350 kcm	4-900 kcm	97	1600	6400
250 kcm	5-600 kcm	78	400 kcm	5-800 kcm	79	2000	8000
350 kcm	6-600 kcm	*	600 kcm	6-900 kcm	*	2500	10,000
400 kcm	8-500 kcm	*	600 kcm	8-750 kcm	*	3000	15,000
500 kcm	8-1000 kcm	*	800 kcm	8-1500 kcm	*	4000	16,000
700 kcm	10-1000 kcm	*	1200 kcm	10-1500 kcm	*	5000	20,000
800 kcm	12-1000 kcm	*	1200 kcm	12-1500 kcm	*	6000	24,000

* Calculations necessary
** 60°C for 20- and 30-ampere devices
*** Based on *NEC*® Chapter 9 Table 8

Based on 1994 Georgia Tech Model

Figure 9–13 *(Table derived from software [SCA] and testing developed at the Georgia Institute of Technology and sponsored by the producers of Steel EMT, IMC, and rigid steel conduit.)*

protected by a listed AFCI combination type installed to protect the branch circuit.

Three exceptions to permit the AFCI overcurrent device to be omitted. These circuits must be installed in metal conduit or cable. conduit is permitted where enclosed in concrete. Exception 1 permits the AFCI device to be installed at the first outlet.

Note: These exceptions were inserted to address high-rise and multifamily dwelling construction where steel wiring methods are commonly utilized. AFCI protection is still required at the first outlet since many of the substantiated fires occur inside the dwelling due to undersized and damaged cords and to faulty appliances. Exception 3 allows the AFCI overcurrent device to be omitted feeding a fire alarm system (note: this does not apply to smoke alarms) where installed in a steel wiring method.

Conclusions—Why Ground Circuits and Systems?

There are many reasons to ground electrical circuits. One reason would be savings in wire cost. For instance, four wires can be used to serve the same load that a 6-wire single-phase would need. A grounded wye (3-phase) system can serve three 120-volt, single-phase loads utilizing a common neutral. Two voltages are available where a grounded circuit is employed. With a 240-volt line-to-line utilizing a grounded system, you then have 120-volt to ground. Thus, a 3-wire system can provide both 230-volt motor and appliance loads; the 120 volts can be applied to smaller appliances, convenience outlets, and lighting in all locations, including residential occupancies.

A good electrical system properly grounded and maintained limits the voltage due to lightning and line surges on the system. A properly grounded system protects us from unintentional contact with high-voltage lines or the influence from higher-voltage lines run in close proximity to the lower-voltage lines. A good grounded system stabilizes the voltage-to-ground during normal operations.

Finally, and most importantly, a properly grounded system provides a low-impedance path to facilitate overcurrent device operation. It makes no difference whether you use a properly sized fuse or circuit breaker. In the case of a line-to-ground fault, the overcurrent device will open. This will prevent needless loss of equipment, fires, and the loss of life and property.

An example of the horrors of ungrounded systems and what can happen was related at a recent meeting. At a manufacturing plant in the southern part of the United States operating on a 480-volt ungrounded system, production was suddenly disrupted. The first evidence came as a motor failure on the 480-volt system, then another, and, in close succession, a third. The maintenance electricians made a quick inspection of the switchboard voltmeter, which measured line-to-line volts and amperes. No unusual conditions were indicated. The system equipment continued to fail. A test voltmeter was rigged up having a full-scale calibration of 1200 volts. Upon connecting it phase-to-ground, the pointer went off the scale. A phase-to-ground potential on the 480-volt system of more than 1200 volts existed! At once, the incoming service transformers were suspected of internal breakdown through high- and low-voltage windings, primary and secondary. As the last of these transformers was isolated and individually tested, it became evident they were not the problem. System equipment continued to fail, and the situation was desperate.

The frantic group went into a huddle and called in outside engineering. It was decided the only way out was to open (turn off) the main incoming service circuit breaker, which would deenergize the entire system. At this point, one of the workmen noticed a small wisp of smoke coming from a motor starting autotransformer; upon approaching it, he could hear a buzzing noise inside. This circuit was switched clear of the system, and the overvoltages disappeared. During the 2-hour period that this arcing fault existed, between 40 and 50 motors had failed.

It was found that the autotransformer enclosing case had been bashed in and was practically in contact with the coil. The spot where arcing had taken place was evident, although not badly burned. An attempt was made to show the plant engineer the source of the trouble. A solid connection was made between the frame and the burned spot on the coil. Much to the bewilderment of the operating men, but according to the expectations of the plant engineer, no more than a 73% increase in voltage-to-ground on the other two phases occurred. The main ingredient of the overvoltage had been omitted (discontinuous conduction). This story is spectacular because of the magnitude of the disturbance and consequential damage. Similar occurrences of lesser extent are not uncommon; there is evidence that they occur more frequently than realized. It is characteristic of an ungrounded system to be subject to relatively severe transient overvoltages. This trouble can be avoided by properly grounding the system. Many other important benefits, as discussed in this chapter, are also obtained.

The term "ungrounded system" is used to identify a system where there is no intentional connection between the system conductors and ground. However, in any practical system, there always exists a capacity coupling between the system conductors and ground. Consequently, the so-called "ungrounded system" is in reality a capacitance grounded system, by virtue of distributed capacitance from the system conductor to ground. When the neutral of a system is not grounded, it is possible for destructive transient overvoltages several times the normal voltage to appear from line-to-ground during normal switching of a circuit having a line-to-ground fault. Tests have shown that over-voltages may be developed by repeated restriking of the arc during interruption of the line-to-ground fault, particularly in low-voltage systems. Experience has proven that these overvoltages may cause failure of insulation at locations other than at the point of fault in the system; thus, line-to-ground fault on one circuit may result in damage to equipment and interruption of service on other circuits. The same condition will result from repeated strike of an arcing fault from line to ground.

In an ungrounded system, the second ground fault on another phase may occur before the first fault is removed. The second fault may be on the same circuit as the original fault or on another. In any event, the resulting line-to-line fault will actuate relays or circuit breakers and trip either one or both circuits. Thus, a single ground fault of relative unimportance may eventually result in considerable damage because of the relatively high line-to-line fault current and interruption of one or both circuits. When the neutral of the system is not grounded, a ground fault on one phase causes a full line-to-line voltage to appear throughout the system between the ground and the two phases, which were not-faulted phases. The voltage is 73% higher than normal. Usually, the insulation between the line and ground is adequate to withstand line-to-line voltage. However, when this voltage is applied for long periods, it may result in failure of insulation that may have already deteriorated because of age or severe service conditions. Line-to-ground faults on ungrounded neutral systems cause a very small ground-fault current to flow through the capacitance of the cables, transformers, and other electrical equipment on the system. This current may have a magnitude of a few amperes to as much as 25 amperes or more on larger systems. This is not, in general, enough to actuate protective devices, but it may do considerable damage if allowed to flow for a long period.

The grounded system, which has been studied, is the topic of many important meetings among the leaders of our industry over the years at IEEE meetings, National Fire Protection Association code development meetings, Canadian Electrical Code meetings, and seminars of all kinds. Proper grounding has been determined to be the safest and most economical practice to follow. The *NEC*® and *Canadian Electrical Code* (*CEC*) provide minimum standards for making the proper grounding connections on services and equipment. IEEE standards provide engineering guidelines that may be of benefit to those with unique requirements for their facility or particular installation. There are many other good documents also available.

Example: The following example illustrates how the size of the grounded (neutral) wire may be selected. Note: To simplify this exercise calculated loads are not considered A new 3-phase, 4-wire feeder is connected in an existing junction box to three single-phase circuits, each consisting of two-phase conductors and a grounded conductor. Two of these circuits are in one conduit; the third is in another. The single-phase circuit conductors are all 2 AWG THW copper. Determine the ampacity of each single-phase circuit and the required size of new service if electric discharge lighting ballasts are in the circuits, but the load is considered noncontinuous.

Step 1. Neutrals in these circuits must be counted as current-carrying conductors. (*NEC*® Article 310, Table 310.15(B)(3)(a). The ampacity of 2 AWG THW conductors from Table 310.15(B)(16) is 115 amperes.

Step 2. Ampacity adjustment is not required for the circuit ampacity in the conduit with the single circuit with only three conductors. (*NEC*® Article 310.) Thus, the allowable ampacity is 115 amperes.

Step 3. In the conduit with two circuits, an 80% ampacity adjustment factor must be used because all six wires are considered as current-carrying conductors (*NEC*® Article 310, Table 310.15(B)(3)(a). Each of these two circuits has a design ampacity of 80% × 115 = 92.

Step 4. The sum of the ampacities of all three circuits is the required ampacity of the service conductors and the service neutral (220.61).

Step 5. The value is 115 + 92 + 92 = 299. Each conductor and neutral requires not less than a 350-kcmil 75°C conductor ampacity. Reference Article 310, Table 310.15(B)(16).

QUESTION REVIEW

Each lesson is purposely designed to require the student to apply the entire *NEC*® text and not specific chapters or articles. It has been found that when studying for a timed, open-book examination, the student must gain proficiency in the Table of Contents, the Index, and the ability to move quickly from cover to cover to find the correct answer to each question in a timely fashion.

1. Required grounding conductors and bonding jumpers shall not be connected solely by _____ connections.

   Answer: _____

   _____

   _____

   _____

   Reference: _____

2. The conductor that bonds together all metal parts of an outline lighting system must be at least size _____ AWG copper.

   Answer: _____

   _____

   _____

   _____

   Reference: _____

3. A connection to a concrete-encased, driven, or buried grounding electrode shall be _____.

   Answer: _____

   _____

   _____

   _____

   Reference: _____

4. The grounded conductor of an electrical branch circuit is identified by the color _____.

   Answer: _____

   _____

   _____

   _____

   Reference: _____

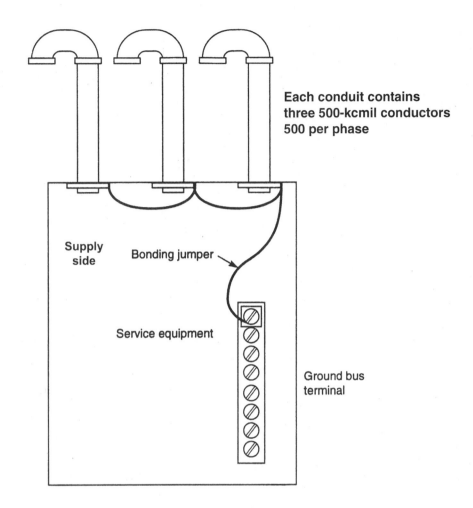

Each conduit contains
three 500-kcmil conductors
500 per phase

Supply side

Bonding jumper

Service equipment

Ground bus terminal

5. *Refer to the figure above.* Each of the supply-side EMT conduits contains three 500-kcmil copper THW service conductors in parallel. As shown, only one copper bonding jumper is used to bond all the conduits to the grounded bus terminal. This bonding jumper must be at least size _____ AWG copper.

Answer: _____

_____

_____

_____

Reference: _____

6. A rigid metal conduit contains three circuits: two 150-ampere, 3-phase circuits and one 300-ampere, single-phase circuit. The equipment bonding jumper for this conduit must be at least size _____ AWG copper.

Answer: _____

_____

_____

_____

Reference: _____

7. Are EMT set-screw connectors permitted for bonding of service raceways, or are jumpers required to bond around them?

   Answer: _____

   _____

   _____

   _____

   Reference: _____

8. Where feeder conductors are paralleled and routed in separate nonmetallic raceways, is an EGC required in each nonmetallic raceway or only in one nonmetallic raceway?

   Answer: _____

   _____

   _____

   _____

   Reference: _____

9. Are submersible deep-well pumps required to have an EGC run with the other conductors and the submersible pump motor housing grounded?

   Answer: _____

   _____

   _____

   _____

   Reference: _____

10. Is it permissible to ground the secondary of a separately derived system to the grounded terminal in the service switchboard instead of the nearest metal frame of a building or structure or nearest water pipe?

    Answer: _____

    _____

    _____

    _____

    Reference: _____

11. Where installing the grounding-electrode conductor in a ferrous metal raceway for physical protection, is the metal raceway required to be bonded to the grounding-electrode conductor? If the answer is yes, is it sufficient to bond it at the service panel only or must bonding be done at all terminations?

    Answer: _____

    _____

    _____

    _____

    Reference: _____

12. Where installing the electrical hookups for RV sites in a recreational vehicle park, can a ground rod be driven at each RV site instead of pulling an EGC with a circuit conductor from the source?

Answer: _____

_____

_____

_____

Reference: _____

13. You have recently installed a separately derived system: a dry-type transformer in a commercial building. No structural steel meeting 250.52(A)(2) was near the transformer location. The transformer secondary is grounded to a nearby metal water pipe. Is a supplemental grounding electrode required?

Answer: _____

_____

_____

_____

Reference: _____

14. When bonding the non–current-carrying metal parts of a recreational vehicle, what is the minimum size copper or equivalent bonding conductor?

Answer: _____

_____

_____

_____

Reference: _____

15. Does the *Code* require interior gas piping systems to be bonded to the grounding-electrode system?

Answer: _____

_____

_____

_____

Reference: _____

16. Two parallel runs of four 500-kcmil conductors are each run in rigid nonmetallic conduits to feed a 600-ampere panel. Is an EGC required in each conduit? What size EGC is required in each conduit?

Answer: _____

_____

_____

_____

Reference: _____

17. When terminating the EGC in an outlet box supplied by nonmetallic sheath cables, can the EGC be twisted together or must they be connected with a wire connector?

    Answer: _____

    _____

    _____

    _____

    Reference: _____

18. What size aluminum EGC will be required to be run with the circuit conductors fed from a 60-ampere fusible switch?

    Answer: _____

    _____

    _____

    _____

    Reference: _____

19. In a run of conduit supplying a motor, it is necessary to install 4 feet (1.2 m) of flexible metal conduit where the raceway leaves the panel to get around a column. When the conduit reaches the motor, an additional 3 feet (900 mm) of flexible metal conduit is employed for convenience and flexibility. Is an EGC required in this metal raceway?

    Answer: _____

    _____

    _____

    _____

    Reference: _____

20. When making an installation of a high-impedance grounded neutral system, how is the grounded system conductor to be sized?

    Answer: _____

    _____

    _____

    _____

    Reference: _____

21. When making an installation of a hydromassage tub in a residence, how are the hydromassage bathtubs and associated electrical components to be protected?

    Answer: _____

    _____

    _____

    _____

    Reference: _____

22. Underwriters Laboratory Standard UL 67 requires that backfed plug-in-type circuit breakers be secured. Does the *NEC*® require that backfed circuit breakers also be secured?

    Answer: _____

    _____

    _____

    _____

    Reference: _____

23. Where several types of grounding electrodes are present on the premises, such as an underground metal water pipe, building steel, and a concrete-encased electrode, is it permissible to jump from one electrode to another with a single properly sized grounding-electrode conductor?

    Answer: _____

    _____

    _____

    _____

    Reference: _____

24. Many contractors use metal 90° elbows in underground nonmetallic raceways for ease of pulling the conductors and to prevent burn-through. Are these 90° bends required to be bonded? If so, how can this be accomplished from a practical standpoint when they are underground?

    Answer: _____

    _____

    _____

    _____

    Reference: _____

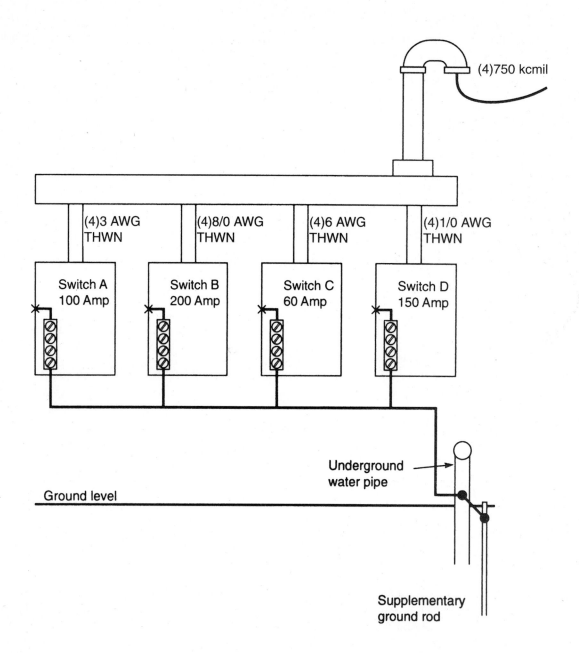

25. In the figure above, what is the required size of the common grounding-electrode conductor and the grounding-electrode conductor taps to Switch A, Switch B, Switch C, and Switch D?

Answer: _____

_____

_____

_____

Reference: _____

CHAPTER 10

OBJECTIVES

Studying this chapter along with the 2011 *NEC®* gives the reader a basic introduction to the branch-circuit and feeder requirements in the *Code* that apply to all installations above and below 600 volts.

After studying this chapter, you should know:

General Wiring Methods

■ The definitions and the application of terms that are applicable and that are unique to the 2011 *NEC®*

■ Identical section numbering scheme now used in Chapter 3 of the *NEC®* 2011

■ The use of acronyms applying to raceways and cables

■ The importance of proper selection of the wiring method raceway type or cable type for the application

Cable Types

■ The *Code* articles in which cables are addressed as they apply to the 2011 *NEC®*

■ Unique requirement for individual metal-sheathed cables types

■ Unique requirement for individual non–metallic-sheathed cables types

■ Common requirements for all cable types

(Objectives, continued)

- The unique requirements for individual metallic-sheathed cable types
- The unique requirements for individual non–metallic-sheathed cable types
- The common requirements for all cable types

Raceway Types

- The unique requirements for individual raceway types
- The common requirements for metal types
- The common requirements for nonmetallic types
- The calculation methods for determining expansion
- The calculation methods for determining the conductor fill and raceway sizing

Enclosures, Panelboards, Switchboards, and Switches

- The *Code* article numbers assigned to enclosures and equipment as they apply to the 2008 *NEC®* (see Figure 10–1 and Figure 10–2)

Cable Tray Systems

- The new *Code* section numbers assigned to cable tray systems as applicable to the 2011 *NEC®*
- The calculation methods for determining conductor fill and sizing

WIRING METHODS

This chapter deals primarily with Chapter 3, which covers the general installation requirements and all the general wiring methods. (See Figure 10–3.) In addition, we reference Chapter 9. Chapter 9 contains Tables 1 through 12(B). Table 10 has been added in the 2011 *NEC®*. Table 4 contains a table for each conduit and tubing type. These tables are necessary because each conduit and tubing type has a unique internal diameter (ID), with intermediate metal conduit (IMC) being the largest and PVC conduit Schedule 80 (PVC) being the smallest standard building wiring method.

Conduit and tubing diameters make a difference.
Trade size 1/2 (16) IMC intermediate metal conduit internal diameter 0.0060 inch

Allowable 40% fill would permit three 8 AWG Type THWN conductors.
Trade size 1/2 (16) Schedule 80 PVC conduit internal diameter 0.526 inch

Allowable 40% fill would permit one 8 AWG Type THWN conductor.

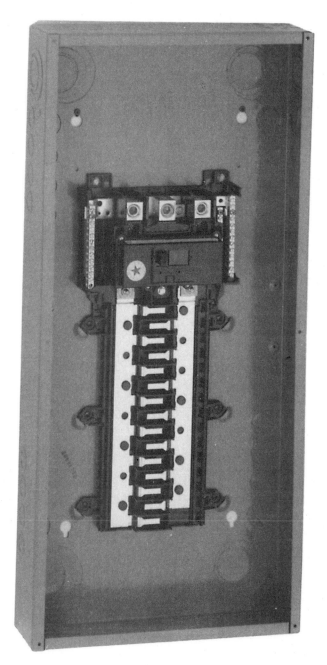

Figure 10–1 Main breaker panelboard (see NEC article 408) and enclosure (see 2011 *NEC*® Article 312). (*Courtesy of Square D Company*)

Figure 10–2 Typical safety switch with view safety glass (see 2011 *NEC*® Article 404). (*Courtesy of Square D Company*)

Table 5 contains the conductor dimensions.

Annex C is for conduit and tubing fill tables of fixture wire and conductors of the same size.

Tables 1, 4, 5, and 8 contain conduit, tubing dimensions, (conductor) wire dimensions, and notes to determine the total cross-sectional area, including insulation, to determine the allowable trade size of conduit or tubing

applicable. Tables 11(A) and 11(B) provide power source limitations for Class 2 and 3 alternating current and direct current power sources. Tables 12(A) and 12(B) provide power source limitations for fire alarm power-limited circuits both alternating current and direct current PLFA power sources. These tables are used in conjunction with the appropriate articles in the *Code*.

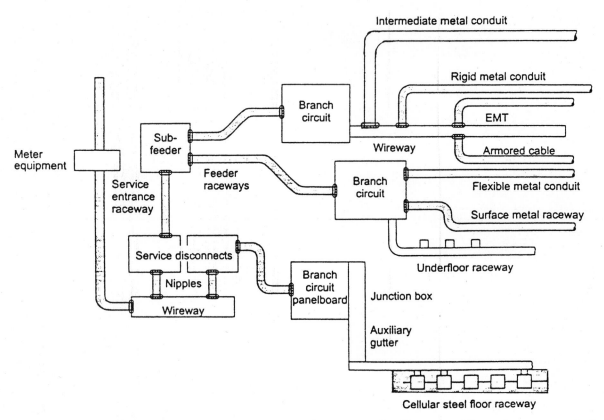

Figure 10–3 Examples of service raceways per *NEC®* Article 230. *Note: Location of overcurrent devices not shown.*

Example: A typical 480/277V, 400-A feeder is run from a distribution unit substation under a concrete slab exposed above a suspended ceiling turning down into a panelboard. Based on the length (200 feet [61 m]) of the run and the calculated load, the installer chooses to pull three 600-kcmil THWN ungrounded circuit conductors and a full-sized 600-kcmil THWN neutral (grounded conductor). If he chooses a nonmetallic raceway for a portion of this run, he recognizes that he will also have to add an equipment grounding conductor sized in accordance with Table 250.122 or a 3 AWG THWN copper. However, when he applies the tables in Chapter 9, he finds that he will need the following sizes of conduit or tubing to make this installation:

Solution:
Table 5. 600-kcmil THWN =
  0.8676 sq. in.   4 × 0.8676 = 3.47 sq. in.

Table 4.
EMT electrical metallic tubing
  40% fill as per Table 1   = 3 (78)

IMC intermediate metal conduit
  40% fill as per Table 1   = 3½ (91)
GRC galvanized rigid conduit
  40% fill as per Table 1   = 3½ (91)
Table 5. 600-kcmil THWN =
  0.8676 sq. in. 4 × 0.8676   = 3.47   sq. in.
Plus an equipment-grounding
  conductor 3 AWG THWN   = 0.0973 sq. in.
                          = 3.5673 sq. in.

Table 4.
ENT electrical nonmetallic
  tubing max. size 2 (63)   = N/A
Sch. 40 rigid nonmetallic conduit
  40% fill as per Table 1   = 3½ (91)
Sch. 80 rigid nonmetallic conduit
  40% fill as per Table 1   = 4 (103)

Using these same calculations with 500-kcmil for the ungrounded and grounded conductors would require a trade size 3 (78) conduit for EMT, IMC, GRC, and PVC Schedule 40, and a trade size 3½ (91) for PVC Schedule 80. As one can see, calculations will be necessary to determine the most efficient wiring method.

The general requirements for wiring methods are found in Article 300. As you may recall in earlier chapters of this book, you have been urged several times to learn the contents of Article 300 thoroughly because it is used in some application for every installation.

The general wiring methods in Article 300 include requirements that apply to the specific wiring methods, such as non–metallic-sheathed cable, which is covered in Article 334 (see Figure 10–4), as specific support requirements and installation criteria, such as found in 300.4(A) through (A) and the Exceptions. These requirements are mandatory and must be adhered to when making an installation of any of the wiring methods and are in Chapter 3 as they apply. Chapter 3 also covers burial depths (in Table 300.5 for conductors under 600 volts) and many conditions that apply to underground installations, such as splices, taps, and the backfill material used so as not to damage the

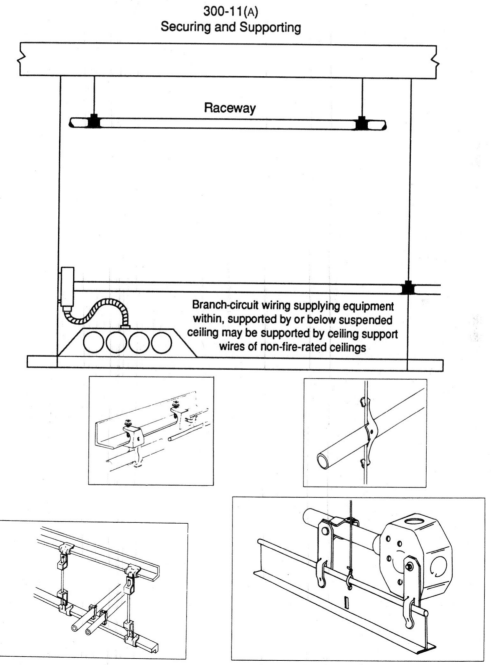

Figure 10–4 Various types of hangers and supporting methods and hardware for raceways. (*Courtesy of Cooper B-Line, Inc.*)

raceway or cable that you have selected as a wiring method for your installation. When a wiring method is installed in earth that is not suitable as physical protection, select fill material should be used, such as sand or soil without rocks.

This article also covers such things as sealing and protection against corrosion, indoor wet locations, raceways exposed to different temperatures, and expansion joints. We will be talking in this chapter about the expansion joint requirements for non-metallic raceways. However, 300.7 covers all wiring methods and where they are required to compensate for expansion or contraction (see Figure 10–5, Figure 10–6, and Figure 10–7). The electrical and mechanical continuity for metallic systems—both the raceway and the enclosures are discussed in 360.10–12, *NEC*® 300.11—requires that these methods, the raceways, cable assemblies, boxes, cabinets, and fittings all be securely fastened where supported on support wires above suspended ceilings (refer to Figure 10–4).

For branch-circuit wiring, read this section very carefully. The wiring systems for feeders or services are not permitted to be secured in this

manner; only certain branch circuits are permitted. Specifically, the junction box out of the box section in Article 314 places more stringent requirements for supporting these box systems. All must be adhered to. Article 300 also tells us where a box or fitting is required and where only a box is permitted. 300.15(A) states that all fittings and connectors must be designed for the specific wiring method for which they are used and also must be listed for that purpose. You are urged to look up the definition for a branch circuit in Article 100 at this point. A very important requirement for all wiring methods is found in 300.21. The requirement aims to limit the spread of fire or products of combustion and requires that the wiring system be installed so as to minimize the spread of fire and products of combustion and that all penetrations through floors, walls, and ceilings be sealed in an approved manner to accomplish this purpose. 300.22, which contains the provisions for the installation and uses of electrical wiring and equipment in ducts, plenums, and other air-handling spaces, deals with safety concerns in determining the wiring method to be used in electrical installations and should be studied carefully before making the final selections. Part II of Article 300 explains the requirements for installations over 600 volts, nominal. 300.50 covers the underground installations and references, which are the general wiring requirements for installations over 600 volts in the table and section in 300.50. Other general articles that relate to wiring methods are found in Chapter 3. However, temporary wiring is covered in Chapter 5. Article 590 covers the temporary wiring requirements needed to provide power for the workers making the installation and constructing the facility during the construction phases of a job. Article 590 also covers other types of temporary wiring, such as temporary wiring concerned with holiday lighting, trade fairs, emergencies and tests, and experimental and developmental work. Be aware of Article 590. Article 310 names the conductors to be used for general wiring and all of the requirements related to these conductors. The requirements do not apply to conductors that form an integral part of equipment, such as motors, motor controllers, and similar equipment, or conductors specifically provided for elsewhere in the *Code*. Examples would be Article 400 and Article 402 for flexible cords and fixture wires. Other sections dealing with conductors would be Chapter 7 and Chapter 8 of the *NEC*®, which deal with limited

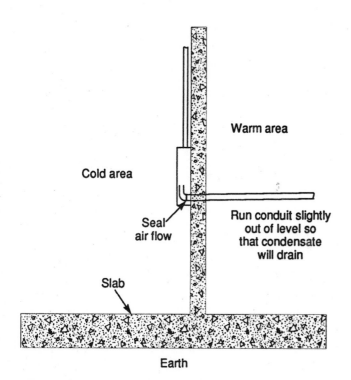

Figure 10–5 Seals are required to comply with *NEC*® 300.7 where the raceway is subjected to more than one temperature, such as shown. The system enters the building from outside.

Table 352.44(A) Expansion Characteristics of PVC Rigid Conduit Coefficient (PVC) of Thermal Expansion = 6.084 × 10⁻⁵ mm/mm/°C (3.38 × 10⁻⁵ in./in./°F)

Temperature Change (°C)	Length Change of PVC Conduit (mm/m)	Temperature Change (°F)	Length Change of PVC Conduit (in./100 ft)	Temperature Change (°F)	Length Change of PVC Conduit (in./100 ft)
5	0.30	5	0.20	105	4.26
10	0.61	10	0.41	110	4.46
15	0.91	15	0.61	115	4.66
20	1.22	20	0.81	120	4.87
25	1.52	25	1.01	125	5.07
30	1.83	30	1.22	130	5.27
35	2.13	35	1.42	135	5.48
40	2.43	40	1.62	140	5.68
45	2.74	45	1.83	145	5.88
50	3.04	50	2.03	150	6.08
55	3.35	55	2.23	155	6.29
60	3.65	60	2.43	160	6.49
65	3.95	65	2.64	165	6.69
70	4.26	70	2.84	170	6.90
75	4.56	75	3.04	175	7.10
80	4.87	80	3.24	180	7.30
85	5.17	85	3.45	185	7.50
90	5.48	90	3.65	190	7.71
95	5.78	95	3.85	195	7.91
100	6.08	100	4.06	200	8.11

Table 355.44 Expansion Characteristics of Reinforced Thermosetting Resin Conduit (RTRC) Coefficient of Thermal Expansion = 2.7 × 10⁻⁵ mm/mm/°C (1.5 × 10⁻⁵ in./in./°F)

Temperature Change (°C)	Length Change of RTRC Conduit (mm/m)	Temperature Change (°F)	Length Change of RTRC Conduit (in./100 ft)	Temperature Change (°F)	Length Change of RTRC Conduit (in./100 ft)
5	0.14	5	0.09	105	1.89
10	0.27	10	0.18	110	1.98
15	0.41	15	0.27	115	2.07
20	0.54	20	0.36	120	2.16
25	0.68	25	0.45	125	2.25
30	0.81	30	0.54	130	2.34
35	0.95	35	0.63	135	2.43
40	1.08	40	0.72	140	2.52
45	1.22	45	0.81	145	2.61
50	1.35	50	0.90	150	2.70
55	1.49	55	0.99	155	2.79
60	1.62	60	1.08	160	2.88
65	1.76	65	1.17	165	2.97
70	1.89	70	1.26	170	3.06
75	2.03	75	1.35	175	3.15
80	2.16	80	1.44	180	3.24
85	2.30	85	1.53	185	3.33
90	2.43	90	1.62	190	3.42
95	2.57	95	1.71	195	3.51
100	2.70	100	1.80	200	3.60

Figure 10–6 (Reprinted with permission from NFPA 70-2011, the *National Electrical Code®*, Copyright, National Fire Protection Association, Quincy, MA 02269. This reprinted material is not the complete and official position of the National Fire Protection Association on the referenced subject, which is represented only by the standard in its entirety.)

Example

380 ft. of conduit is to be installed on the outside of a building exposed to the sun in a single straight run. It is expected that the conduit will vary in temperature from 0°F in the winter to 140°F in the summer (this includes the 30°F for radiant heating from the sun). The installation is to be made at a conduit temperature of 90°F. From the table, a 140°F temperature change will cause a 5.7 in. length change in 100 ft. of conduit. The total change for this example is 5.7″ × 3.8 = 21.67″ which should be rounded to 22″. The number of expansion couplings will be 22 ÷ coupling range (6″ for E945, 2″ for E955). If the E945 coupling is used, the number will be 22 ÷ 6 = 3.67 which should be rounded to 4. The coupling should be placed at 95 ft. intervals (380 ÷ 4). The proper piston setting at the time of installation is calculated as explained above.

$$0 = \left[\frac{140 - 90}{140} \right] 6.0 = 2.1 \text{ in.}$$

Insert the piston into the barrel to the maximum depth. Place a mark on the piston at the end of the barrel. To properly set the piston, pull the piston out of the barrel to correspond to the 2.1 in. calculated above.

See the drawing below.

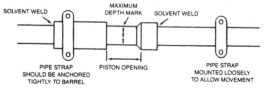

Figure 10–7 PVC conduit. For aluminum raceway, 21.7 × 0.1 = 2.17 in. an expansion fitting may be required. (*Courtesy of Carlon, a Lamson & Sessions Company*)

voltage wire, such as in Articles 725, 760, and so on. Article 310 must be used in every installation made. The requirements in 310.1 through 310.15 are general requirements for the use and installation of conductors.

> *Note:* When using the applicable tables, read each heading and each column carefully. Read the ambient temperature correction factors below the column, and remember that the notes (footnotes) under each table are a part of those tables and are mandatory requirements.

This is important when studying for a test because many test questions will require you to apply these additional parts of the table to get the correct answer. For an example, the obelisk note under each table applies to general wiring methods. However, notice that it says unless specifically otherwise provided for or permitted in this *Code.* Motors, for instance, are generally not required to comply with this note because Article 430 has specific requirements for branch-circuit conductors for motors. Study these tables carefully.

The next important part of Article 310 is 310.15 (B)(7). Table 310.15(B)(7) has less restrictive ampacity requirements for dwelling services and feeders.

> *Note:* These only apply to single-phase, 120/240 volt, 3-wire main power feeders and do not apply to any other type of systems a circuits.

Example: *NEC®* 392.22

Width selection for cable tray containing 600-volt multiconductor cables, sizes 4/0 AWG and larger only. Cable installation is limited to a single layer. The sum of the cable diameters (*Sd*) must be equal to or less than the usable cable tray width.

Cable tray width is obtained as follows:

*A—*Width required for 4/0 AWG and larger multiconductor cables [*see diagram top of page 190*]:

Item No.	List Cable Sizes	(D) List Cable Outside Diameter	(N) List Number of Cables	Multiply (D) × (N) Subtotal of Sum of Cable Diameters (Sd)
1	3/C–500 kcmil	2.25 in	4	9.00 in
2	3/C–250 kcmil	1.76 in	3	5.78 in
3	3/C–4/0 kcmil	1.55 in	10	15.50 in

The sum of the diameters (*Sd*) of all cables (add total *Sd* for items 1, 2, and 3):

9.00 in + 5.78 in + 15.50 in = 30.28 in (Sd)

A cable tray with a usable width of 30 in is required. For approximately 15% more, a 36-in wide cable tray could be purchased that would provide for some future cable additions.

> *Notes:*
> 1. Cable sizes used in this example are a random selection.
> 2. Cables—copper conductor with cross-linked polyethylene insulation and a PVC jacket. (These cables could be ordered with or without an equipment grounding conductor.)
> 3. Total cable weight per foot for this installation:
>   61.4 lb/ft (without equipment-grounding conductors)
>   69.9 lb/ft (with equipment-grounding conductors)

NEC® 392.22(A)(1)

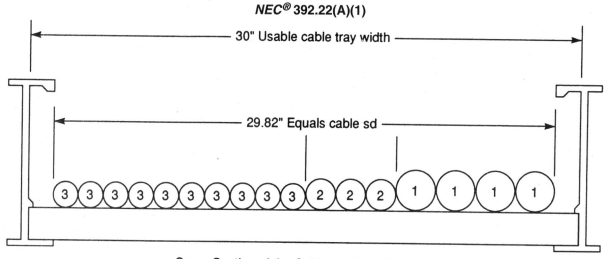

Cross Section of the Cables and the Cable Tray

(Courtesy of Cooper B-Line, Inc.)

This load can be supported by a load symbol "B" cable tray–75 lb/ft.

Example: *NEC® 392.22(A)(2)*

Width selection for cable tray containing 600-volt multiconductor cables, sizes 3/0 AWG and smaller. Cable tray allowance fill areas are listed in Column 1 of Table 392.22(A).

Method 1:

The sum of the total areas for items 1, 2, 3, and 4:

3.34 sq. in. + 3.04 sq. in. + 6.02 sq. in. + 16.00 sq. in. = 28.40 sq. in

From Table 392.9, Column 1, a 30-in (750 mm) wide tray with an allowable fill area of 35 sq in

must be used. The 30-in (750 mm) cable tray has the capacity for additional future cables (6.60 sq. in. of additional allowable fill area can be used).

Method 2:

The sum of the total areas for items 1, 2, 3, and 4 multiplied by (6 sq. in./7 sq. in.) = cable tray width required:

3.34 sq. in. + 3.04 sq. in. + 6.02 sq. in. + 16.00 sq. in. = 28.40 sq. in.

$$\frac{28.40 \text{ sq. in.} \times 6 \text{ sq. in.}}{7 \text{ sq. in.}} = \begin{array}{l}24.34 \text{ inch} \\ \text{cable tray width} \\ \text{required}\end{array}$$

Use a 30-in (750 mm) cable tray.

NEC® 392.22(A)(2)

Cross Section of the Cables and the Cable Tray

(Courtesy of Cooper B-Line, Inc.)

Notes:

1. The cable sizes used in this example are a random selection.
2. Cables—copper conductors with cross-linked polyethylene insulation and a PVC jacket. These cables could be ordered with or without an equipment-grounding conductor.
3. Total cable weight per foot for this installation: 31.9 lb/ft. (Cables in this example do not contain equipment-grounding conductors.) This load can be supported by a load symbol "A" cable tray—50 lb/ft.

Example: NEC® 392.22(A)(1)(c)

Width selection for cable tray containing 600-volt multiconductor cables, sizes 4/0 AWG and larger (single layer required) and 3/0 AWG and smaller. These two groups of cables must have dedicated areas in the cable tray.

Cable tray width is obtained as follows:

A—Width required for 4/0 AWG and larger multiconductor cables:

Item No.	List Cable Sizes	(D) List Cable Outside Diameter	(N) List Number of Cables	Multiply (D) × (N) Subtotal of Sum of Cable Diameters (Sd)
1	3/C–500 kcmil	2.26 in	3	6.78 in
2	3/C–4/0 kcmil	1.55 in	4	6.20 in

Total cable tray width required for items 1 and 2:

6.78 in + 6.20 in = 12.98 in

B—Width required for 3/0 AWG and smaller multiconductor cables:

Total cable tray width required for items 3, 4, and 5:

(3.20 sq. in. + 4.00 sq. in. + 3.20 sq. in.) (6 sq. in. / 7 sq. in.)1 = (10.4 sq. in.) (6 sq. in. / 7 sq. in.)1 = 8.92 in

Actual cable tray width is A "width" (12.98 in) + B "width" (8.92 in) = 21.90 in

A 24-in (600 mm) wide cable tray is required. The 24-in (600 mm) cable tray has the capacity for additional future cables (3.1 in or 3.6 sq. in allowable fill can be used).

Notes:

1. This ratio is the inside width of the cable tray in inches divided by its maximum fill area in square inches from Table 392.22(A), Column 1.
2. The cable sizes used in this example are a random selection.
3. Cables—copper conductor with cross-linked polyethylene insulation and a PVC jacket.
4. Total cable weight per foot for this installation: 40.2 lb/ft. (Cables in this example do not contain equipment grounding conductors.) This load can be supported by a load symbol "A" cable tray—50 lb/ft.

NEC® 392.22(A)(1)(C)

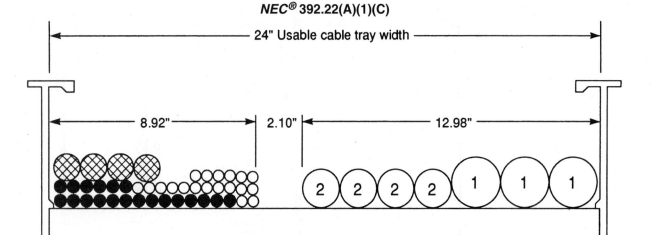

Cross Section of the Cables and the Cable Tray

(Courtesy of Cooper B-Line, Inc.)

Example: NEC® Section 392.22(A)(2)

50 % of the cable tray usable cross-sectional area can contain type PLTC cables.

4 in. × 6 in. × 0.500 = −12 sq. in. allowable fill area

2/C 16 AWG 300-volt shielded instrumentation cable O.D. = 0.224 in.

Cross-Sectional Area = 0.04 sq. in.

$$\left(\frac{12\,sq.in.}{0.04\,sq.in.\,cable}\right) = 300\ \text{cables and be installed in the cable tray}$$

$$\left(\frac{300\,cables}{26\,cables\,/\,row}\right) = 11.54\ \text{rows can be installed in the cable tray}$$

Notes:

1. The cable sizes used in this example are a random selection.
2. Cable—copper conductors with PVC insulation, aluminum/mylar shielding, and PVC jacket.

Following these general application articles (300 and 310), the remaining articles in Chapter 3 are specific articles. Article 392 covers cable trays (again see Figure 10–4). Cable trays are a support system and not a wiring method.

Note: As you begin to study this part of Chapter 3, refer to the definitions in Article 100 for a raceway. An informational note will alert you to the types of raceways. There is a distinct difference between raceways and cables and other wiring methods. Study that definition or refer to it as needed. A cable tray is neither a raceway nor a cable. It is merely a support system for other wiring methods.

392.10 Uses Permitted. Cable tray shall be permitted to be used as a support system for service conductors, feeders, branch circuits, communications circuits, control circuits, and signaling circuits. Cable tray installations shall not be limited

NEC® 392.22(A)(2)

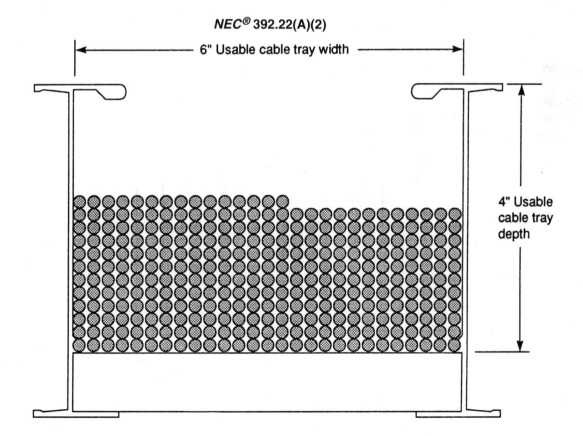

Cross Section of the Cables and the Cable Tray

(Courtesy of Cooper B-Line, Inc.)

to industrial establishments. Where exposed to direct rays of the sun, insulated conductors and jacketed cables shall be identified as being sunlight resistant. Cable trays and their associated fittings shall be identified for the intended use.

(A) Wiring Methods. The wiring methods in **Table 392.10(A)** shall be permitted to be installed in cable tray systems under the conditions described in their respective articles and sections.

(B) In Industrial Establishments. The wiring methods in **Table 392.10(A)** shall be permitted to be used in any industrial establishment under the conditions described in their respective articles. In industrial establishments only, where conditions of maintenance and supervision ensure that only qualified persons service the installed cable tray system, any of the cables in **392.10(B)**(1) and **(B)(2)** shall be permitted to be installed in ladder, ventilated trough, solid bottom, or ventilated channel cable trays.

(1) Single-conductor cables shall be permitted to be installed in accordance with (B)(1)(a) through (B)(1)(c).

(a) Single-conductor cable shall be 1/0 AWG or larger and shall be of a type listed and marked on the surface for use in cable trays. Where 1/0 AWG through 4/0 AWG single-conductor cables are installed in ladder cable tray, the maximum allowable rung spacing for the ladder cable tray shall be 225 mm (9 in.).

(b) Welding cables shall comply with the provisions of Article **630**, Part IV.

(c) Single conductors used as equipment grounding conductors shall be insulated, covered, or bare, and they shall be 4 AWG or larger.

(2) Single- and multiconductor medium voltage cables shall be Type MV cable. Single conductors shall be installed in accordance with **392.10(B)**(1).

(C) Hazardous (Classified) Locations. Cable trays in hazardous (classified) locations shall contain only the cable types and raceways permitted by other articles in this *Code*.

(D) Nonmetallic Cable Tray. In addition to the uses permitted elsewhere in **392.10**, nonmetallic cable tray shall be permitted in corrosive areas and in areas requiring voltage isolation.

392.12 Uses Not Permitted. Cable tray systems shall not be used in hoistways or where subject to severe physical damage.

Example: Application of 392.80(B). Ampacity of Type MV and Type MC Cables (2001 Volts or Over) in Cable Trays—(B) Single Conductor cables. These single conductor cables can be installed in a cable tray cabled together (triplexed, quadruplexed, and so on) if desired. Where the cables are installed according to the requirements of 392.22(C), the ampacity requirements are as shown in the accompanying chart.

Refer to the articles in Chapter 3 for the specific types of raceways and cables as needed. In addition to the references for raceways in Chapter 3, you will find references related to Chapter 9 tables and examples, such as the reference found in 342.22, the numbers of conductors in conduit. When sizing the number of conductors in a raceway, conduit, tubing, and so on, it is necessary to refer to the permitted percent fills

Sec. No.	Cable Sizes	Solid Unventilated Cable Tray Cover	Applicable Ampacity Tables*	Amp. Table Values By	Special Conditions
(1)	1/0 AWG and Larger	No Cover Allowed**	310.60(C)(69) and 310.60(C)(70)	0.75	—
(1)	1/0 AWG and Larger	Yes	310.60(C)(69) and 310.60(C)(70)	0.70	—
(2)	1/0 AWG and larger in Single Layer	No Cover Allowed**	310.60(C)(69) and 310.60(C)(70)	1.00	Maintained Spacing of 1 Cable Diameter
(2)	Single Conductor in Triangle Config. 1/0 AWG	No Cover Allowed**	310.60(C)(70) and 310.60(C)(72)	1.00	Maintained Spacing of 2.15 × 1 Conductor O.D.

*The ambient ampacity correction factors must be used.

**At a specific position where it is determined that the tray cables require mechanical protection, a single cable tray cover of 6 feet (1.8 m) or less in length can be installed.

specified in the tables in Chapter 9, Table 1, for percentage fill and Table 4 for the conduit dimensions. Similar references are found in 344.22 for rigid metal conduit, 352.22 for rigid PVC conduit, 358.22 for electrical metallic tubing, and 348.22 for flexible metallic conduit. Other similar references are found for other raceway types. The number of conductors allowed in a given raceway must be calculated in square inches based on the insulation and wire size listed in Chapter 9, Table 5 and Table 5-A, for compact strand aluminum. For bare conductors, Chapter 8 dimensions are used. Where conductors of the same size are installed in a single raceway, Tables C1 through C12A in Annex 3 may be used.

The following example will help you learn to apply the provisions of Chapter 9 where necessary to calculate the proper size raceway for the enclosed conductors.

Example: What size PVC conduit is required for a multiple branch circuits supplying a cooling tower located adjacent to a hospital containing three 6 AWG THWN stranded copper conductors, three 4 AWG THWN copper conductors, four 1/0 AWG THW copper conductors, and one 2 AWG bare copper conductor?

Step 1. Rigid PVC conduit, Article 352. Number of conductors, 352.22; Ref. Table 1, Chapter 9.

Step 2. Chapter 9, Notes 1, 2, Table 1 and Notes. Read the notes to the table carefully because they are a part of the table and, where applicable, they are mandatory.

Step 3. Chapter 9 Table 1. Over two conductors permit a 40 % fill of the raceway.

Step 4. Table 4. Dimensions and percent are of conduit and tubing in square inches. Therefore, select square-inch dimensions from Tables 5 and 8 for the correct calculations.

Step 5.
Table 5.
3 6 AWG THWN = 0.0507 × 3 = 0.1521 sq. in.
3 4 AWG THWN = 0.0824 × 3 = 0.2472 sq. in.
4 6 AWG THWN = 0.2223 × 4 = 0.8892 sq. in.

Table 8.
1 2 AWG BARE = 0.067 × 1 = <u>0.067</u> sq. in.

Total = 1.3555 sq. in

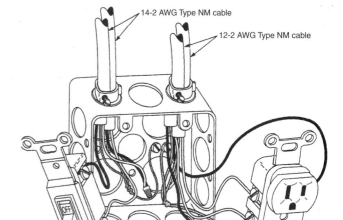

NEC® 314.16(A)(2)

14-2 AWG Type NM cable
12-2 AWG Type NM cable

Table 4.
Over two conductors in a raceway 40 %. Trade size 2 (53) Rigid PVC Schedule 40 raceway is 1.316 sq. in. Trade size 2½(63) raceway is 1.878. Therefore, 2 (53) is not large enough, and it is necessary to install 2½ (63) PVC conduit for this installation.

Note: Trade size 2 (53) (IMC) intermediate metal conduit (Article 342) would permit a 40 % fill of 1.452 sq. in. Therefore, trade size 2 (53) IMC could be used.

Article 314 covers outlet boxes, pull boxes, junction boxes, conduit bodies, and fittings. A conduit body is what is normally referred to in the trade as a Tee, LB, or as a condulet. Use Article 314 carefully, and as you determine the number of conductors for a given box, the sizing requirements will be found in 314.16.

Note: A deduction must be added for boxes with the integral-type clamps, such as nonmetallic outlet boxes. In a test, normally the test material will tell you whether there are clamps to be calculated. Study 314.16(A) and (B) and the tables carefully.

Example: What size outlet box with integral clamps is required to accommodate a 15-ampere snap switch to control lighting and a 20-ampere duplex receptacle? The switch has a 2-wire w/GRD 14 AWG non–metallic-sheathed cable feeding it with another 14 AWG NM 2-wire

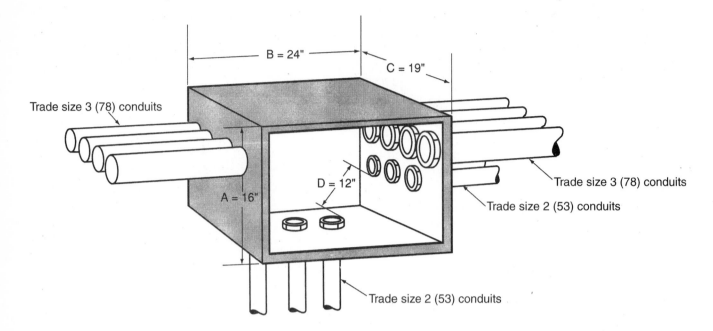

w/GRD feeding the luminaire. The receptacle is fed with a 2-wire w/GRD 12 AWG non–metallic-sheathed cable, and a 2-wire w/GRD 12 AWG non–metallic-sheath cable goes on to feed additional duplex receptacles in the same area.

Solution: 314.16(A)(1) and (2): The minimum size of box in cubic inches is based on the following calculations:

The required deduction for each type fitting (i.e., clamps, hickeys, and the like) and all equipment-grounding conductors require one deduction for each based on the largest conductor entering the box; 314.16(A) (1), (2), and Table 314.16(B).

In this example, a 12 AWG is the largest; therefore, a 2.25 cubic inch deduction is required for each.

Equipment grounding conductors = 2.25
Cable clamps = 12 AWG 2.25 × 1 = 2.25

Each device requires two deductions each based on the size of conductor connected to the device.

Receptacle = 12 AWG 2 × 2.25 = 4.50
Switch = 14 AWG 2 × 2.00 = 4.00

Conductors entering the box—one deduction each.

12 AWG 2.25 × 4 = 9.00
14 AWG 2.00 × 4 = 8.00

Answer: The minimum size box permitted would be a 30 cubic in. box.

NEC® 314.16(B)(4) also requires that devices or utilization equipment wider than a 2-inch (50 mm) device box must have double volume allowances for each gang required for mounting even if it will fit in a single gang box. A dryer receptacle is an example of this.

In addition, sizing of pull boxes and junction boxes will be included in most examinations. These requirements are found in 314.28.

Example: Four trade size 3 (78) rigid steel conduits or EMT, straight pull. Three trade size 2 (53) rigid steel conduits or EMT, angle pull.

See the figure above. Recommended minimum spacing for trade size 3 (78) conduits, 4¾ in.; trade size for 2 (53) conduits, 3⅜ in.; and between a trade size 3 (78) and a trade size 2 (53), conduit, 4 inches.

For depth of box (A):

Three trade size 2 (53) conduits, angle pull:

$$A = (6 \times 2) + 2 + 2 \qquad = 16 \text{ in.}$$

Spacing for two rows of conduit:

Edge of box to center of trade size 3 (78) conduit	= 2⅜ in.
Center to center, trade size 3 (78) to trade size 2 (53) conduit	= 4
Minimum distance, *D*	= 8½
Total	= 14⅞ in.

Required depth: Use 16 in.

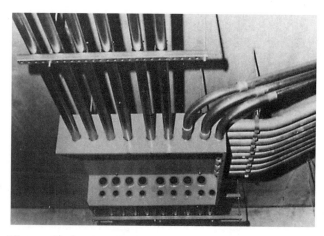

Figure 10–8 A typical pull box (junction manufactured and sized in accordance with Article 314). (*Courtesy of Allied Tube and Conduit*)

Figure 10–9 Typical cabinet manufactured in accordance with Article 312. (*Courtesy of Unity Manufacturing*)

For length of box (B):

Four trade size 3 (78) conduits, straight pull:

$B = 8 \times 3$ = 24 in.

Trade size 2 (53) conduit, angle pull:

$B = A$ = 16 in.

Required length: Use 24 in.

For width of box (C):

Recommended spacing for
trade size 3 (78) conduits   = 4¾ in.

$C = 4 \times 4¾$ in.   = 19 in.

Required width: Use 19 in.

Angle pull distance (D):

$D = 6 \times 2$ in.   = 12 in.

Required distance: Use 12 in.

This section should also be studied, and you must know where to find it quickly so that you can determine the size of pull of junction box required (see Figure 10–8).

Article 312 covers cabinets, cutout boxes, and meter socket enclosures. Most panelboards, switches, disconnecting switches, conductors, and so on are enclosed in a cabinet or cutout box (see Figure 10–9). When determining the requirements for cabinets and cutout boxes, refer to Article 312.

Article 366 covers auxiliary gutters. Auxiliary gutters are not raceways or cables. An auxiliary gutter is an extension used to supplement wiring spaces in meter enclosures, motor control centers, distribution centers, switchboards, and similar wiring systems and may enclose conductors or bus bars but is not permitted for switches, overcurrent devices, appliances, or other similar equipment. An auxiliary gutter, although very similar in appearance to a wireway, has a different application and use. Be aware of Article 366 for auxiliary gutters (see Figure 10–10).

Article 404 covers switches. In this article, you may find requirements for all types of switches and some installation requirements, such as accessibility and grouping. *NEC®* Article 406 covers receptacles, cord conductors, and attachment caps. This article now requires all 125- and 250-volt non–locking-type receptacles installed in a wet or damp location to be listed as "weather-resistant" type.

Section 406.11 requires that all 125-volt, 15- and 20-ampere receptacles installed in dwellings as specified per 210.52 must be listed as "tamper-resistant receptacles."

New receptacle rules for the 2011 NEC

406.13 now requires all non–locking-type, 125-volt, 15- and 20-ampere receptacles located in guest rooms and guest suites be listed tamper-resistant receptacles and 406.14 requires all child care facilities, all non–locking-type, 125-volt, 15- and 20- ampere receptacles be listed tamper-resistant receptacles.

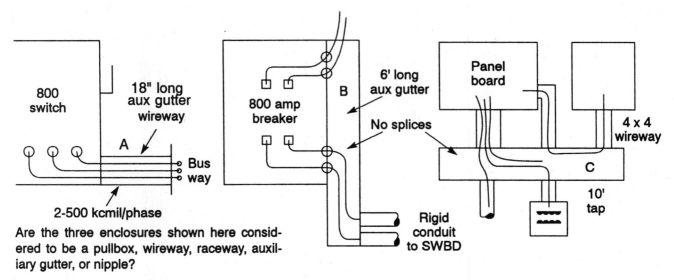

Are the three enclosures shown here considered to be a pullbox, wireway, raceway, auxiliary gutter, or nipple?

Figure 10–10 This figure was given to a panel of *NEC®* experts and the question was asked, "Which of these sketches are auxiliary gutters?" The consensus of that group was that only the center sketch is an auxiliary gutter (Article 366); the other two are wireways (Article 376).

New replacement rules

406.4(D)(5) requires Listed tamper-resistant receptacles be used where replacements are made at receptacle outlets that are required to be tamper-resistant elsewhere in this *Code.*

406.4(D)(6) Weather-Resistant Receptacles. Weather-resistant receptacles shall be provided where replacements are made at receptacle outlets that are required to be so protected elsewhere in this *Code.*

Article 408 covers switchboards and panelboards. This article will undoubtedly be included as a part of most examinations and includes many of the requirements for panelboards and switches. Of particular importance is 110.26(F), which applies to equipment covered in Articles 408 and 404 for the dedicating working space above and below equipment. This section will also undoubtedly be referred to in most examinations. You need not try to memorize this because it contains specific requirements for this equipment, and your proficiency in using this as a reference document will adequately guide you through these wiring method articles. Be familiar with the article names as listed in the Table of Contents, which will greatly assist you in using this book as a reference or as a tool in the field and will greatly increase your proficiency in preparing for an examination.

QUESTION REVIEW

Each lesson is purposely designed to require the student to apply the entire *NEC®* text and not specific chapters or articles. It has been found that when studying for a timed, open-book examination, the student must gain proficiency in the Table of Contents, the Index, and the ability to move quickly from cover to cover to find the correct answer to each question in a timely fashion.

1. When installing Type SER service entrance cable in a building from the service panel to a dryer outlet, what are the support requirements for this cable?

   Answer: _____

   _____

   _____

   _____

   Reference: _____

2. When installing conduit bodies that are legibly marked with their cubic inch capacity, how can one determine the number of conductors, splices, and taps permitted?

   Answer: _____

   _____

   _____

   _____

   Reference: _____

3. Is trade size ⅜(12) liquidtight flexible metal conduit permitted for enclosing motor leads in accordance with 430.245(B)?

   Answer: _____

   _____

   _____

   _____

   Reference: _____

4. Structural bar joists are often spaced at up to 5-foot (1.5 m) intervals. Support is difficult to achieve when installing metal raceways on these bar joists. Is it permitted to mount an outlet box on a bar joist and extend intermediate metal conduit (IMC) 5 feet (1.5 m) to the first support?

   Answer: _____

   _____

   _____

   _____

   Reference: _____

5. Is it permissible to install electrical nonmetallic tubing (ENT), through metal studs without supporting the ENT every 3 feet (900 mm)?

Answer: _____

_____

_____

_____

Reference: _____

6. What are the support requirements for Type MC cable?

Answer: _____

_____

_____

_____

Reference: _____

7. What are the support requirements and methods of securing Type NM cable?

Answer: _____

_____

_____

_____

Reference: _____

8. Can nonmetallic wireways be installed in hazardous Class I or Class II locations?

Answer: _____

_____

_____

_____

Reference: _____

9. What wiring methods are permitted for installing a branch circuit routed through the bar joists of a metal building for a 4160-volt, 3-phase motor?

Answer: _____

_____

_____

_____

Reference: _____

10. A contract to install a branch circuit under a bank president's desk to serve a calculator is awarded to a contractor. The bank has asked that there be no drilled holes in the walls. Can the conductors be laid on the floor under the carpet? If so, what type of wiring method should be used?

    Answer: _____

    _____

    _____

    _____

    Reference: _____

11. When installing a wireway metallic or nonmetallic, either type, how do you calculate the ampacity adjustment for the conductors when exceeding the number permitted?

    Answer: _____

    _____

    _____

    _____

    Reference: _____

12. When installing Type AC cable on wood studs, routing them parallel to the framing members, what distance clearance must be maintained where the Type AC cable is likely to be penetrated by screws or nails?

    Answer: _____

    _____

    _____

    _____

    Reference: _____

13. Is it required to support electrical metallic tubing (EMT) within 3 feet (900 mm) of each coupling or fitting?

    Answer: _____

    _____

    _____

    _____

    Reference: _____

14. How do the installation requirements for IMC and that of rigid metal conduit (RMC) differ?

    Answer: _____

    _____

    _____

    _____

    Reference: _____

15. What is the minimum size conductor that can be run in parallel, generally? (Do not consider exceptions.)

Answer: _____

_____

_____

_____

Reference: _____

16. Where installing electrical nonmetallic tubing for the branch-circuit conductors in a five-story building, does the material from which the walls, floors, and ceilings are constructed need to be considered?

Answer: _____

_____

_____

_____

Reference: _____

17. Holiday tree lights have been installed on the local courthouse. The mayor suggested that they be left up throughout the year. Is this permitted? How long are they permitted to remain installed?

Answer: _____

_____

_____

_____

Reference: _____

18. When installing a run of conduit, it is necessary to install a piece of liquidtight flexible metal conduit 6 feet (1.8 m) long through a plenum. Is this permitted by the *Code*? How many feet of liquidtight flexible metal conduit can you install in a plenum used for environment air?

Answer: _____

_____

_____

_____

Reference: _____

19. Where making an installation of PVC conduit, Schedule 80, along the outside of a building to feed outdoor floodlights and studying the *Code*, you find that in 300.7(B), expansion joints must be provided where necessary. How much expansion or contraction must PVC conduit have before an expansion fitting is required?

Answer: _____

_____

_____

_____

Reference: _____

20. A column panel is a narrow panel that is installed inside or flush with an I-beam. It is primarily used in warehouses and areas where the panel is subjected to moving traffic and warehousing storage, such as forklifts and palletizing. When installing a column-type panel, an auxiliary gutter is installed above the panel, and the neutral connections are made up at the top of the auxiliary gutter and not brought down into the panelboard. Is this an acceptable installation method?

Answer: _____

_____

_____

_____

Reference: _____

21. When wiring a single-family dwelling to save space, a trade size 2 (53) rigid conduit is installed from the top of the panelboard into the attic space and non–metallic-sheathed cables are pulled within that trade size 2 (53) conduit. Is this a permitted wiring method?

Answer: _____

_____

_____

_____

Reference: _____

22. A small lake cabin is to be wired. This cabin will contain wood heating, no air conditioning, and only six single-pole branch circuits for lighting and receptacle outlets. Using only six switches in this panel, are you exempt from using a main circuit breaker?

Answer: _____

_____

_____

_____

Reference: _____

23. A location in a small commercial office building appears to be dangerous. The light switches located on the wall are installed in a standard three-gang switch box. Each light switch controls a row of fixtures, and each is fed from a different phase in the panelboard. The light switch on the right is fed from B-phase of a 480-volt, 277-volt panelboard. The center switch is fed from B-phase, and the switch on the right is fed from C-phase, making the potential between each switch at 480 volts. This appears to be dangerous. Which section in the *Code* prohibits this installation?

Answer: _____

_____

_____

_____

Reference: _____

24. Two new circuits are to be added in an existing old house with knob-and-tube wiring. The owner has asked that this extension be made with the same wiring method that is now in the house: concealed knob-and-tube. The attic space has had 6 in. (150 mm) of blown insulation installed; the wires are covered by the insulation. The owner has stated that you should pull the insulation back, make the extension, and then relocate the insulation over the wiring. Is this wiring method still acceptable in the *Code*?

Answer: _____

_____

_____

_____

Reference: _____

25. What is the minimum size single conductor cable permitted within cable trays to be used as ungrounded circuit conductors?

Answer: _____

_____

_____

_____

Reference: _____

CHAPTER 11

OBJECTIVES

Studying this chapter along with the 2011 *NEC*® gives the reader a basic introduction to the branch-circuit and feeder requirements in the *Code* that apply to all installations above and below 600 volts.

After studying this chapter, you should know:

■ The definitions and the application of terms that are applicable and that are unique to the 2011 *NEC*®

■ The utilization equipment requirements found in this chapter

■ The flexible cords and fixture wire requirements and applications

■ The switch types, requirements, and applications

■ The requirements for receptacles, cord connectors, and attachment plugs (caps)

■ The switchboard and panelboard requirements and applications

■ The basic requirements for luminaires, motors, transformers, and other general-use equipment commonly used in many installations

GENERAL-USE UTILIZATION EQUIPMENT

The equipment covered in Chapter 4 is equipment found generally in all occupancies—residential, commercial, and industrial. These are things that utilize electricity to operate and include items such as luminaires, motors, and appliances. They are not considered as special equipment, which is dealt with in Chapter 6. Chapter 4 starts out with two articles, Articles 400 and 402, dealing with wire. One might think that these articles are misplaced and should be found in Chapter 3. However, flexible cords, cables, and fixture wires are most generally associated with general-purpose equipment, such as flexible cords for appliances and motors and fixture wiring for luminaires. The application for these articles is not the same as general wiring. *NEC®* 400.7 and 400.8 cover the uses permitted and not permitted. Cords are not generally used as a fixed-wiring method; there are exceptions for limited length and use applications. It would not be acceptable to connect many types of equipment with a flexible cord—for example, water heaters, gas furnaces, luminaires—except as permitted in 410.24(A). Many appliances, however, are listed with a flexible cord attached or are designed to have a flexible cord installed—for example, dishwashers, disposals, washing machines, dryers, and ranges. Article 402 is used even less in normal applications than is Article 400. Apply the rules carefully and in the proper application.

Article 410 covers luminaires, portable luminaires, lampholders, pendants, incandescent filament lamps, arc lamps, electric-discharge lamps, decorative lighting products, lighting accessories for temporary seasonal and holiday use, portable flexible lighting products, and the wiring and equipment forming part of such products and lighting installations. Article 410 does not cover those lighting systems that consist of an isolating power supply with low-voltage luminaires and associated equipment identified for the use where the output circuits of the power supply are rated for not more than 25 amperes and operate at 30 volts (42.4 volts peak) or less under all load conditions. These systems are covered in Article 411.

The term "luminaire(s)" replace the terms "lighting fixture," "lighting fixtures," and "fixture(s)." The reason for this change was that luminaire(s) has a broader use internationally. Article 406—Receptacles, Cord Connectors, and Attachment Plugs (Caps) is a compilation of requirements that were formerly found in Article 210 and Part L of Article 410. This was done

in an attempt to make the *Code* more user-friendly. Many had believed that receptacles and plugs should be a part of Article 404 (switches). They are not utilization equipment, and those just beginning to study the *Code* should mark this article or make notes to remind themselves where this information is located. Notes can be beneficial to quickly find items in the *Code*. However, do not mark up your *Code* book so that there are so many notes that they are difficult to distinguish from each other. Some testing agencies do not permit the book used for the examination to be marked up. (See Figure 11–1, Figure 11–2, and Figure 11–3.) In Article 406, the terms "tamper-resistant" and "weather-resistant" and requirements for their use have been added. Section 406.12 requires listed tamper-resistant 125-volt, 15- and 20-ampere receptacles for all areas specified in 210.52. This section applies to all dwellings. Sections 406.9(A) and 406.9(B) for damp and wet locations require that all 15- and 20-ampere, 125- and 250-volt nonlocking receptacles shall be a listed weather-resistant type. Sections 406.19 and 406.14 extends these requirements to quest rooms, guest suites, and child care facilities.

Figure 11–1 Compliance with 406.9. *(Courtesy of TayMac Corporation)*

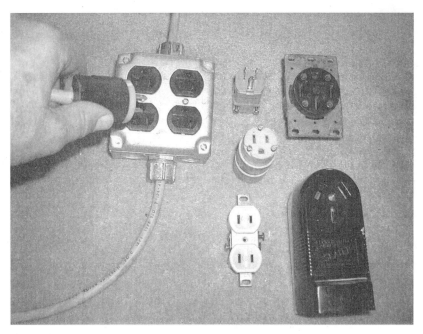

Figure 11–2 Examples of common receptacles, cord connectors, and attachment plugs (caps) covered in Article 406.

As you study Article 410, it will quickly become apparent that this article is divided into sections that cover only specific types of luminaires—that is, incandescent, high discharge (including fluorescent), and recessed fixtures and lampholders. 410.10 covers the requirements for installing fixtures in specific locations. This section contains the only requirements found in the *NEC®* for locating fixtures in bathrooms and clothes closets. Many believe that the *Code* contains requirements for protecting luminaires in bathrooms with ground-fault protection. Although it is not prohibited, the *NEC®* contains no such requirements.

---

1. The maximum wattage permitted for a medium-base incandescent lamp (standard lightbulb) is _____ watts.
2. The maximum wattage permitted for a mogul-base incandescent lamp is _____ watts.
3. A special base or other means must be used for incandescent lamps of _____ watts and larger.

Answers: 1. 300;   2. 1500;   3. 1501.
Reference: Section 410.103

Section 410.130(G), requires a disconnecting means either internal or external to each fluorescent luminaires that utilize double-ended lamps and contain ballast(s) that can be serviced in place. The line side terminals of the disconnecting means shall be guarded.

The disconnecting means shall simultaneously break all the supply conductors to the ballast, including the grounded conductor in a multiwire circuit.

The disconnecting means shall be accessible to qualified persons before servicing or maintaining the ballast. Where the disconnecting means is external to the luminaire, it shall be a single device and shall be attached to the luminaire or the luminaire shall be located within sight of the disconnecting means.

This new rule does not apply in dwellings or outdoors, and is retroactive when a ballast is replaced. There are five exceptions where the new rule does not apply: hazardous (classified) location(s); for emergency illumination; for cord-and-plug-connected luminaries; in industrial establishments with restricted public access where conditions of maintenance and supervision ensure that only qualified persons service the installation by written procedures; and in locations where more than one luminaire is installed and supplied by other than a multiwire branch circuit, a disconnecting means shall not be required for every luminaire when the design of the installation includes disconnecting means, such that the illuminated space cannot be left in total darkness.

Article 411—Lighting Systems Operating at 30 Volts or Less: This is a very short article with only seven sections, and the requirements are very easy to understand. However, one must be aware that it exists; otherwise, misapplication of Article 410 may result.

LANGUAGE *NEC®* 406.9 Receptacles in Damp or Wet Locations*

406.9 Receptacles in Damp or Wet Locations.

(A) Damp Locations. A receptacle installed outdoors in a location protected from the weather or in other damp locations shall have an enclosure for the receptacle that is weatherproof when the receptacle is covered (attachment plug cap not inserted and receptacle covers closed).

An installation suitable for wet locations shall also be considered suitable for damp locations.

A receptacle shall be considered to be in a location protected from the weather where located under roofed open porches, canopies, marquees, and the like, and will not be subjected to a beating rain or water runoff. All 15- and 20-ampere, 125- and 250-volt nonlocking receptacles shall be a listed weather-resistant type.

Informational Note: The types of receptacles covered by this requirement are identified as 5-15, 5-20, 6-15, and 6-20 in ANSI/NEMA WD 6-2002, National Electrical Manufacturers Association Standard for Dimensions of Attachment Plugs and Receptacles.

(B) Wet Locations.

(1) 15- and 20-Ampere Receptacles in a Wet Location. 15- and 20-ampere, 125- and 250-volt receptacles installed in a wet location shall have an enclosure that is weatherproof whether or not the attachment plug cap is inserted. For other than one- or two-family dwellings, an outlet box hood installed for this purpose shall be listed, and where installed on an enclosure supported from grade as described in **314.23(B)** or as described in **314.23(F)** shall be identified as "extra-duty." All 15- and 20-ampere, 125- and 250-volt non–locking-type receptacles shall be listed weather-resistant type.

Informational Note No. 1: Requirements for extra-duty outlet box hoods are found in ANSI/UL 514D-2000, *Cover Plates for Flush-Mounted Wiring Devices.*

Informational Note No. 2: The types of receptacles covered by this requirement are identified as 5-15, 5-20, 6-15, and 6-20 in ANSI/NEMA WD 6-2002, National Electrical Manufacturers Association Standard for Dimensions of Attachment Plugs and Receptacles.

PURPOSE To ensure receptacles remain weatherproof when equipment is used in damp or wet locations.

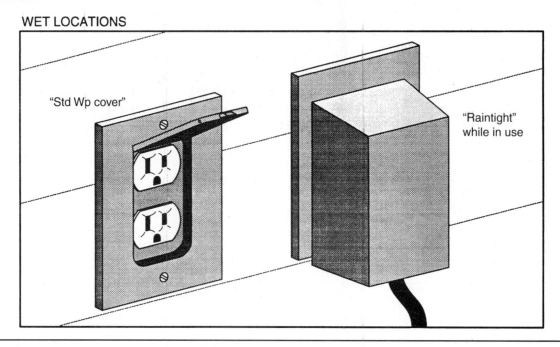

WET LOCATIONS

"Std Wp cover"

"Raintight" while in use

Figure 11–3 Weatherproof cover on the left meets 406.9(A). The weatherproof cover on the right meets the general rule 406.9(B). (*Courtesy of TayMac Corporation*) (*Reprinted with permission from NFPA 70-2011, the *National Electrical Code®* Copyright 2007, National Fire Protection Association, Quincy, MA 02269. This reprinted material is not the complete and official position of the National Fire Protection Association on the referenced subject, which is represented only by the standard in its entirety.)

Exception: 15- and 20-ampere, 125- through 250-volt receptacles installed in a wet location and subject to routine high-pressure spray washing shall be permitted to have an enclosure that is weatherproof when the attachment plug is removed.

(2) Other Receptacles. All other receptacles installed in a wet location shall comply with (B)(2)(a) or (B)(2)(b).

(a) A receptacle installed in a wet location, where the product intended to be plugged into it is not attended while in use, shall have an enclosure that is weatherproof with the attachment plug cap inserted or removed.

(b) A receptacle installed in a wet location, where the product intended to be plugged into it will be attended while in use (e.g., portable tools), shall

have an enclosure that is weatherproof when the attachment plug is removed.

(C) Bathtub and Shower Space. Receptacles shall not be installed within or directly over a bathtub or shower stall.

(D) Protection for Floor Receptacles. Standpipes of floor receptacles shall allow floor-cleaning equipment to be operated without damage to receptacles.

(E) Flush Mounting with Faceplate. The enclosure for a receptacle installed in an outlet box flush-mounted in a finished surface shall be made weatherproof by means of a weatherproof faceplate assembly that provides a watertight connection between the plate and the finished surface.

Figure 11–3 *(Continued)* *(Courtesy of TayMac Corporation)* (*Reprinted with permission from NFPA 70-2011, the *National Electrical Code®* Copyright 2007, National Fire Protection Association, Quincy, MA 02269. This reprinted material is not the complete and official position of the National Fire Protection Association on the referenced subject, which is represented only by the standard in its entirety.)

Article 422 covers appliances. "*Appliance*" is defined in Article 100 as *Utilization equipment, generally other than industrial, that is normally built in standardized sizes or types and is installed or connected as a unit to perform one or more functions such as clothes washing, air conditioning, food mixing, deep frying, and so forth.* This article gives some special requirements for calculating the branch circuits supplying appliances, but Articles 210 and 220 govern generally. 422.11 should be studied carefully because specific overcurrent protection requirements are contained in this section for appliances.

NEC® 422.12 states that central heating equipment must be supplied by an "**individual branch circuit.**" What is an individual branch circuit?

Answer: An individual branch circuit is defined in Article 100 as a branch circuit that supplies only one utilization equipment.

Would it be permissible to connect a waste (garbage) disposal and a built-in dishwasher with flexible cords? Both cords would be a minimum of 42 in. (1.1 m).

Answer: *NEC*® 422.16(B)(2) would permit the built-in dishwasher to be connected with a cord 3 feet (900 mm) to 4 feet (1.2 m) long. 422.16(B)(1),

however, would prohibit the 42-in. (1.1 m) cord on the waste disposal. This section limits the cord for disposals to 18 to 36 inches (450 to 900 mm).

Note: Portable dishwashers are supplied by the manufacturer with a cord and plug attached. This cord is evaluated as a part of the listing, and length is not limited by 422.16.

Sections 422.51 and 422.52 requires for both cord-and-plug-connected vending machines and electric drinking fountains to be protected with accessible ground-fault circuit-interrupter (GFCI) protection.

422.51 requires all cord-and-plug-connected vending machines manufactured or re-manufactured to include a GFCI as an integral part of the attachment plug or be located within 12 in. (300 mm) of the attachment plug. Older vending machines manufactured or remanufactured prior to January 1, 2005, must be connected to a accessible GFCI-protected outlet. 422.52 requires electric- drinking fountains shall be protected with GFCI protection whether they are cord-and plug-connected or direct wired.

Fixed electric space heating equipment is covered in Article 424. It is important to understand that this article covers electrical resistance furnaces, baseboard heaters, heating cable, duct heaters, and boilers, resistance or electrode types. It does not cover the typical gas-fired furnace common in many residential

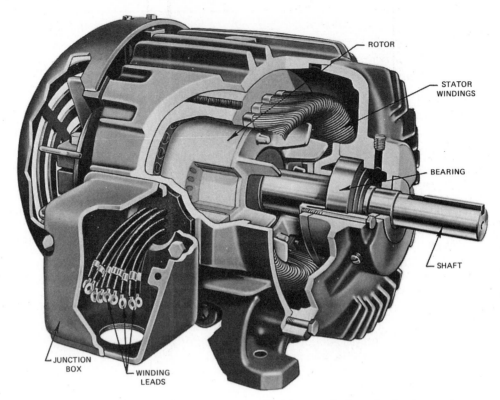

Figure 11–4 Cutaway of a standard totally enclosed fan-cooled squirrel-cage motor.

occupancies throughout the United States, which are appliances and covered in Article 422. It also does not cover heat pumps; they are covered in Article 440.

Article 430 covers motors, motor circuits, and controllers (see Figure 11–4). This article technically amends several articles in Chapters 1–4—210, 220, and 310. Although Article 90 states that Chapter 4 is a part of the general requirements, each of these articles contains a special allowance to which Article 430 applies.

Example: Motor branch-circuit conductors ampacities are in accordance with Article 310. However, 240.4(D) limiting 14 AWG to a 15-ampere overcurrent device, 12 AWG to a 20-ampere device, and 10 AWG to a 30-ampere device does *not* apply. 430.6 only references the table and not 240.4(D).

Read *NEC*® 430.6 carefully.

The motor nameplate current is *not* to be used; the values in Tables 247–250 are to be used instead.

Motor-circuit conductors are determined in Article 430, Part II; branch-circuit and overcurrent protection in Part III. Article 430 also contains special tap rules for motors in 430.28. Where the values for branch-circuit, short-circuit, and ground-fault protective devices do not correspond to standard sizes or ratings, the next higher standard size, rating, or possible setting is permitted.

Careful study is required for Part IX of Article 430, which covers the disconnecting means, the location, type, size, and so on.

Article 440 covers air-conditioning and refrigeration equipment (see Figure 11–5). The rules for hermetically sealed units are covered in this article; belt-driven units are referenced to Article 430 for the rules.

Example: In a steel mill, a number of small 3-horsepower, 3-phase, 480-volt motors are supplied by a 400-A feeder. What is the minimum size 10-foot (3.0 m) tap conductor permitted to be used to feed each motor?

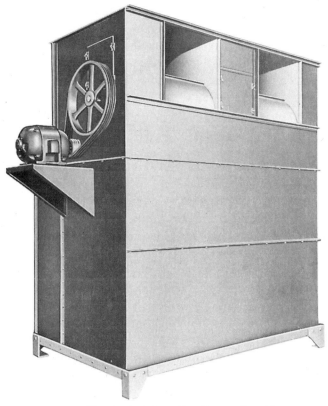

Figure 11–5 Typical air handler (fan coil unit).

Step 1. 40.4(E), which permits tap conductors to be protected against overcurrent in accordance with 210.19(A)(3), 240.5 (B)(2), 240.21, 368.17(B), 368.17(C), and 430.53(D).

Step 2. 240.21(F) references 430.28 and 430.53 (D) for motor taps.

Step 3. *NEC®* 430.28, 400 amperes ÷ 1000% = 40 amperes

Step 4. Table 310.15(B)(16)

Answer: 8 AWG THWN

Article 409—Industrial Control Panels
These control panels are for general use; they operate at 600 volts or less. They are general manufactured or shop fabricated to UL 508A, the safety standard for industrial control panels. They are defined as an assembly of two or more components consisting of one of the following:

(1) Power circuit components only, such as motor controllers, overload relays, fused disconnect switches, and circuit breakers

(continued)

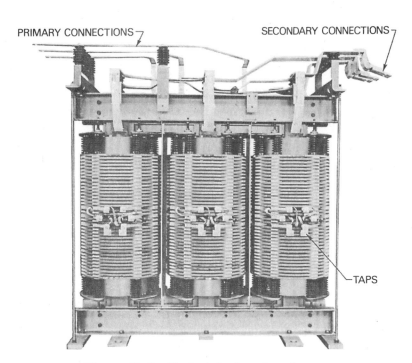

Figure 11–6 Typical dry-type transformer.

(2) Control circuit components only, such as push-buttons, pilot lights, selector switches, timers, switches, control relays

(3) A combination of power and control circuit components.

These components, with associated wiring and terminals, are mounted on or contained within an enclosure or mounted on a subpanel. The industrial control panel does not include the controlled equipment.

Note: Some may incorrectly permit a 10 AWG conductor to be used in this calculation as taken from the 90° column because a 3-horsepower motor (Table 430.250) would only draw 4.8 amperes, and the temperature would not be exceeded.

Article 450 covers transformers and transformer vaults (see Figure 11–6). The secondary conductor rules for protecting found in 240.21(C). Autotransformers (often referred to as buck or boost transformers) are covered in this article; however, branch-circuit and feeder rules are found in 210.9.

NEC® Chapter 4 also contains several articles for special-use types of general equipment, such as phase converters (Article 455), which are used often in rural farm-type installations; capacitors (Article 460) now generally found in many types of equipment and used to improve the efficiency of many services as a power factor correction; and generators (Article 445) often used to supplement or provide emergency or standby power; and others.

As your studies take you through Chapter 4 of the *NEC®*, be aware that the requirements in Article 410 cover luminaires, lampholders, and lamps. However, Article 410 does not apply to signs. Article 422, which covers appliances, generally has specific requirements for branch circuits, installation of appliances, and the control and protection of appliances. Space heating, motors, and air-conditioning and refrigeration equipment articles also have general provisions and installation provisions within those articles. Study them carefully for correct application. Many examination questions are derived from Chapter 4.

QUESTION REVIEW

> Each lesson is purposely designed to require the student to apply the entire *NEC®* text and not specific chapters or articles. It has been found that when studying for a timed, open-book examination, the student must gain proficiency in the Table of Contents, the Index, and the ability to move quickly from cover to cover to find the correct answer to each question in a timely fashion.

THREE-PHASE MOTOR AS FOLLOWS:

Type:	Squirrel cage, high reactance	Duty:	Continuous
Voltage:	460 volts, 3-phase	Start:	High reactance
HP:	30	Code letter:	None
Current:	Nameplate is the same as *NEC®* Tables 430.247–250	Service factor:	None

Refer to the description of the 3-phase motor above for Questions 1–3.

1. Where in the *NEC* is the requirement that Tables 430.247–250 be used instead of the actual current rating marked on the motor nameplate?

   Answer: _____

   _____

   _____

   _____

   _____

   Reference: _____

2. What size inverse-time circuit breaker is needed to provide the maximum branch-circuit, short-circuit, ground-fault protection?

   Answer: _____

   _____

   _____

   _____

   _____

   Reference: _____

3. A dual-element (time-delay) fuse can be sized a maximum of _____ % when used as branch-circuit protection.

   Answer: _____

   _____

   _____

   _____

   _____

   Reference: _____

4. A 120-volt single-phase store sign panel supplies only the sign as follows: Hours-on 24 hours a day, three ¼ horsepower motors, 24-0.80 ampere ballasts. Each ungrounded conductor in the feeder to the sign panel must be at least _____ AWG *aluminum* THW.

Answer: _____

_____

_____

_____

Reference: _____

5. Where small conductors are tapped to larger conductors as permitted in the *NEC*®, the tap cannot be larger than the _____ size of the conductor being tapped.

Answer: _____

_____

_____

_____

Reference: _____

In Questions 6–9, determine the ampere rating of the fuse size needed to protect the following copper conductors where no motors are connected.

6. 12 AWG THHN _____ ampere-rated fuse

Answer: _____

_____

_____

_____

Reference: _____

7. 8 AWG THW _____ ampere-rated fuse

Answer: _____

_____

_____

_____

Reference: _____

8. 3 AWG THW _____ ampere-rated fuse

Answer: _____

_____

_____

_____

Reference: _____

9. 3/0 AWG THWN _____ ampere-rated fuse

   Answer: _____

   _____

   _____

   _____

   Reference: _____

10. For a transformer 600 volts or less that has no secondary overcurrent protection, the primary overcurrent protection shall not exceed _____ % of the primary current.

    Answer: _____

    _____

    _____

    _____

    Reference: _____

11. The maximum branch-circuit ground-fault short circuit protection using time-delay fuses for typical motors are sized at _____ % of the motor's full-load current.

    Answer: _____

    _____

    _____

    _____

    Reference: _____

12. A motor pulls 22 amperes at 230 volts, and the feeder circuit is 150 feet (45.7 m) in length. If a 10 AWG copper wire is installed, what would be the voltage drop?

    Answer: _____

    _____

    _____

    _____

    Reference: _____

13. What is the percentage of voltage drop on the circuit described in Question 12?

    Answer: _____

    _____

    _____

    _____

    Reference: _____

14. What is the correct size conductor to maintain 3% or less voltage drop in Question 12?

    Answer: _____

    _____

    _____

    _____

    Reference: _____

15. Overload protection (heaters, relays, time-delay fuses, thermal overloads) for typical motors having a service factor of 1.15 is generally sized at _____ % of the motor's full-load current.

    Answer: _____

    _____

    _____

    _____

    Reference: _____

16. A motor controller that is installed with the expectation of its being submerged in water occasionally for short periods shall be installed in a rated enclosure type _____.

    Answer: _____

    _____

    _____

    _____

    Reference: _____

17. Where recessed high-intensity discharge fixtures are installed _____ and operated by remote ballasts, both the fixture and the ballast require thermal protection.

    Answer: _____

    _____

    _____

    _____

    Reference: _____

18. Any pipe or duct system foreign to the electrical installation must not enter a transformer vault. The _____ piping is not considered foreign to the vault.

    Answer: _____

    _____

    _____

    _____

    Reference: _____

19. A 10 kVA dry-type transformer rated at 480 volts can be installed on a building column and shall not be required to be _____.

    Answer: _____

    _____

    _____

    _____

    _____

    Reference: _____

20. A capacitor is located indoors. It must be enclosed in a vault or outdoor fenced enclosure if it contains more than _____ gallon(s) of flammable liquid.

    Answer: _____

    _____

    _____

    _____

    _____

    Reference: _____

21. A single-phase hermetic refrigerant motor-compressor has a rated load current of 24 amperes and a branch-circuit selection current of 30 amperes. The branch-circuit conductors are copper, Type TW. They operate at 80°F, and they are the only conductors in the conduit to this compressor. The smallest possible branch-circuit conductors must be at least size _____ AWG.

    Answer: _____

    _____

    _____

    _____

    _____

    Reference: _____

22. A 50-horsepower, 3-phase, squirrel-cage motor with full-voltage reactor starting has no code letter. The rating of the fuse to provide the branch circuit ground-fault and short-circuit protection of the motor would be sufficient for the starting current of the motor. The non–time-delay fuse to protect the circuit of this motor shall be sized at _____ % of full-load current.

    Answer: _____

    _____

    _____

    _____

    _____

    Reference: _____

23. A dry-type transformer is to be installed indoors. If rated more than _____ kilovolt amperes, the transformer must be installed in a fire-resistant room.

    Answer: _____

    _____

    _____

    _____

    Reference: _____

24. An electric resistance heater is rated for 2400 watts at 240 volts. What power is consumed when the heater is operated at 120 volts?

    Answer: _____

    _____

    _____

    _____

    Reference: _____

25. A 230-volt, single-phase circuit has 10 kilowatts of load and 50 amperes of current. The power factor is _____ %.

    Answer: _____

    _____

    _____

    _____

    Reference: _____

CHAPTER 12

OBJECTIVES

Studying this chapter along with the 2011 *NEC*® gives the reader a thorough review of the 2011 *NEC*®.

After studying this chapter, you should know:

- The definitions and the application of terms that are applicable and that are unique to the 2011 *NEC*®

- The *Code* requirements in which special equipment and occupancies are addressed as they apply to the *NEC*®

- The differences between Class I, Class II, and Class III locations as they relate to hazardous occupancies

- The differences between U.S. Division classification and the International Zone classification methods

- The wiring methods allowed in special occupancies

- Healthcare facilities and the requirements for those occupancies

- The definitions and the application of terms that are applicable and that are unique to Article 517 of the 2011 *NEC*®

SPECIAL EQUIPMENT AND OCCUPANCIES

Special occupancies are covered in Chapter 5, Articles 501 through 590. It is very important to be aware that the *NEC®* covers only installations of wiring and equipment. The classification of areas and operational rules are covered in other standards.

Articles 500 through 517 cover hazardous (classified) locations (see figure 12–1). When an electrical system, or any portion of it, is to be installed in an area containing a hazardous atmosphere, protection against possible explosions becomes a major design factor. All the basic design considerations still apply to determination of conductor sizes and general layout of the overall system. In addition, the portion of the system in or passing through an area where flammable gases, combustible dusts, or ignitable fibers and flyings may exist must have equipment enclosures specially designed to prevent electrical sparks and arcs from igniting the surrounding atmospheres and causing serious fires and/or explosions. For this portion of the design, the designers must be doubly careful to select equipment specifically permitted for the conditions encountered.

The *NEC®* describes three classifications of hazardous locations where an unconfined spark could result in explosion or fire. Each of these classifications identifies conditions that may exist either under normal operations—under occasional but frequent conditions resulting from maintenance operations—or under conditions of accidental breakdown that may cause the release of flammable or readily combustible materials.

> **Warning:** *NEC®* 90.3, Chapter 5, only modifies and augments Chapters 1–4. Corrosion, capacity, suitability, and service are not considered. Therefore, it is the responsibility of the designer to take these factors into account along with the hazardous location requirements that are covered in Chapter 5.

NEC® Articles 500 through 517 contain rules covering the type of wiring and equipment to be installed in hazardous locations. Because the actual degree of hazard is frequently evaluated by local inspection and fire department authorities, they should be consulted during the design stage of the wiring system that may be in the hazardous area.

Figure 12–1 Special occupancy hazardous (classified) fuel storage facility. (*Courtesy Crouse Hinds, Division of Cooper Industries*)

Class I

Those areas in which flammable gases, flammable liquid-produced vapors, or combustible liquid-produced vapors are or may be present in the air in quantities sufficient to produce explosive or ignitible mixtures. Class I locations include those specified in 500.5(B)(1) and (B)(2).

Class II

Those areas that are hazardous because of the presence of combustible dust. Class II locations include those specified in 500.5(C)(1) and (C)(2).

Class III

Those areas that are hazardous because of the presence of easily ignitible fibers or materials producing combustible flyings are handled, manufactured, or used but in which such fibers/flyings are not likely to be in suspension in the air in quantities sufficient to produce ignitible mixtures. Class III locations include those specified in 500.5(D)(1) and (D)(2).

Division I in the Normal Situation

A hazard would be expected to be present in everyday production operations or during frequent repair and maintenance activity (see Figure 12–2, Figure 12–3, and Figure 12–4).

Division II in the Abnormal Situation

The hazardous material is expected to be confined within closed containers or closed systems and will be present only through accidental rupture, breakage, or unusual faulty operation (see Figure 12–2, Figure 12–3, and Figure 12–4).

Groups

The gases and vapors of Class I locations are broken into four groups A, B, C, and D. These materials are grouped according to their explosion pressures and other flammable characteristics. Class II dust locations are groups E, F, and G. These groups are classified according to the ignition temperature and the conductivity of the hazardous substance (see Figure 12–2, Figure 12–3, and Figure 12–4). The equitable fibers/flyings are not broken down into groups.

Seals

Special fittings are required in Class I locations to minimize the passage of gases and vapors or, in the case of an explosion, prevent the passage of flames. They are also required in Class II locations to prevent the passage of combustible dusts (see Figure 12–5).

Articles 500 through 503

These articles explain in detail the requirements for the installation of wiring or electrical equipment in hazardous locations. These articles, along with other applicable regulations, local governing inspection authorities, insurance representatives, and qualified engineering/technical assistance, should be your guides to the installation of wiring or electrical equipment in any hazardous location.

Class I Wiring Methods

The minimum wiring methods are specified in *NEC®* 501.10(A) for Class I Division 1 locations. *NEC®* 501.10(B) covers Class I Division 2 locations.

Class I, Division 1 Wiring Methods

NEC® 501.10(A) requires wiring methods of *NEC®* Article 344—Threaded Rigid Metal Conduit (RMC); Article 342—Threaded Steel Intermediate Metal Conduit (IMC); or Article 332—Type MI cable (see Figure 12–6) with termination fittings listed for the location. All boxes, fittings, and joints shall be threaded for a connection to conduit or cable terminations and shall be approved for Class I, Division 1. Threaded joints must be made up with at least five threads fully engaged.

> MI cable must be installed and supported in a manner to avoid tensile stress at the terminations.

When it is necessary to use flexible connections as provided in *NEC®* 501.10, flexible fittings listed for Class I locations must be used. *NEC®* 501.140, however, permits flexible cord or connection between portable lighting equipment or other portable utilization equipment and a fixed portion of its supply circuit with rigid requirements: Extra-hard usage cord must be employed and must, in addition to the conductors in the circuit, employ a grounding conductor complying with *NEC®* 400.23. It must be an equipment

Hazardous(Classified) Locations in Accordance with *Article 500, National Electrical Code*® – 2011

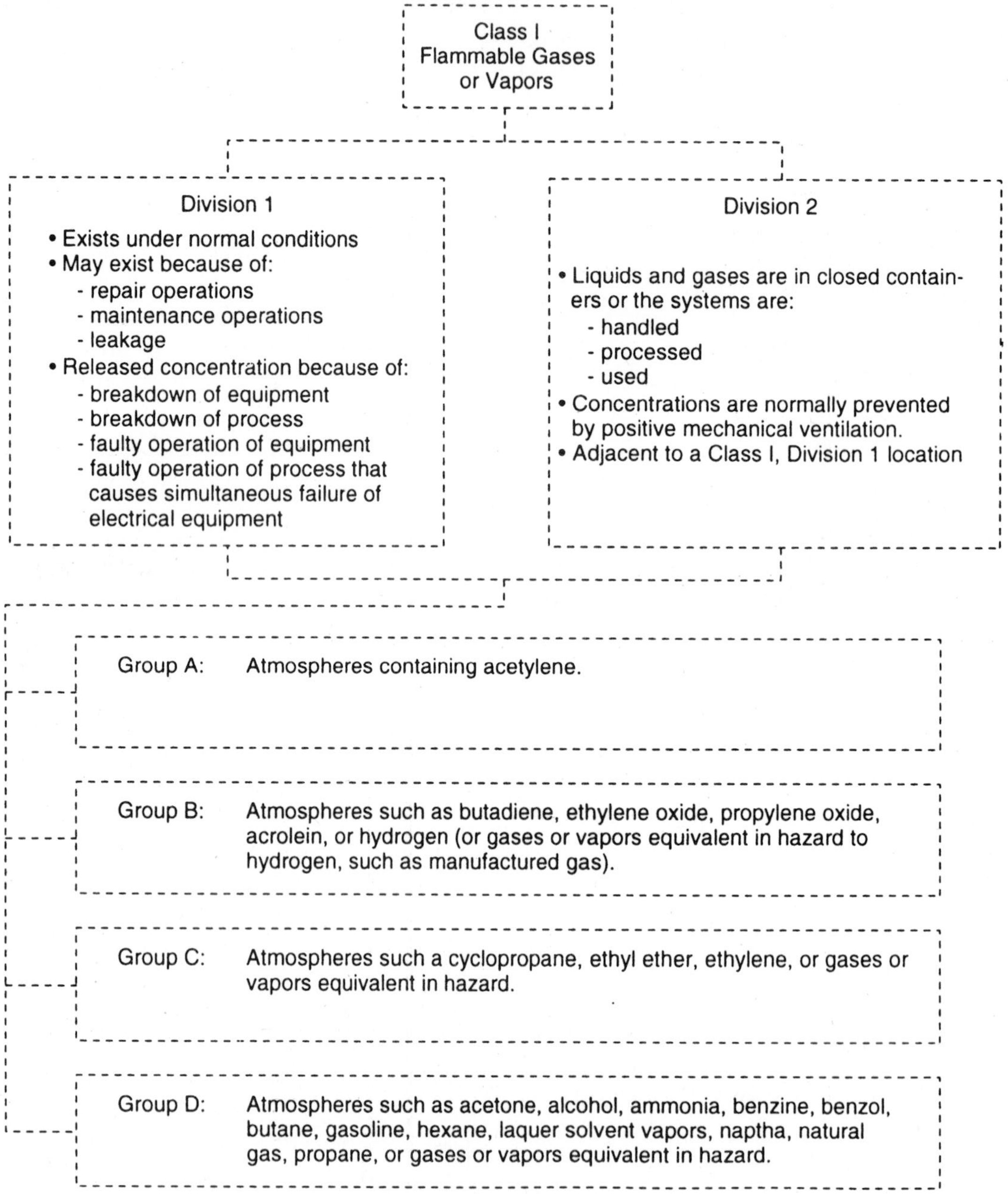

Class I
Flammable Gases
or Vapors

Division 1

• Exists under normal conditions
• May exist because of:
  - repair operations
  - maintenance operations
  - leakage
• Released concentration because of:
  - breakdown of equipment
  - breakdown of process
  - faulty operation of equipment
  - faulty operation of process that causes simultaneous failure of electrical equipment

Division 2

• Liquids and gases are in closed containers or the systems are:
  - handled
  - processed
  - used
• Concentrations are normally prevented by positive mechanical ventilation.
• Adjacent to a Class I, Division 1 location

Group A: Atmospheres containing acetylene.

Group B: Atmospheres such as butadiene, ethylene oxide, propylene oxide, acrolein, or hydrogen (or gases or vapors equivalent in hazard to hydrogen, such as manufactured gas).

Group C: Atmospheres such a cyclopropane, ethyl ether, ethylene, or gases or vapors equivalent in hazard.

Group D: Atmospheres such as acetone, alcohol, ammonia, benzine, benzol, butane, gasoline, hexane, laquer solvent vapors, naptha, natural gas, propane, or gases or vapors equivalent in hazard.

Figure 12–2   Class I flammable gases or vapors. See 500.5(B) and 500.6(A) for precise rules.

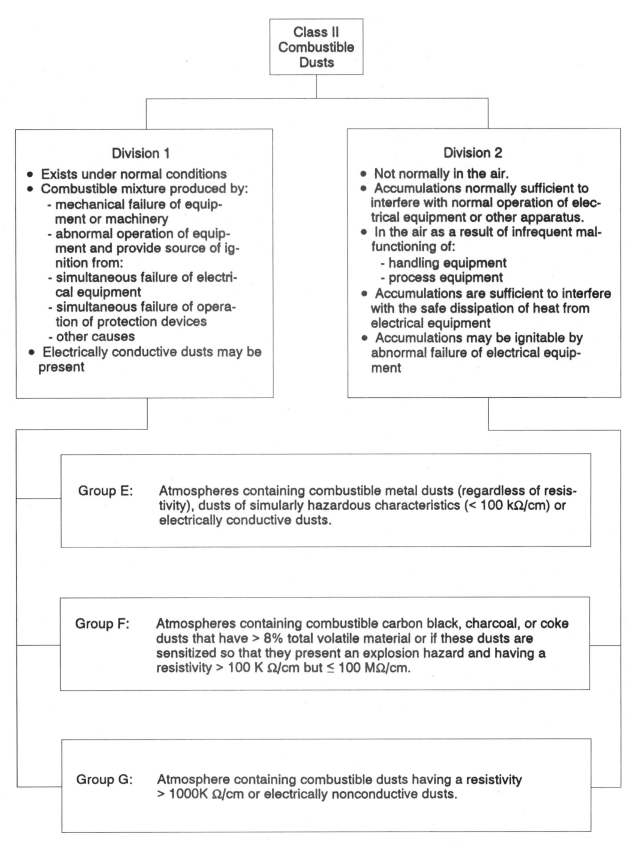

Figure 12–3 Class II combustible dusts. See 500.5(C) and 500.6(B) for precise rules.

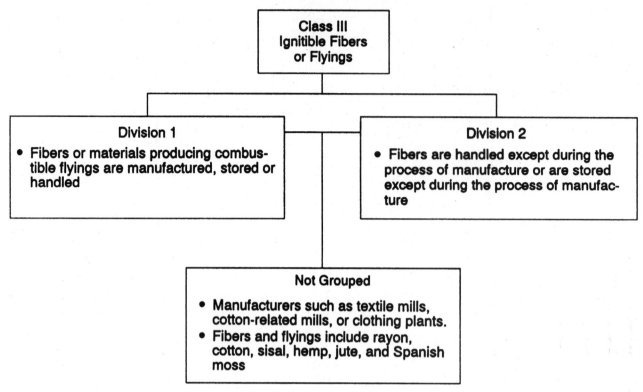

Figure 12–4 Class III ignitable fibers or flyings. See 500.5(D) for precise rules.

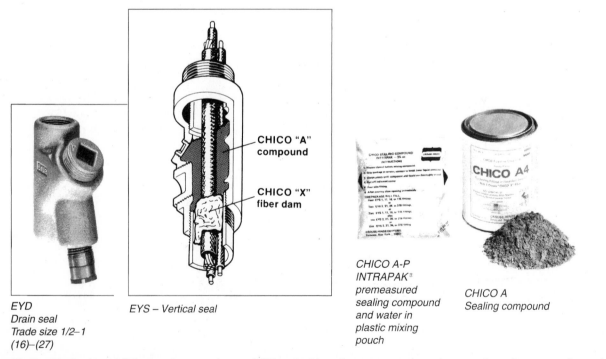

EYD
Drain seal
Trade size 1/2–1
(16)–(27)

EYS – Vertical seal

CHICO "A"
compound

CHICO "X"
fiber dam

CHICO A-P
INTRAPAK
premeasured
sealing compound
and water in
plastic mixing
pouch

CHICO A
Sealing compound

Figure 12–5 Typical seal fitting, cutaway of a seal fitting that has been poured, and examples of approved pouring compound. (*Courtesy of Crouse Hinds, Division of Cooper Industries*)

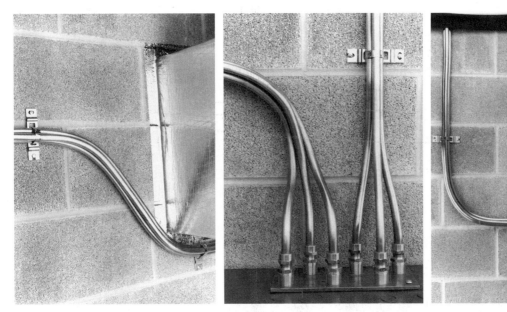

Figure 12–6 MI cable installed in accordance with *NEC®* Article 332. (*Courtesy of Pyrotenax USA Inc., photos by Robert Stewart Associates*)

supported by clamps or other suitable means in such manner that there will be no tension on the terminal connections and must be provided with seals where the cord enters boxes, fittings, or enclosures of explosion-proof type. Also permitted are Type PVC conduit and Type RTRC conduit encased in a concrete envelope a minimum of 2 in. (50 mm) thick and provided with not less than 24 in. (600 mm) of cover measured from the top of the conduit to grade. The concrete encasement shall be permitted to be omitted where subject to the provisions of 514.8, Exception No. 2, and 515.8(A). Threaded rigid metal conduit or threaded steel intermediate metal conduit shall be used for the last 24 in. (600 mm) of the underground run to emergence or to the point of connection to the aboveground raceway. An equipment-grounding conductor shall be included to provide for electrical continuity of the raceway system and for grounding of non–current-carrying metal parts. In industrial establishments with restricted public access, where the conditions of maintenance and supervision ensure that only qualified persons service the installation, Type MC-HL and Type ITC-HL cable cable, listed for use in Class I, Zone 1, or Division 1 locations, with a gas/vaportight continuous corrugated metallic sheath, an overall jacket of suitable polymeric material.

The restrictions in 330.12 apply to MC-HL and 250.122. See 727.4 and 727.5 for restrictions on use of Type ITC cable.

Class I, Division 2 Wiring Methods

NEC® 501.10(B) allows somewhat less stringent wiring requirements. However, it is equally as important to wire areas classified as Class I, Division 2 correctly. The Class I, Division 2 minimum wiring methods are *NEC®* Article 342—Threaded Steel Intermediate Conduit; *NEC®* Article 344—Threaded Rigid Metal Conduit; *NEC®* Article 368—Enclosed Gasketed Busways; *NEC®* Article 376—Enclosed Gasketed Wireways; PLTC cable in accordance with the provisions of *NEC®* Article 725 and Types MI, MC, or TC, or SNM cable with listed fittings and terminations.

> Types PLTC, MI, MC, MV, and TC are permitted to be installed in cable trays and shall be installed in a manner to avoid tensile stress at the termination fittings.

In industrial establishments with restricted public access where the conditions of maintenance and supervision ensure that only qualified persons service the installation and where metallic conduit does not provide sufficient corrosion resistance, reinforced thermosetting resin conduit (RTRC), factory elbows, and associated fittings, all marked with the suffix -XW, and Schedule 80 PVC conduit, factory elbows, and associated fittings shall be permitted.

Boxes and joints are not required to be explosion-proof in these areas, except as required in *NEC®* 501.105(B)(1), 501.115(B)(1), and 501.150(B)(1). When provisions must be made for limited flexibility, such as at motor terminals, flexible metal conduit with approved fittings, *NEC®* Article 320—Type AC Armored Cable with approved fittings, *NEC®* Article 350—Liquidtight Flexible Metal Conduit with approved fittings, and Article 356—Liquidtight Flexible Nonmetallic Conduit with approved fittings, or flexible cord listed for hard uses provided with approved bushed fittings may be used. The additional conductor for grounding must be included in the flexible cord unless other accessible means for grounding are provided.

Wiring that cannot release sufficient energy to ignite a specific ignitable atmospheric mixture by opening, shorting, or grounding may be permitted in this area using any of the methods suitable for general wiring applications.

Sealing and Drainage Requirements

NEC® 501.15 covers the requirements for sealing and drainage in Class I, Divisions 1 and 21 locations. Fittings approved for the location and purpose must be used with approved sealing compounds. Sealing compounds approved for Type MI cable must also exclude moisture and other fluids from the cable insulation. In addition, see the requirements of *NEC®* 501.15(D).

> Seals are provided in conduit and cable systems to minimize the passage of gases and vapors and prevent the passage of flames through the wiring method from one electrical system or enclosure to another. Type MI cable provides this inherently and prevents the passage of gases and vapors through the insulation.

The specific requirements for installing the seals are found in *NEC®* 501.15(A) through (F) and the manufacturer's instructions. Installation of these seals is a critical part of this installation and if improperly made will not provide the protection needed to isolate accidental explosions. Raceways in Class I hazardous locations are required to be provided with sealing fittings to prevent the passage of gases, vapors, or flames through the raceway from one portion of an electrical installation to another and to prevent pressure piling or cascading. Pressure piling may occur in these locations when an explosion in some part of the system, such as

a switch enclosure, is not effectively blocked or sealed off. The first explosion may be made harmless if properly confined by a seal to the switch enclosure. If allowed to enter the raceway, it builds up great pressure and another more powerful explosion may follow. The second explosion builds up additional pressures in the gas-filled conduit, and a greater explosion results. This may continue in very long raceway runs not provided with intermittent seals until some part of the system ruptures. The hot or burning gases could then escape and ignite flammable gases in the surrounding atmosphere, causing an explosion.

Proper sealing is one of the most important requirements for safety in a Class I location. Explosive vapors are prevented from passing to an unclassified area where they could be ignited and carry the flame back to the classified location. In the manner of a pilot light on a gas range, arcs, sparks, and hot particles from an arcing device are also prevented from entering the raceway by the seal. Sealing fittings are designed to permit the pouring of an approved water-base sealing compound into the fitting around and between the conductors (refer to Figure 12–5).

> Conduit seals are limited to a 25% conductor fill unless specifically identified for a higher percentage fill. Where raceways are filled to 40%, check with the manufacturer for conductor fill limits.

The raceway should first be blocked at the bottom of the sealing fitting with a packed material to prevent the sealing mixture from draining out of the fitting. The liquid mixture solidifies after a few hours and becomes very hard. Proper mixing of the compound and liquid is essential; otherwise, the seal would not prevent the pressure piling. Even though it may prevent the passage of vapors, arcs, and hot particles, all voids under, between, and over conductors must be filled. If the sealing mixture is to perform its function of preventing the passage of gases, vapors, or flames through the raceway, complete elimination of the pressure piling can be achieved by installing effective seals in every raceway regardless of size where it enters or leaves any junction box at regular intervals, in all raceway runs, and at points at which they presently are required by the *NEC®*. In the absence of tests for various gases, determine the proper spacing of seals. In long conduit runs, it is not unreasonable to specify that they be installed at 50-foot (15.25 m) intervals in all conduit runs within a Class I Division 1 location.

If installed in accordance with *NEC*® requirements, a rigid steel conduit system with suitable enclosure and seals will effectively contain an internal explosion and prevent the escape of burning gases until they are cooled sufficiently to prevent the ignition of flammable vapors in the surrounding atmosphere. The gases are cooled by expansion as they escape through the machine-grounded joints of predetermined width and through threaded joints. Ground joints, unless properly maintained, are not as reliable as threaded joints. A single loose screw in a cover of a ground joint or scratches in ground surfaces will render an otherwise explosion-proof enclosure ineffective. Threaded joints, on the other hand, are reliable because the gases escape only by following the threads. The *NEC*® requires that at least five full threads be engaged at each coupling. Effective seals must be placed in a raceway as required by the *NEC*® within 18 in. (450 mm) of an arcing device or at or near the point where the raceway leaves or enters a classified area. In the case of a raceway 2 in. (50 mm) in diameter or larger, seals are required within 18 in. (450 mm) of any junction box containing splices, taps, or terminal connections. No coupling or fitting, except listed reducers, is permitted between the seal and the boundary between the hazardous and nonhazardous areas. In many cases, special panelboards and controls are designed with factory seals that are an integral part of the equipment designed for Class I locations. Because such seals effectively isolate the arc-producing segments of the assembly, additional seals are not generally required in the conduit leaving these panelboards. Incandescent and fluorescent fixtures in Class I locations are also available with integral seals.

> Seals are not intended to prevent the passage of gases or vapors at a continuous pressure. Even at differences in pressure across the seal, there may be a slow passage of gases or vapors.

> Leakage and the passage or propagation of flames through the interstices of stranded conductors can be reduced through the use of compact stranding or sealing of the individual stranding.

Although wiring methods are less restrictive in Class I, Division 2 locations, sealing is still required as specified in *NEC*® 501.15(B). In each conduit run passing from a classified location into an unclassified area, a sealing fitting is required to be on either side of the boundary, designed and installed to minimize the amount of gas or vapor that may enter the conduit system within the Class II location from being communicated to the conduit beyond the seal. Rigid metal conduit and threaded steel intermediate conduit shall be used between the sealing fitting and the point at which the conduit leaves the classified location, and a threaded connection shall be used at the sealing fitting. No fitting or coupling is permitted between the sealing fitting and the point in which the conduit leaves the Division 2 location. Conduit systems terminating to an open raceway in an outdoor unclassified area are not required to be sealed between the point at which the conduit leaves the classified location and enters the open raceway. The requirements for the location of sealing fittings are found in *NEC*® 501.15.

Class II, Division 1 Wiring Methods

Class II, Division 1 502.10(A)(1) and (2) permit threaded rigid metal conduit, threaded steel intermediate metal conduit, or Type MI cable with termination fittings listed for the location and installed to avoid tensile stress at the fittings. The fittings and boxes with threaded bosses and close-fitting covers shall have no openings through which dust might enter or sparks or burning material might escape. In locations where combustible dusts of an electrically conductive nature are likely, the fittings and boxes that are designed to contain taps, joints, or terminal connections must be identified for Class II locations. In industrial establishments with limited public access, where the conditions of maintenance and supervision ensure that only qualified persons service the installation, Type MC-HL cable, listed for use in Class II, Division 1 locations, with a gas/vaportight continuous corrugated metallic sheath, an overall jacket of suitable polymeric material, a separate equipment grounding conductor(s) in accordance with 250.122, and provided with termination fittings listed for the application, is permitted. Flexible connections shall be made with dusttight flexible connectors; liquidtight flexible metal conduit with listed fittings; liquidtight flexible nonmetallic conduit with listed fittings; or flexible cord listed for extra-hard usage and provided with bushed fittings. Flexible cords must comply with 502.140.

> FPN: See 502.30(B) for grounding requirements where flexible conduit is used. [502.10(A)(2), reprinted with permission from NFPA 70-2011.]

Class II, Division 2 Wiring Methods

Class II, Division 2 502.10(B)(1) and (2) permit the use of rigid metal conduit, intermediate metal conduit, electrical metallic tubing, dusttight wireways, or Types MI and MC, cable with listed termination fittings. ITC, PLTC, or TC cable installed in ventilated channel-type cable trays in a single layer with a space of not less than the larger cable diameter between the two adjacent cables is also permitted. Wiring in nonincendive circuits is permitted using any of the methods suitable for wiring in ordinary locations.

Sealing Requirements for Class II, Divisions 1 and 2

Combustible and explosive dusts constitute a hazard in Class II locations. The term "dust-ignition proof" is defined in *NEC®* 500.2. Such dusts do not travel for great distance within a raceway; therefore, seals are not required if the raceway between a dust-ignition proof enclosure and an enclosure that is not dust-ignition proof consists of either a horizontal section not less than 10 feet (3.0 m) in length or a vertical section not less than 5 feet (1.5 m) in length extending downward from the dust-ignition-proof enclosure through the raceway. Where such raceways are not provided, seals must be used to prevent the entrance of dust into ignition-proof enclosures through the raceway. Sealing fittings and poured sealing compound may be used in both Division 1 and 2. They are not required to meet the same stringent requirements of those in 501.5.

Class III Divisions 1 and 2
Live uninsulated parts shall not be exposed.
[503.25]

Wiring Methods for Class III, Division 1 Locations

The permitted wiring methods in class III, Division 1 are rigid metal conduit, intermediate metal conduit (IMC), electrical metallic tubing (EMT), PVC rigid conduit, Type MC, PVC rigid or MI cable with listed termination fittings or dusttight wireways. Flexible connections must be made with dusttight flexible fittings and installed in accordance with 503.10(A)(2) using liquidtight flexible metal conduit with approved fittings, liquidtight flexible nonmetallic conduit with approved fittings, or flexible cord in accordance with 503.140. Nonincendive field wiring in accordance with control drawings.

Boxes and fittings must be dusttight.

Class III, Division 2 Locations

Wiring methods approved for Class III Division 1 locations are acceptable and, in addition, open wiring on insulators may be used in accordance with all of the provisions of Article 398, not subject to physical damage in compartments or locations solely for storage and containing no machinery.

Seals are not required in Class III locations.

Grounding and Bonding Class I, II, and III, Divisions 1 and 2

Bonding jumpers with proper fittings or other approved means of bonding shall be used at all intervening raceways, fittings, boxes, enclosures, and so on between classified locations and the point of grounding for service equipment or point of grounding of a separately derived system, except as otherwise permitted.

The locknut-bushing and double-locknut types shall not be depended upon for bonding purposes.

Equipment-grounding conductors shall be installed in accordance with 250.120. Where flexible conduit is used, it must be installed with internal or external bonding jumpers with each conduit and complying with 250.102, with exceptions in Chapter 5.

Article 504 covers "intrinsic safe systems." Intrinsic safety is an explosion-prevention technique applied to electrical equipment and the wiring of hazardous locations. The technique is based on limiting electrical and thermal energy levels below that which is required to ignite a specific flammable or combustible atmospheric mixture in its most easily ignitable concentration.

The entire intrinsically safe system must be installed in accordance with control drawings, as required in 504.10. The control drawing identification must be marked on all equipment and apparatus. Generally, intrinsically safe systems may use any wiring method suitable for unclassified locations. However, sealing shall be provided in accordance with 504.70, and separation of intrinsically safe systems shall be provided in accordance with 504.30.

Warning: Articles 500 through 517 do not consider corrosion and physical protection, only circuit safety. It is the designer's responsibility to consider all facets of the installation for performance, safety, and life cycle issues.

It should be noted that intrinsically safe systems always fail in the safe position. Should a wire or cable be crushed or cut accidentally, the operation of the equipment must fail safe in every instance. No arcing or sparking will occur due to the design and nature of intrinsically safe systems. Designers generally require physical protection, such as a metal raceway, to protect intrinsically safe conductors and to provide assurance from costly accidental shutdowns and operational stoppages. *NEC*® 504.70 requires that conduits and cables be sealed in accordance with the hazardous location in which they are installed, such as Article 501 for Class I locations. Sealing of intrinsically safe systems must be provided as specified in 501.25. For flammable or combustible dusts, sealing is required as specified in 502.25. Seals are not required for enclosures that contain only intrinsically safe apparatus.

Article 505 introduces the European zone system as an alternate method for designers additional. It provides a level of the classification of Class I hazardous (classified) locations. This alternate method uses the zone concept to meet the requirements for electrical equipment and wiring for all voltages in locations where fire or explosive hazards may exist due to flammable gases, vapors, and flammable liquids.

Article 505 modifies the general rules of the *Code* as it applies to the electric wiring and equipment in locations classified as Class I, Zone 0, Zone 1, and Zone 2.

Article 506—Zones 20, 21, and 22 provide the alternative classification scheme for Class II or Class III locations, where flammable dust and fibers, and flyings are likely to be present. The article covers the requirements for the zone classification system as an alternative to the division classification system covered in Articles 500, 502, and 503 for electrical and electronic equipment and wiring for all voltages. Article 506 divides hazardous (classified) locations where fire and explosion hazards may exist due to combustible dusts, or ignitable fibers, or flyings into Zone 20, Zone 21, and Zone 22. Zone 20, Zone 21, and Zone 22 area classifications are based on a modified IEC area classification system.

Combustible metallic dusts are not covered by the requirements of Article 506.

The varying rules for hazardous classified locations as covered in Articles 500 through 504 are applied to the specific locations covered in Articles 510 through 517, with amendments to those rules. For example, Article 511 covers commercial garages. The standard for classifying these areas is NFPA 88. The hazardous concerns are the presence of Class I liquids, gases, and vapors in dangerous concentrations.

The difficulty in interpreting the rules is that Chapters 1–4 apply generally. Article 511 amends those general rules but references Articles 500 and 501.

Article 517 covers the installation requirements for healthcare facilities. This article contains many extracts from NFPA 99, which covers equipments other than installation requirements, for these locations.

Chapter 5 also contains many other articles related to special occupancies from places of assembly (Article 518) to mobile homes (Article 550).

Chapter 6 covers special equipment. This equipment is not found in all occupancies, such as Article 600—Electric Signs and Outline Lighting. They are not required nor are they found on all occupancies. However, when installed, they do introduce a hazard if not safely installed; therefore, the *NEC*® contains specific requirements for sign installations.

Examination Study

Questions related to Chapter 6, Special Equipment, can generally be answered quickly because these articles are usually short and cover very specific requirements. The most important study advice is to learn the names of these articles for quick reference.

- Article 600—Signs and Outline Lighting
- Article 604—Manufactured Wiring Systems
- Article 605—Office Furnishings
- Article 610—Cranes and Hoists
- Article 620—Elevators, Dumbwaiters, Escalators, Moving Walks, Wheelchair Lifts, and Stairway Chairlifts
- Article 625—Electric Vehicle Charging System Equipment.
- Article 626—Electrified Truck Parking Spaces
- Article 630—Electric Welders
- Article 640—Sound Recording and Similar Equipment
- Article 645—Information Technology Equipment (formerly Computer/Data Processing Equipment)
- Article 647—Sensitive Electronic Equipment
- Article 650—Pipe Organs
- Article 660—X-ray Equipment
- Article 665—Induction and Dielectric Heating Equipment
- Article 668—Electrolytic Cells
- Article 669—Electroplating
- Article 670—Industrial Machinery
- Article 675—Electrical-Driven or Controlled Irrigation Machinery
- Article 680—Swimming Pools, Fountains, and Similar Equipment
- Article 682—Natural and Artificially Made Bodies of Water
- Article 685—Integrated Electrical Systems
- Article 690—Solar Photovoltaic Systems
- Article 692—Fuel Cell Systems
- Article 694—Small Wind Electric Systems
- Article 695—Fire Pumps. This article has added installation requirements into the *Code* that were formerly found in NFPA 20.

QUESTION REVIEW

> Each lesson is purposely designed to require the student to apply the entire NEC® text and not specific chapters or articles. It has been found that when studying for a timed, open-book examination, the student must gain proficiency in the Table of Contents, the Index, and the ability to move quickly from cover to cover to find the correct answer to each question in a timely fashion.

1. Name three types of optical fiber cable.

   Answer: _____

   _____

   _____

   _____

   Reference: _____

2. Fixed wiring over a Class I location in a commercial garage shall not be enclosed in non–metallic-sheathed cables. True or False?

   Answer: _____

   _____

   _____

   _____

   Reference: _____

3. Are seals required for an outdoor propane-dispensing unit located 50 feet (15.24 m) from the office where the branch circuit supplying the unit originates? Where are the seals required to be located?

   Answer: _____

   _____

   _____

   _____

   Reference: _____

4. Emergency electrical systems are those systems legally required and classed as emergency by Article 700. True or False?

   Answer: _____

   _____

   _____

   _____

   Reference: _____

5. In a major repair garage without dispensing, Class _____, Division _____ wiring methods must be met for an electrical outlet installed 12 inches (300 mm) above the floor if there is no mechanical ventilation.

Answer: _____

_____

_____

_____

Reference: _____

6. Transfer equipment for emergency systems shall be designed and installed to prevent _____ of normal and emergency sources of power.

Answer: _____

_____

_____

_____

Reference: _____

7. The sealing compound in a completed conduit seal for a Class I location must be at least _____ inch(es) (_____ mm) thick.

Answer: _____

_____

_____

_____

Reference: _____

8. Bare conductors are field connected to fixed terminals of different phases on an outdoor, 13.8-kV circuit. According to the *NEC*®, the air separation between these conductors shall be a minimum of _____ inch(es) (_____ mm).

Answer: _____

_____

_____

_____

Reference: _____

9. In a hospital's general care areas, the number of receptacles required to be in a patient-bed location is a minimum of _____.

Answer: _____

_____

_____

_____

Reference: _____

10. A 120-volt, single-phase store sign panel supplies only the sign as follows: Hours-on 24 hours a day, three ⅓ horsepower motors and thirty-six 0.80 ampere ballasts. Each ungrounded conductor in the feeder to the sign panel must be at least _____ AWG copper THWN.

    Answer: _____

    _____

    _____

    _____

    Reference: _____

11. Portable gas tube signs for interior use can be connected with a supply cord having a maximum length of _____ feet (_____ m).

    Answer: _____

    _____

    _____

    _____

    Reference: _____

12. Electric discharge (neon) tubing shall be a _____ and _____ so as not to cause steady overvoltage on the transformer.

    Answer: _____

    _____

    _____

    _____

    Reference: _____

Answer Questions 13–20 about Type S fuses with True or False.

13. The S indicates "size rejection."

    Answer: _____

    _____

    _____

    _____

    Reference: _____

14. A 15-ampere Type S fuse will fit into a 20-ampere Type S adapter.

    Answer: _____

    _____

    _____

    _____

    Reference: _____

15. A 20-ampere Type S fuse will fit into a 20-ampere Type S adapter.

Answer: _____

_____

_____

_____

Reference: _____

16. A 20-ampere Type S fuse will fit into a 15-ampere Type S adapter.

Answer: _____

_____

_____

_____

Reference: _____

17. A 30-ampere Type S fuse will fit into a 20-ampere Type S adapter.

Answer: _____

_____

_____

_____

Reference: _____

18. A 25-ampere Type S fuse will fit into a 30-ampere Type S adapter.

Answer: _____

_____

_____

_____

Reference: _____

19. A 6½-ampere Type S adapter will accept a 6¼-ampere Type S fuse.

Answer: _____

_____

_____

_____

Reference: _____

20. A 6¼-ampere Type S adapter will accept a 3-, a 4-, a 4½-, a 5-, a 5⁶⁄₁₀-, and a 6¼-ampere Type S fuse.

Answer: _____

_____

_____

_____

Reference: _____

21. The *NEC®* requires that the combination rating be marked when tested series rated systems are installed. (a) Who is required to install that marking? (b) What must the marking state? (c) Who is responsible for providing this information?

Answer: _____

_____

_____

_____

Reference: _____

22. A 1½-horsepower, single-phase motor that has an efficiency of 80% operates at 230 volts and has an input current of _____ amperes. (1 hp = 746 watts)

Answer: _____

_____

_____

_____

Reference: _____

23. A 30-horsepower, wound-rotor induction motor with no code letter is to be installed with 460-volt, 3-phase alternating current. Disregarding all the exceptions, the non–time-delay fuse for short-circuit protection of the motor branch circuit must be rated at a maximum _____ amperes.

Answer: _____

_____

_____

_____

Reference: _____

24. The following 480-volt, 3-phase, 3-wire, intermittent-use equipment is in a commercial kitchen: two 5000-watt water heaters, four 3000-watt fryers, and two 6000-watt ovens. Each ungrounded conductor in the feeder circuit for this kitchen equipment must be sized to carry a minimum calculated load of _____ amperes.

Answer: _____

_____

_____

_____

Reference: _____

25. A 277/480-volt, 3-phase panelboard supplies only one 15,000 VA, 480-volt, 3-phase balanced resistive load. Each ungrounded conductor in the feeder to this panel has calculated load of _____ amperes.

Answer: _____

_____

_____

_____

Reference: _____

CHAPTER 13

OBJECTIVES

Studying this chapter along with the 2011 *NEC*® gives the reader a basic introduction to the requirements in the *Code* that apply to special systems, all limited voltage, and communications installations.

After studying this chapter, you should know:

Special Systems (Conditions)

- The definitions and the application of terms that are applicable and that are unique to the 2011 *NEC*®

- The *Code* requirements in which special systems or conditions are addressed as they apply to the *NEC*®

- The differences between emergency systems and legally required or optional systems

- The differences between power-limited and non—power-limited circuits and equipment

- The basics of optical fiber cables and raceways

- The interconnected electrical power production sources (commonly called "cogeneration")

- The definitions and the application of terms that are applicable and that are unique to Chapter 8 of the 2011 *NEC*®

(Objectives, continued)

Communications Systems

■ The *Code* requirements in which special systems or conditions are addressed as they apply to Chapter 8 systems

■ The network-powered and premises-powered broadband communications systems

---

SPECIAL CONDITIONS AND COMMUNICATION SYSTEMS
Special Conditions

The first three articles in Chapter 7 cover Emergency Systems in Article 700, Legally Required Standby Systems in Article 701, and Optional Standby Systems in Article 702. The determination for which type system is installed in a building or facility is made by others—fire marshal, building official, or the like. The *NEC*® does not make these requirements; it does not require exit lights in buildings; it does not require smoke detectors. However, when these items are required to be installed, the *NEC*® governs and the installation requirements must conform to these articles as appropriate.

Article 708 for critical operations power systems addresses facilities that are involved with disasters and emergencies. The article applies to the installation, operation, monitoring, control, and maintenance of the portions of the premises wiring system intended to supply, distribute, and control electricity to designated critical operations areas (DCOA) in the event of disruption to elements of the normal system. Critical operations power systems (COPS) are those systems so classified by municipal, state, federal, or other codes by any governmental agency having jurisdiction or by facility engineering documentation establishing the necessity for such a system. These systems include but are not limited to power systems, HVAC, fire alarm, security, communications, and signaling for designated critical operations areas.

The source of power shall be such that in the event of failure of the normal supply to the DCOA, critical operations power is available within the time required for the application. The supply system for critical operations power, in addition to the normal services

to the building, shall be one or more of the types of systems described in 708.20(D) through (H) which includes surge protection devices, storage battery, generators, and fuel cells. The rules for using these power sources are included in Section 708.20. The wiring methods must provide protection from physical damage and be kept entirely independent of all other wiring and equipment.

725.121 Power Sources for Class 2 and Class 3 Circuits.

(A) Power Source. The power source for a Class 2 or a Class 3 circuit shall be as specified in **725.121(A)**(1), (A)(2), (A)(3), (A)(4), or (A)(5):

> Informational Note No. 1: **Figure 725.121** illustrates the relationships between Class 2 or Class 3 power sources, their supply, and the Class 2 or Class 3 circuits.

> Informational Note No. 2: **Table 11(A)** and **Table 11(B)** in Chapter **9** provide the requirements for listed Class 2 and Class 3 power sources.

(1) A listed Class 2 or Class 3 transformer
(2) A listed Class 2 or Class 3 power supply
(3) Other listed equipment marked to identify the Class 2 or Class 3 power source

Exception No. 1 to (3): Thermocouples shall not require listing as a Class 2 power source

Exception No. 2 to (3): Limited power circuits of listed equipment where these circuits have energy levels rated at or below the limits established in Chapter 9, Table 11(A) and Table 11(B).

> Informational Note: Examples of other listed equipment are as follows:

(1) A circuit card listed for use as a Class 2 or Class 3 power source where used as part of a listed assembly

(2) A current-limiting impedance listed for the purpose, or part of a listed product, used in conjunction with a non–power-limited transformer or a stored energy source, for example, storage battery, to limit the output current

(3) A thermocouple

(4) Limited voltage/current or limited impedance secondary communications circuits of listed industrial control equipment

(4) Listed information technology (computer) equipment limited-power circuits.

Informational Note: One way to determine applicable requirements for listing of information technology (computer) equipment is to refer to UL 60950-1-2003, *Standard for Safety of Information Technology Equipment*. Typically such circuits are used to interconnect information technology equipment for the purpose of exchanging information (data).

(5) A dry cell battery shall be considered an inherently limited Class 2 power source, provided the voltage is 30 volts or less and the capacity is equal to or less than that available from series connected No. 6 carbon zinc cells.

(B) Interconnection of Power Sources. Class 2 or Class 3 power sources shall not have the output connections paralleled or otherwise interconnected unless listed for such interconnection.

Except for dry-cell batteries and thermocouples, **725.121(B)** requires listed power sources for Class 2 and Class 3 circuits.

725.124 Circuit Marking. The equipment supplying the circuits shall be durably marked where plainly visible to indicate each circuit that is a Class 2 or Class 3 circuit.

725.127 Wiring Methods on Supply Side of the Class 2 or Class 3 Power Source. Conductors and equipment on the supply side of the power source shall be installed in accordance with the appropriate requirements of Chapters **1** through **4**. Transformers or other devices supplied from electric light or power circuits shall be protected by an overcurrent device rated not over 20 amperes.

Exception: The input leads of a transformer or other power source supplying Class 2 and Class 3 circuits shall be permitted to be smaller than 14 AWG, but not smaller than 18 AWG if they are not over 12 in. (305 mm) long and if they have insulation that complies with **725.49(B)**.

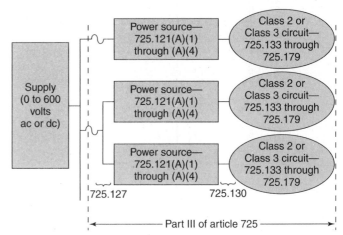

Figure 725.121 Class 2 and Class 3 Circuits.

Article 720 covers the *few* circuits and equipment operating at less than 50 volts that are not already covered in the *Code* by Articles 411, 551, 650, 669, 690, 725, and 760.

Article 725 covers circuits of Class 1, 2, and 3. It is very important that one understand the parameters of each class of wiring system. For Class 2 and Class 3 the power is limited by the power source, generally a transformer.

NEC® 725.24 covers the mechanical execution of work as follows:*

These circuits can often be run as open wiring without physical protection with specific cable types for the location. They can also be run with standard types of building wire when placed in the proper wiring method.

For example: Class II conductors that run through a plenum space must be run as Type CL2P, CMP, or CL3P or with appropriate building wire run in RMC, IMC, EMT Type MI cable, or specific types of solid-sheathed MC cable. (See Section 300.22.)

Article 760 covers fire alarm systems, both the power-limited and the non–power-limited. This article is very complete in its requirements for these systems. The Non–power-Limited Fire Alarm (NPLFA) wiring methods are in accordance with Chapters 1–4 and 300.15(A), with the exceptions provided in 760.41 through 760.53. Power-Limited Fire Alarm (PLFA) wiring methods are covered in 760.130 and are similar to those in Article 725 as shown previously.

*Reprinted with permission from NFPA 70-2011.

For example: PLFA conductors that run through a plenum must be run as Type FPLP, MPP, or CMP or with appropriate building wire run in RMC, IMC, EMT Type MI cable, or specific types of solid-sheathed MC cable. (See Section 300.22.)

NEC® 760.24 covers the mechanical execution of work for fire alarm circuits, and equipment shall be installed in a neat and professional manner. Cables shall be supported by the building structure in such a manner that the cable will not be damaged by normal building use.

Article 840 covers premises-powered optical fiber-based broadband communications systems that provide any combination of voice, video, data, and interactive services through an optical network terminal A typical premises powered system consists of optical fiber cable run to the premise structure to supply a broadband signal which is converted to component electrical signals, such as telephone, television and high-speed internet accomplished through a power supply unit with battery backup connected to AC at the premises. The optical fiber cable is unpowered and may be nonconductive or conductive.

Article **770** permits the orderly development and use of optical fiber technology in conjunction with electrical conductors for communications, signaling, and control circuits in lieu of metallic conductors. Because they are not affected by electrical noise, optical fiber cables may be desirable in some circumstances to transmit data or other communications where electrical noise is a problem. This article covers the installation of optical fiber cables, raceways, and cable routing assemblies. It provides applications and listing requirements. Optical fiber cables are commonly nonconductive; however, they may be composite, containing electrical conductors. Cable Routing assemblies are U-shaped wiring troughs that may or may not have covers. (If they always had covers, effectively they would be raceways.) The significant difference between optical fiber or communications cable routing assemblies and optical fiber raceways is that the routing assemblies are larger and open, and therefore may present a greater fire load.

The question always comes up: Why are optical fiber installation requirements needed in the *NEC®* this is not an electrical system? The concern is for static and resistance to the spread of fire. There are different types of fiber-optic cables. They are grouped into three types, as defined in 770.2. The composite types are

hybrid and can carry both current and noncurrent. All types are capable of controlling current flow.*

NEC® 770.24 covers mechanical execution of work for optical fiber cables and raceways, and equipment shall be installed in a neat and professional manner. Cables shall be supported by the building structure in such a manner that the cable will not be damaged by normal building use.

This chapter is divided into five articles.

Chapter 8 covers communications systems and are not subject to the requirements of Chapter 1 through 7 except where the requirements are specifically referenced in Chapter 8.

Article 800 covers communications circuits and equipment.

Informational Note No. 1: See **90.2(B)**(4) for installations of communications circuits and equipment that are not covered.

Informational Note No. 2: For further information for remote-control, signaling, and power-limited circuits, see Article **725**.

Informational Note No. 3: For further information for fire alarm systems, see Article **760**.

Article 810 covers antenna systems for radio and television receiving equipment, amateur and citizen band radio transmitting and receiving equipment, and certain features of transmitter safety. This article covers antennas such as wire-strung type, multi-element, vertical rod, and dish and also covers the wiring and cabling that connects them to equipment. This article does not cover equipment and antennas used for coupling carrier current to power line conductors.*

Article 820 covers coaxial cable distribution of radio frequency signals typically employed in community antenna television (CATV) systems.

Informational Note: See **90.2(B)**(4) for installations of CATV and radio distribution systems that are not covered.

Article 830 article covers network-powered broadband communications systems that provide any combination of voice, audio, video, data, and interactive services through a network interface unit.

Informational Note No. 1: A typical basic system configuration includes a cable supplying power and broadband signal to a network interface unit that converts the broadband signal to the component signals. Typical cables are coaxial cable with both broadband signal and power on the center conductor, composite metallic cable with a coaxial member for the broadband signal and a twisted pair for power, and composite optical fiber cable with a pair of conductors for power. Larger systems may also include network components such as amplifiers that require network power.

Informational Note No. 2: See **90.2(B)**(4) for installations of broadband communications systems that are not covered.

Article 840 covers premises-powered optical fiber-based broadband communications systems that provide any combination of voice, video, data, and interactive services through an optical network terminal (ONT).

Informational Note No. 1: A typical basic system configuration consists of an optical fiber cable to the premises (FTTP) supplying a broadband signal to an ONT that converts the broadband optical signal into component electrical signals, such as traditional telephone, video, high-speed internet, and interactive services. Powering of the ONT is typically accomplished through an ONT power supply unit (OPSU) and battery backup unit (BBU) that derive their power input from the available ac at the premises. The optical fiber cable is unpowered and may be nonconductive or conductive.

Informational Note No. 2: See **90.2(B)**(4) for installations of premises-powered broadband communications systems that are not covered in this article.

Articles 800, 820, and 830, 840 do not reference Table 300.5 (wiring method burial requirements); therefore, the table cannot be applied. There are no burial depth requirements for these communications cables. The reason is that the *NEC* is a safety document. Communications generally pose no safety hazards by interruption of service. Therefore, the installers often bury these cables for service and economic reasons. However, they pose no safety hazards, so the *NEC* does not set any requirements. Local requirements or amendments may apply.

NEC 800.24, 820.24, 830.24, and 840.24 cover the mechanical execution of work for communications circuits, and equipment shall be installed in a neat and professional manner. Cables shall be supported by the building structure in such a manner that the cable will not be damaged by normal building use.

Article 830 covers requirements for "Network-Powered Broadband Communications Systems." This is fairly a new generation communication system and provides multimedia services in any combination of audio, voice, video, data, and interactive services. This new system may operate at higher levels of power and recognizes two levels of power (see *NEC* Table 830.15).

Network Power Source	Low	Medium
Maximum power rating (volt-amperes)	100	100
Maximum voltage rating (volts)	100	150

It is because of these higher operating voltages that Article 830 contains burial depth requirements. *NEC* 830.47(A) through (D) and Table 830.47 contain requirements similar to those in Table 300.5 and generally require cables to be buried at 18 in. (450 mm) below grade. When the cables are enclosed in a nonmetallic raceway 12 in. of burial depth is permitted (300 mm) and in RMC or IMC, the burial depth is 6 in. (150 mm). The notes at the bottom of the table provide additional information that can be applied if necessary.

Chapter 8 is unique. Study it carefully. Study the scope, as shown throughout this chapter, of each article carefully to be sure that you are answering a question from the right article and section of this chapter.

QUESTION REVIEW

Each lesson is purposely designed to require the student to apply the entire *NEC*® text and not specific chapters or articles. It has been found that when studying for a timed, open-book examination, the student must gain proficiency in the Table of Contents, the Index, and the ability to move quickly from cover to cover to find the correct answer to each question in a timely fashion.

1. Does the *Code* address grounding CATV service to a dwelling unit?

    Answer: _____

    _____

    _____

    _____

    _____

    Reference: _____

2. After you install a satellite dish at a local residence, the inspector informs you that you must bury the cable from the satellite dish to the house in accordance with Table 300.5 of the *NEC*®. Was the inspector correct?

    Answer: _____

    _____

    _____

    _____

    _____

    Reference: _____

3. Does the *Code* require grounding the metal sheath of CATV cable entering a dwelling?

    Answer: _____

    _____

    _____

    _____

    _____

    Reference: _____

4. According to the *NEC*®, open conductors for communication equipment on a building shall be separated at least _____ feet (_____ m) from *lightning* conductors, where practicable.

    Answer: _____

    _____

    _____

    _____

    Reference: _____

5. In a classified location where propane is used, the customer states that the wiring to the control circuit is of sufficient low voltage that a spark cannot be generated from this circuit. The devices and wiring have a control drawing number on them. What type wiring is most likely employed in this installation where there is not enough voltage to generate a spark sufficient to cause an explosion?

Answer: _____

_____

_____

_____

_____

Reference: _____

6. A motor control center is to be installed in a local factory. Does the *NEC®* cover this type of installation? If so, which article is applicable?

Answer: _____

_____

_____

_____

_____

Reference: _____

7. You have been asked to install a small transformer that will transform 480 volts to 240/120 volts. The inspector states that this must be wired as a separately derived system. What is a separately derived system? How do you wire this transformer?

Answer: _____

_____

_____

_____

_____

Reference: _____

8. You recently encountered a receptacle with an orange triangle on the face of it. You were told that this was an isolated ground receptacle. Where can this be found in the *NEC®*?

Answer: _____

_____

_____

_____

_____

Reference: _____

9. You recently encountered a rental storage facility that is an area in a building to store old movie films. Where would the wiring requirements for this area be found in the *NEC*®?

Answer: _____

_____

_____

_____

Reference: _____

10. The inspector informed you that a bathroom area was defined in the *NEC*®. Where is this definition located?

Answer: _____

_____

_____

_____

Reference: _____

11. In wiring a large church, an installation of a pipe organ is required. Are pipe organs covered in the *NEC*®?

Answer: _____

_____

_____

_____

Reference: _____

12. In making an installation, it was noted that the clearance between two large switchboards was rather tight—only about 3 feet (900 mm). With the doors open, there was no room to pass. Are there any requirements in the *NEC*® to cover the requirements for working clearances?

Answer: _____

_____

_____

_____

Reference: _____

13. After you completed wiring a new home, the inspector came and told you to cover the panelboard—that if the sheetrockers got their spackling material or paint on the panel, it was a code violation. Is this correct?

Answer: _____

_____

_____

_____

Reference: _____

14. You recently mounted a new service on a new mobile home on a privately owned lot. The inspector turned it down and said this was not permitted. Which section of the *NEC®* was violated?

    Answer: _____

    _____

    _____

    _____

    Reference: _____

15. What is an optional standby system?

    Answer: _____

    _____

    _____

    _____

    Reference: _____

16. Is color coding required in the *NEC®*? You know that the neutral conductor is required to be white and the grounding conductor is required to be green, but what about the ungrounded conductors?

    Answer: _____

    _____

    _____

    _____

    Reference: _____

17. You recently installed 14 AWG THHN conductors to supply a general purpose branch circuit supplying duplex receptacles in an office area. You connected the 14 AWG THHN to a 20-ampere breaker. Table 310.15(B)(16) allows 25 amperes on a 14 AWG conductor. Was this in accordance with the *NEC®*?

    Answer: _____

    _____

    _____

    _____

    Reference: _____

18. In making an installation to an outdoor ball field, you proposed a common neutral to be run with multiple branch-circuit conductors. How many branch-circuit conductors can be supplied with a common neutral?

    Answer: _____

    _____

    _____

    _____

    Reference: _____

19. You have been asked to wire a dairy barn. Are there any specific requirements in the *NEC®* for this type of installation?

Answer: _____

_____

_____

_____

Reference: _____

20. You recently observed a local electrician installing wiring under a carpet in a bank. Is this wiring method permitted? Which article covers wiring under carpets?

Answer: _____

_____

_____

_____

Reference: _____

21. More buildings are increasingly requiring lightning arrestors (surge arrestors or surge protection devices) to be installed on the service. Which article of the *NEC®* covers these installations?

Answer: _____

_____

_____

_____

Reference: _____

22. Are temporary wiring methods covered in the *NEC®*? You regularly install a temporary meter loop to supply power to the construction crews until the permanent power system has been installed. Where is this covered in the *NEC®*?

Answer: _____

_____

_____

_____

Reference: _____

23. A 480-volt, 3-phase circuit with a rated load of 150 kW draws 210 amperes. The power factor is _____ %.

Answer: _____

_____

_____

_____

Reference: _____

24. A _____-horsepower, 208-volt, 3-phase motor operating at 91% efficiency has an input current of 141 amperes. (1 hp = 746 watts)

Answer: _____

_____

_____

_____

Reference: _____

25. A feeder supplying a second building requires a maximum conductor resistance of 0.02 ohm. The minimum size conductor must have a resistance of no more than _____ ohm(s) per 1000 feet (305 m). The total length of the circuit conductor is 190 feet (58 m).

Answer: _____

_____

_____

_____

Reference: _____

CHAPTER 14

OBJECTIVES

Studying this chapter along with the 2011 *NEC®* gives the reader a thorough review of the 2011 *NEC®*.

To complete this chapter, you should:

■ Know the basics of preparing for any examination

■ Know how to read, understand, and simplify examination questions

■ Know how to locate the answer to each question by using the Table of Contents and the Index

■ Take the 50-question practice examination to evaluate your progress before beginning the practice exams

■ Take practice exams until you are confident that you are ready to be tested on your knowledge of the 2011 *NEC®*

■ Take the final examination to assure yourself that you are ready to take the actual examination

ELECTRICAL AND *NEC®* QUESTION REVIEW

Now you are about to begin the final chapter of this book. You should have a good understanding of the *Code* arrangement and its content. You should now understand that the chapters are arranged in the order in which you use them in most common installations, from Article 90—Introduction; Chapter 1—general requirements; Chapter 2—wiring and protection; Chapter 3—wiring methods and materials; Chapter 4—general use equipment; Chapters 5–7—special occupancies, uses, and equipment; and Chapter 8—communication systems. The final chapter in the *NEC®* is Chapter 9, which has the tables. Furthermore, you should now understand this arrangement and know the articles contained in each chapter. For instance, you should understand that the article covering rigid metal conduit (RMC) would automatically be found in Chapter 3 because RMC is a wiring method. You should understand that the conditions for installation of a luminaire would be found in Chapter 4 because luminaires are general-use equipment. Or you should understand that hospital installation requirements would be found in Chapter 5 because a hospital is a special occupancy and those are found in Chapter 5, as are mobile homes, hazardous locations and spray booths, commercial garages, and so on. You should automatically remember that a sign would be found in Chapter 6, special equipment, because, although signs are common, they are not general equipment found in all types of installations. Signs are special equipment and are found in Chapter 6. These things must be understood before you take the examination. It is necessary so that you might use the *NEC®* efficiently as a reference document. Memorization is not recommended. However, with a good, clear understanding, one can refer to the Table of Contents in the front of the book and quickly find the article number if you have an understanding of where that article should be found in the *Code*. You should then be able to go to the Index and find the proper reference so that you might quickly look up the question and prove the answer. Once this has been done, you should have no problem passing any competency exam given throughout the country that is based on your knowledge of the *NEC®*.

With that competency achieved and your understanding of the *NEC®* arrangement ensured, the next important task you must improve is the understanding of what each question is asking. You must read the question through, then determine exactly what the question is asking. You will find key words, and with those key words you can simplify the question generally so it can be quickly researched using the Table of Contents, the Index, and your knowledge of the *Code* arrangement.

Example: As an example, in the following question, select the key words and simplify the question:

"May single insulated conductors cable be used for a 120/240-volt branch circuit as temporary wiring in a building under construction where the cable is supported on insulators at intervals of not more than 10 feet (3 m)?"

Now, you may rephrase the question slightly differently in your mind or you may look for the key words. Single insulated conductors is a key word, temporary wiring is a key, and supported on insulators at intervals of not more than 10 feet (3 m) is a key. So,

"Are Single insulated conductors permitted as temporary wiring where supported on insulators?"

Now, with the three key words in mind, we must remember that they are not asking where this wiring method is acceptable; they are telling you that the installation would be temporary wiring. If we go to the Table of Contents in the front of the *NEC®*, we find that temporary wiring is found in Article 590. As we turn to Article 590 and we look in the scope under 590.1, we find out that, yes, lesser methods are acceptable. As we research that article quickly, we find in 590.4(C) that this is an acceptable method, provided that it is supported on insulators at intervals of not more than 10 feet (3 m), so the answer would be yes. References would be *NEC®* 590.1 and 590.4(C). As you see in this case when selecting the key words or phrases, using the Table of Contents was sufficient and it was not necessary to go to the Index.

Example: Let us look at another example. The question is:

"Is liquidtight flexible nonmetallic conduit (LFNC) permitted to be used in circuits in excess of 600 volts?"

As we look at that question, we see that LFNC is the key phrase. Then, we look at permitted use as a key phrase and 600 volts as the final key phrase.

Again, we can go to the Table of Contents and find that LFNC is found in Article 356. As we turn to Article 356, we see uses permitted in 356.10 and uses not permitted in 356.12. Immediately, as we scan down the uses permitted and the uses not permitted, we find in Section 356.12(4) that LFNC is not permitted for circuits in excess of 600 volts. However, you will note that there is an exception for electrical signs, so the Answer: is: "Yes, for electrical signs by exception. No, generally."

If you will take each question and break it down so that you are looking at the key phrase or key word, you will find it much easier to reference the correct *Code* article and section, and you will not waste the valuable time needed to complete your examination. All questions will not be as easy as these two examples to identify the key phrases, but in all questions, there are key phrases or key words, and with the practice you will receive during the following exercises in Chapter 14, you should greatly improve your competency and speed when preparing for an examination, which will allow you ample time to get through the examination and still have some time to do additional research on the tough questions. What you will find by following these examples is that there will be fewer tough questions in the exam. Good luck!

2011 Highlighted Code Changes

There were approximately 5,016 proposals and 2,910 public comments submitted for modifications to the 2011 edition of the *NEC*. "FPNs" "Fine Print Notes" have appeared in all recent editions of the NEC. They have now been changed and are now referred to as "Informational Notes" throughout the Code to clarify that they are not enforceable or mandatory.

The term "grounding conductor" was replaced with the term "grounding electrode conductor," "bonding jumper," or "bonding conductor" in several places throughout the NEC to clarify the purpose of the conductor referenced.

Three new articles added:

- Article 399—Outdoor, Overhead Conductors, Over 600 volts

- Article 694—Small Wind Electric Systems

- Article 840—Premises-Powered Broadband Communications Systems

All annexes are now "informative annexes" and are not enforceable but they do provide excellent material to assist users in the application of the enforceable text in the NEC.

PRACTICE EXAMINATION

The following test was developed and administered by a nationally known seminar presenter and author at a meeting of *Code* authorities. Although this test is not representative of national electricians' tests, it does provide a good exercise in researching the *NEC®*. You will find the questions to be somewhat tricky in some cases. However, if you can pass this examination within one hour, you should be able to pass most national examinations. It will exercise your knowledge of the *NEC®*, your ability to use the document as a reference, and your ability to research difficult questions.

1. Equipment enclosed in a case or cabinet that is provided with a means of sealing or locking so that live parts cannot be made accessible without opening the enclosure is said to be
   A. guarded.
   B. protected.
   C. sealable equipment.
   D. lockable equipment.

   Answer: _____ Reference: _____

2. The letter(s) _____ indicate(s) two insulated conductors laid parallel within an outer nonmetallic covering.
   A. D
   B. M
   C. T
   D. II

   Answer: _____ Reference: _____

3. Class 1 circuit conductors shall be protected against overcurrent
   A. in accordance with the values specified in Table 310.15(B)(16) through Table 310.15(B)(21) for 14 AWG and larger.
   B. not exceeding 7 amperes for 18 AWG.
   C. not exceeding 10 amperes for 16 AWG.
   D. and ampacity adjustment and correction factors do not apply.
   E. all of the above.

   Answer: _____ Reference: _____

4. Enclosures for switches and overcurrent devices shall be permitted to contain feed through conductors and splices provided _____.
   A. The total of all conductors installed at any cross section of the wiring space does not exceed 40% of the cross-sectional area of that space.
   B. The total area of all conductors, splices, and taps installed at any cross section of the wiring space does not exceed 75% of the cross-sectional area of that space.
   C. A warning label is applied to the enclosure that identifies the closest disconnecting means for any feed-through conductors.
   D. All of the above

   Answer: _____ Reference: _____

5. A plug fuse of the Edison base has a maximum rating of _____ amperes.
   A. 20
   B. 30
   C. 40
   D. 50

   Answer: _____   Reference: _____

6. Non–power-limited fire-protective alarm circuit conductors shall be
   A. solid copper.
   B. stranded copper.
   C. 18 AW6 or larger.
   D. all of the above.

   Answer: _____   Reference: _____

7. When conduit or tubing nipples having a maximum length not to exceed _____ inches (_____ mm) are installed between boxes and similar enclosures, the fill shall be permitted to 60%.
   A. 6 (150)
   B. 12 (300)
   C. 24 (600)
   D. 30 (750)

   Answer: _____   Reference: _____

8. Circuit breakers shall be
   A. capable of being opened by manual operation.
   B. capable of being closed by manual operation.
   C. Trip-free.
   D. all of the above.

   Answer: _____   Reference: _____

9. For a circuit operating at less than 50 volts, standard lampholders having a rating not less than _____ watts shall be used.
   A. 300
   B. 550
   C. 660
   D. 770

   Answer: _____   Reference: _____

10. 18 AWG TFF wire has an allowable ampacity of _____.
    A. 14
    B. 10
    C. 8
    D. 6

    Answer: _____   Reference: _____

11. A 10 AWG bare solid copper conductor has a cross-section area of _____ square inches (_____ mm²).
    A. .008 (5.26)
    B. 1.21 (3.984)
    C. .006 (4.25)
    D. .109 (70.41)

    Answer: _____ Reference: _____

12. Conductors of light or power can occupy the same enclosure or raceway with conductors of power-limited fire-protective signaling circuits.
    A. True
    B. False

    Answer: _____ Reference: _____

13. Explanatory material in the *NEC®* is in the form of
    A. footnotes.
    B. Informational notes.
    C. obelisks and asterisks.
    D. red print.

    Answer: _____ Reference: _____

14. Straight runs of trade size 1¼ (35) RMC using threaded couplings can be secured at not more than _____ foot (_____ m) intervals.
    A. 5 (1.5)
    B. 10 (3)
    C. 12 (3.7)
    D. 14 (4.3)

    Answer: _____ Reference: _____

15. An isolating switch is one that
    A. is not readily accessible to persons unless special means for access are used.
    B. is capable for interrupting the maximum operating overload current of a motor.
    C. is intended for use in general distribution and branch circuits.
    D. is intended for isolating an electrical circuit from the source of power.

    Answer: _____ Reference: _____

16. Where damage to remote-control circuits of safety-control equipment would produce a hazard, all conductors the circuit shall be installed in _____ or otherwise suitably protected from physical damage.
     I. mineral-insulated or metal clad cable
    II. rigid metallic conduit
    III. rigid nonmetallic conduit
    A. I only
    B. II only
    C. III only
    D. I, II, or III

    Answer: _____ Reference: _____

17. _____ on equipment to be grounded shall be removed from contact surfaces to assure good electrical continuity.
    A. Conductive coatings
    B. Nonconductive coatings
    C. Manufacturer's instructions
    D. all of the above

    Answer: _____   Reference: _____

18. A _____ conductor is encased within materials of thickness and composition that are not recognized by this *Code*.
    A. noninsulating
    B. bare
    C. covered
    D. none of these

    Answer: _____   Reference: _____

19. To guard live parts, _____ accessible to qualified persons only.
    A. isolate in a room
    B. locate on a balcony
    C. enclose in a cabinet
    D. any of these

    Answer: _____   Reference: _____

20. Conductors sizes are expressed in AWG or _____.
    A. inches.
    B. circular mils.
    C. square inches.
    D. none of the above

    Answer: _____   Reference: _____

21. The metric system used in the 2011 *NEC*® is known as the
    A. English metric system.
    B. International system of units.
    C. SPOT metric conversion.
    D. hard metric conversion.

    Answer: _____   Reference: _____

22. The voltage of a circuit is defined by the *Code* as the _____ root-mean-square (effective) difference of potential between any two conductors in the circuit.
    A. lowest
    B. greatest
    C. average
    D. nominal

    Answer: _____   Reference: _____

23. Doorbell wiring, rated as Class 2, in a residence _____ run in the same raceway with light and power conductors.
    A. is permitted with a 600-volt insulation
    B. shall not be
    C. shall be
    D. is permitted if insulation is equal to the highest-installed.

    Answer: _____   Reference: _____

24. Equipment or materials to which has been attached a symbol or other identifying mark acceptable to the authority having jurisdiction is known as
    A. listed.
    B. labeled.
    C. approved.
    D. rated.

    Answer: _____   Reference: _____

25. The grounded conductor of a branch circuit shall be identified by a _____ color.
    I. gray
    II. continuous green
    III. continuous white
    IV. green with yellow stripe
    A. I or III
    B. II or IV
    C. III only
    D. II only

    Answer: _____   Reference: _____

26. Circuits for lighting and power shall not be connected to any system containing
    A. hazardous material.
    B. trolley wires with ground returns.
    C. poor wiring methods.
    D. dangerous chemicals or gases.

    Answer: _____   Reference: _____

27. The minimum clearance between an electric space-heating cable and an outlet box shall not be less than _____ inches (_____ mm).
    A. 8 (200)
    B. 12 (300)
    C. 18 (450)
    D. 6 (150)

    Answer: _____   Reference: _____

28. The ampacity requirements for a disconnecting means of X-ray equipment shall be based on _____ % of the input required for the momentary rating of the equipment if greater than the long-time rating.
    A. 125
    B. 100
    C. 50
    D. none of the above

    Answer: _____ Reference: _____

29. Open conductors installed outside shall be separated from open conductors of other circuits by not less than _____ inches (_____ mm).
    A. 4 (100)
    B. 6 (150)
    C. 8 (200)
    D. 10 (250)

    Answer: _____ Reference: _____

30. The minimum headroom of working spaces about motor control centers shall be
    A. 3½ feet (1 m).
    B. 5 feet (1.5 m).
    C. 6 feet, 6 inches (2 m).
    D. 6 feet, 3 inches (1.8 m).

    Answer: _____ Reference: _____

31. Where fixed multioutlet assemblies are employed in locations where a number of appliances are likely to be used simultaneously, each foot or fraction thereof shall be considered as an outlet of not less than _____ VA.
    A. 180
    B. 200
    C. 60
    D. 35

    Answer: _____ Reference: _____

32. Soft-drawn or medium-drawn copper lead in conductors for television equipment antenna systems shall be permitted where the maximum span between points of support is less than _____ feet (_____ m).
    A. 35 (11)
    B. 30 (9.1)
    C. 20 (6)
    D. 10 (3)

    Answer: _____ Reference: _____

33. For a one-family dwelling, the service disconnect shall not be less than _____.
    A. 60
    B. 100
    C. 150
    D. 200

    Answer: _____ Reference: _____

34. Which of the following machines shall be provided with speed-limiting devices or other speed-limiting means?
    I. series motors
    II. induction motors
    III. self-excited DC motors
    A. I only
    B. II only
    C. III only
    D. I and III

    Answer: _____   Reference: _____

35. Where within _____ feet (_____ m) of any building or other structure, open wiring on insulators shall be insulated or covered.
    A. 3 (900 mm)
    B. 4 (1.2)
    C. 6 (1.8)
    D. 10 (3)

    Answer: _____   Reference: _____

36. Conductors over 600 volts in tunnels shall be installed in
    A. rigid metal conduit.
    B. MC cable.
    C. other metal raceways.
    D. any of the above.

    Answer: _____   Reference: _____

37. For garages and outbuildings on residential property, a _____ suitable for use on branch circuits shall be permitted as the disconnecting means.
    A. snap switch
    B. set of three-way or four-way snap switches
    C. set of two-way or four-way snap switches
    D. A or B

    Answer: _____   Reference: _____

38. Utilization equipment fastened in place connected to a branch circuit with luminaires shall not exceed _____% of the branch-circuit rating.
    A. 50
    B. 60
    C. 80
    D. 100

    Answer: _____   Reference: _____

39. The maximum amperage rating of a 4 inch × ½ inch copper busbar in an auxiliary gutter is _____ amperes.
    A. 500
    B. 700
    C. 1000
    D. 2000

    Answer: _____   Reference: _____

40. For temporary wiring over 600 volts, nominal, _____ shall be provided to prevent access of other than authorized and qualified personnel.
    I. fencing
    II. barriers
    III. signs
    A. I only
    B. II only
    C. III only
    D. I or II

    Answer: _____   Reference: _____

41. Each continuous-duty motor with _____ horsepower or less, not permanently installed, that is nonautomatically started and is within sight of the controller shall be permitted to be protected against overload by a branch-circuit protective device.
    A. ⅛
    B. ½
    C. ¾
    D. 1

    Answer: _____   Reference: _____

42. The grounded conductor in a mobile home shall be insulated from the equipment grounding conductor to which of the following location(s)?
    A. range and oven
    B. distribution panel
    C. clothes dryer
    D. all of the above

    Answer: _____   Reference: _____

43. The bottom of sign and outline lighting enclosures shall not be less than _____ feet (_____ m) above areas accessible to vehicles.
    A. 12 (3.7)
    B. 14 (4.25)
    C. 16 (4.9)
    D. 18 (5.5)

    Answer: _____   Reference: _____

44. The conductor used to ground the outer cover of a coaxial cable shall be
    I. insulated, covered, or bare
    II. 14 AWG minimum.
    III. guarded from physical damage when necessary.
    A. I only
    B. II only
    C. III only
    D. I, II, and III

    Answer: _____   Reference: _____

45. Where energized live parts are exposed, the minimum clear workspace shall not be less than _____ feet (_____ m) high for over 600 volts.
    A. 3 (900 mm)
    B. 3½ (1)
    C. 5 (1.8)
    D. 6¼ (1.9)
    E. 6½ (2)

    Answer: _____   Reference: _____

46. Conductors _____ AWG or larger supported on solid knobs shall be securely tied thereto by tie wires having an insulation equivalent to that of the open wire.
    A. 14
    B. 12
    C. 10
    D. 8

    Answer: _____   Reference: _____

47. A branch circuit that supplies two or more receptacles or outlets for lighting and appliances is known as a
    A. general-purpose branch circuit.
    B. a multipurpose branch circuit.
    C. a utility branch circuit.
    D. none of the above

    Answer: _____   Reference: _____

48. The minimum size conductor that can be used for an overhead feeder without a messenger wire from a residence to a remote garage when less than 50 feet away is _____.
    A. 10 AWG cu.
    B. 12 AWG cu.
    C. 6 AWG al.
    D. 10 AWG al.

    Answer: _____   Reference: _____

49. Two or more _____, 120-volt small-appliance branch circuits and 15- or 20-ampere receptacle outlets in a dwelling unit, kitchen, dining room, breakfast room, pantry, or similar dining areas are required in all dwelling units.

    A. 15-ampere
    B. 20-ampere
    C. 30-ampere
    D. 15- or 20-ampere

    Answer: _____   Reference: _____

50. Each kitchen countertop surface in a dwelling unit shall be supplied by not fewer than _____ small-appliance branch circuit(s).

    A. one
    B. two
    C. three
    D. no minimum

    Answer: _____   Reference: _____

CODE QUIZ 1

> **Each lesson is purposely designed to require the student to apply the entire *NEC®* text and not specific chapters or articles. It has been found that when studying for a timed, open-book examination, the student must gain proficiency in the Table of Contents, the Index, and the ability to move quickly from cover to cover to find the correct answer to each question in a timely fashion.**

1. Which article in the *NEC®* is applicable when correctly installing the electrical systems for commercial fish ponds and sewer ponds?

   Answer: _____   Reference: _____

2. Where is high-density polyethylene conduit covered in the *Code*?

   Answer: _____   Reference: _____

3. Where are the requirements for installing temporary wiring on a construction job for the workers only during construction?

   Answer: _____   Reference: _____

4. An area including a basin with one or more of the following—a toilet, a tub, shower, a bidet or similar plumbing fixture—is defined in the *NEC®* as a _____.

   Answer: _____   Reference: _____

5. Hazardous locations determined to be neither Class I, Division 1; Class I, Division 2; Class I, Zone 0; Class I, Zone 1; Class I, Zone 2; Class II, Division 1; Class II, Division 2; Class III, Division 1; Class III, Division 2; nor any combination thereof are _____.

   Answer: _____   Reference: _____

6. _____ contains optional administrative and enforcement rules.

   Answer: _____   Reference: _____

7. The main bonding jumper is used to bond the grounded conductor to the grounding electrode and equipment grounding. What is the system bonding jumper used for?

   Answer: _____   Reference: _____

8. The conductor used to connect the grounding electrode(s) to the grounded conductor, at the service, at each building or structure where supplied by a feeder(s) or branch circuit(s) from a common service, or at the source of a separately derived system is called the _____.

   Answer: _____   Reference: _____

9. An enclosure for use in underground systems, provided with an open or closed bottom, and sized to allow personnel to reach into but not enter, for the purpose of installing, operating, or maintaining equipment or wiring or both is defined as a _____.

   Answer: _____   Reference: _____

10. A kitchen is defined as an area with a sink and permanent facilities for food preparation and cooking. What section of the *Code* places special requirements for these areas when not located in dwelling?

    Answer: _____   Reference: _____

11. All 125-volt, 15- or 20-ampere receptacles located in other than dwelling units outdoors must have _____ for personnel.

    Answer: _____   Reference: _____

12. Where is the allowable minimum bending radius for all conduit and tubing found?

    Answer: _____   Reference: _____

13. Article 830 covers network-powered broadband communications systems that are covered by the *NEC*®. Those not covered are described in Section __.

    Answer: _____   Reference: _____

14. Article 820 covers CATV systems and radio distribution systems. How is the maximum number of coaxial cables permitted in conduit calculated?

    Answer: _____   Reference: _____

15. GFCI protection is required for 15 A and 20 A, 125-volt receptacles in aircraft hangers. True or False?

    Answer: _____   Reference: _____

CODE QUIZ 2

1. Abandoned power cables, communication cables, and connecting and interconnection cables are required to be removed under raised floors in information technology (IT) rooms unless contained in _____.

   Answer: _____   Reference: _____

2. Contact conductors and exposed short lengths of conductors at resistors, collectors, and other equipment for cranes or hoists are not required to be _____.

   Answer: _____   Reference: _____

3. For mobile manufactured homes, Outdoor or under-chassis line-voltage (120 volts, nominal, or higher) wiring exposed to moisture or physical damage shall be protected by _____ or _____ unless otherwise permitted.

   Answer: _____   Reference: _____

4. A service for a carnival would have to comply with all the requirements of Article _____ and Sections _____.

   Answer: _____   Reference: _____

5. The wiring methods for Zone 20 are found in Section _____.

   Answer: _____   Reference: _____

6. Resistors commonly found in electrical control cabinets and other electrical apparatus are covered in _____.

   Answer: _____   Reference: _____

7. The disconnecting means for motor circuits rated 600 volts, nominal, or less shall have an ampere rating not less than _____% of the full-load current rating of the motor, except for specific listed nonfused motor-circuit switches.

   Answer: _____   Reference: _____

8. Receptacles shall not be installed within or directly above _____ or _____.

   Answer: _____   Reference: _____

9. Conductors in vertical poles used as raceways shall be supported in accordance with Section _____.

   Answer: _____   Reference: _____

10. Flexible cords and cables must conform to the descriptions provided in _____.

    Answer: _____   Reference: _____

11. Open wiring on insulators shall not be permitted in other than industrial or agricultural installations or where concealed by the building structure, or where the voltage exceeds _____.

    Answer: _____   Reference: _____

12. Do the ampacity adjustment factors apply to conductors in cellular metal raceways?

    Answer: _____   Reference: _____

13. Nonmetallic cable tray shall be permitted only in corrosive areas and in areas requiring voltage isolation. True or False?

    Answer: _____   Reference: _____

14. Busways may not be installed where subject to _____ or severe _____.

    Answer: _____   Reference: _____

15. If I ask for 10 lengths of 27 IMC, what should I receive from the supplier?

Answer: _____   Reference: _____

CODE QUIZ 3

1. Integrated gas spacer cable is not permitted to be used in contact with buildings or _____.

Answer: _____   Reference: _____

2. _____ receptacles or covers are required in psychiatric and pediatric general-care areas where accessible to patients.

Answer: _____   Reference: _____

3. The _____ shall not be used as the sole equipment grounding conductor or as an effective ground-fault current path.

Answer: _____   Reference: _____

4. The processes of purging or pressurizing supplying enclosures in Class 1 hazardous locations to an acceptable level are acceptable protection techniques. What NFPA standard covers these two protection techniques?

Answer: _____   Reference: _____

5. Single conductors specified in Table 310.104(A) must only be installed in a recognized wiring method of Chapter 3 or be part of a Chapter 3 wiring method except in accordance with _____.

Answer: _____   Reference: _____

6. Where grounded conductors of different systems are installed in the same raceway, cable, box, auxiliary gutter, or other type of enclosure, each grounded conductor shall be identified by system. This means of identification shall be permanently _____.

Answer: _____   Reference: _____

7. An attachment plug and multioutlet receptacle assembly where permitted assembly shall consist of an attachment plug, a flexible cord, and a multioutlet receptacle strip. Each assembly must be energized from a receptacle outlet or cord connection body.

Answer: _____   Reference: _____

8. In mobile or manufactured homes, receptacle outlets shall not be permitted to be installed within a _____ or _____ or _____ position in any countertop or _____ electric baseboard heaters, unless provided for in the listing's or manufacturer's instructions.

Answer: _____   Reference: _____

9. Conduit or tubing not smaller than _____ trade size for neon secondary circuit conductors over 1000 volts, nominal, shall contain no more than _____ conductor(s).

   Answer: _____   Reference: _____

10. Which article covers the requirements for electric cutting process equipment?

   Answer: _____   Reference: _____

11. Is a disconnect switch required to be within sight of a motors or can the disconnect be located elsewhere if it can be locked in the off position?

   Answer: _____   Reference: _____

12. How is the size of an equipment bonding jumper determined?

   Answer: _____   Reference: _____

13. Do raceways to be installed by directional boring equipment have to be listed for the purpose?

   Answer: _____   Reference: _____

14. What cables do not have to be marked as "sunlight resistant" when they are used in installations exposed to direct sunlight?

   Answer: _____   Reference: _____

15. Is it permissible to have 240-volt lighting circuits in a dwelling?

   Answer: _____   Reference: _____

16. Would a 125-volt receptacle installed outdoors at a dwelling for a sewer lift pump require GFCI protection?

   Answer: _____   Reference: _____

17. What is the trade size of a 103 intermediate conduit?

   Answer: _____   Reference: _____

18. Is a green or bare equipment grounding conductor required to be installed with the circuit conductors installed in electrical metallic tubing (EMT)?

   Answer: _____   Reference: _____

19. Where are the installation requirements for industrial control cabinets found?

   Answer: _____   Reference: _____

20. Where installed an SPD fated 1kV or less shall be connected to each _____ conductor.

   Answer: _____   Reference: _____

CODE QUIZ 4

1. Is it permissible for a switch in a patient care area, located outside the patient vicinity, to be supplied by Type MC cable?

   Answer: _____

   _____

   _____

   _____

   Reference: _____

2. In the *NEC®*, what in commonly referred to as a lighting fixture is referred to as a _____.

   Answer: _____

   _____

   _____

   _____

   Reference: _____

3. Does the *NEC®* contain adoptive ordinance language?

   Answer: _____

   _____

   _____

   _____

   Reference: _____

4. Has the *Code* been changed or modified in any way to make it a more suitable document for international use?

   Answer: _____

   _____

   _____

   _____

   Reference: _____

5. Does 90.4 address signaling and communications systems as part of the term "electrical installations"?

   Answer: _____

   _____

   _____

   _____

   Reference: _____

6. Are informative product standards referenced in the *NEC®*? If so, where?

   Answer: _____

   _____

   _____

   _____

   Reference: _____

7. Are metric measurements the measurement system of preference?

   Answer: _____

   _____

   _____

   _____

   Reference: _____

8. Where are the requirements for receptacles, cord connectors, and attachment plugs found in the *NEC®*?

   Answer: _____

   _____

   _____

   _____

   Reference: _____

9. Would the definition of a garage in Article 100 include a building where electric vehicles are stored?

   Answer: _____

   _____

   _____

   _____

   Reference: _____

10. Would 110.22 literally not require identification of a disconnecting means that was not required by the *Code*?

    Answer: _____

    _____

    _____

    _____

    Reference: _____

11. Is panic hardware on electrical room doors required?

    Answer: _____

    _____

    _____

    _____

    Reference: _____

12. Where can the requirements for the specific wiring methods for electric discharge luminaires supported independent of the outlet box be found?

    Answer: _____

    _____

    _____

    _____

    Reference: _____

13. What does the term "luminaire" used throughout the *Code* refer to

    Answer: _____

    _____

    _____

    _____

    Reference: _____

14. Does the *Code* prohibit receptacles from being installed in a face-up position in a dwelling-unit laundry room countertop?

    Answer: _____

    _____

    _____

    _____

    Reference: _____

15. Are dwelling unit bedroom lighting circuits required to be AFCI protected?

    Answer: _____

    _____

    _____

    _____

    Reference: _____

16. Receptacles are commonly installed in appliance garages on dwelling-unit kitchen countertops. Can this receptacle be considered as one of the required receptacles for the countertop?

    Answer: _____

    _____

    _____

    _____

    Reference: _____

17. Is GFCI protection required for electrically heated floors in bathrooms, kitchens, hydromassage bathtub, spa, and hot tub locations?

    Answer: _____

    _____

    _____

    _____

    Reference: _____

18. Is it permissible to use a gray-colored conductor as a grounded (neutral) conductor?

    Answer: _____

    _____

    _____

    _____

    Reference: _____

19. Are branch circuits over 600 volts permitted in residential and commercial occupancies where there are no maintenance or supervision provisions?

    Answer: _____

    _____

    _____

    _____

    Reference: _____

20. Is an overhead service conductor permitted to be bare?

    Answer: _____

    _____

    _____

    _____

    Reference: _____

CODE QUIZ 5

1. Are service conductors permitted to be spliced?

   Answer: _____

   _____

   _____

   _____

   Reference: _____

2. Is cable tray a wiring method for all service conductors?

   Answer: _____

   _____

   _____

   _____

   Reference: _____

3. *NEC*® 230.33 permits service conductors to be spliced. Could a crimp connection or exothermic welding be used for this purpose?

   Answer: _____

   _____

   _____

   _____

   Reference: _____

4. Would 230.70 allow the use of a shunt trip switch at the location specified in 230.70(A)(1) as the service disconnecting means if the circuit breaker were located inside the building?

   Answer: _____

   _____

   _____

   _____

   Reference: _____

5. A mobile home is equipped with a chassis bonding connection as required in 550.16(C)(1). The metal gas piping is bonded with a factory-installed bonding jumper connected to the mobile home frame at a point remote from the chassis bonding connection. Does this installation meet all the requirements?

   Answer: _____

   _____

   _____

   _____

   Reference: _____

6. Does the *Code* specifically allow a molded case switch rated in amperes to be used as a motor controller?

Answer: _____

_____

_____

_____

Reference: _____

7. The service-disconnecting means is located outside a building. Can it be located 50 feet (15 m) away or within sight of the building?

Answer: _____

_____

_____

_____

Reference: _____

8. Where can requirements for separately derived systems that operate at 120 volts line-to-line and 60 volts line-to-ground be found?

Answer: _____

_____

_____

_____

Reference: _____

9. What section of the *Code* permits solar photovoltaic systems, fuel cell systems, or interconnected electric power production sources to be connected to the supply side of the service disconnect?

Answer: _____

_____

_____

_____

Reference: _____

10. How are overcurrent devices for fire pumps sized?

Answer: _____

_____

_____

_____

Reference: _____

11. Is GFCI protection required for all single-phase, 15- and 20-ampere receptacles in commercial kitchens?

Answer: _____

_____

_____

_____

Reference: _____

12. Where in the *Code* are the branch-circuit overcurrent devices supplying a motel room required to be readily accessible to the motel room occupant?

Answer: _____

_____

_____

_____

Reference: _____

13. Are circuit breakers permitted to be used as a switch? Are they permitted for switching HID luminaires if they are only marked SWD?

Answer: _____

_____

_____

_____

Reference: _____

14. Are there any *Code* requirements for the removal or termination of abandoned low-voltage cables in buildings or structures?

Answer: _____

_____

_____

_____

Reference: _____

15. Where in the *Code* does it require that temporary electrical power and lighting installations for holiday decorative lighting be limited to 90 days?

Answer: _____

_____

_____

_____

Reference: _____

16. Are underground water pipe supplementary grounding electrodes required to meet the requirements of 250.53(1)(2) if they are rod pipe or plate types of electrodes?

    Answer: _____

    _____

    _____

    _____

    Reference: _____

17. Is it permissible to connect a grounding electrode to equipment such as parking light poles?

    Answer: _____

    _____

    _____

    _____

    Reference: _____

18. Where are the requirements for a surge arresters and a surge protective devices found?

    Answer: _____

    _____

    _____

    _____

    Reference: _____

19. In walls and ceilings constructed of wood or other combustible surface material, boxes shall be flush with the finished surface. Is gypsum board (sheetrock) considered a combustible surface material?

    Answer: _____

    _____

    _____

    _____

    Reference: _____

20. A luminaire that weighs more than _____ shall be supported independently of the outlet box unless the outlet box is listed for the weight to be supported.

    Answer: _____

    _____

    _____

    _____

    Reference: _____

CODE QUIZ 6

1. Type AC cable shall be permitted to be unsupported where the cable is not more than _____ from the last point of support for connections within an accessible ceiling to luminaire(s) or equipment.

   Answer: _____

   _____

   _____

   _____

   Reference: _____

2. Does the *Code* require a raceway to be sealed where it passes from the interior to the exterior of a building where condensation is known to be a problem?

   Answer: _____

   _____

   _____

   _____

   Reference: _____

3. Does the *Code* allow cables to be used as a means of support for other cables?

   Answer: _____

   _____

   _____

   _____

   Reference: _____

4. Is liquidtight flexible metal conduit permitted to be used in ducts or plenums as long as no single run exceeds 6 feet (1.8 m) in length?

   Answer: _____

   _____

   _____

   _____

   Reference: _____

5. Which article of the *Code* covers temporary installations?

   Answer: _____

   _____

   _____

   _____

   Reference: _____

6. What is an arc-fault circuit interrupter (AFCI)?

   Answer: _____

   _____

   _____

   _____

   Reference: _____

7. Do the ampacity adjustment factors for cables apply to all Type MC or AC cable?

   Answer: _____

   _____

   _____

   _____

   Reference: _____

8. Is the messenger-supported wiring method permitted where subject to physical damage?

   Answer: _____

   _____

   _____

   _____

   Reference: _____

9. Does the *Code* allow Type AC cable to be run unsupported in lengths not exceeding 6 feet (1.8 m) between luminaires in accessible ceilings?

   Answer: _____

   _____

   _____

   _____

   Reference: _____

10. Hybrid-type vehicles incorporate gasoline engines and electric motors. Would Article 625 or Article 511 govern repair facilities for these vehicles?

    Answer: _____

    _____

    _____

    _____

    Reference: _____

11. Could intermediate metal conduit (IMC) be used as an exposed vertical riser to support a lighted display case in a department store?

Answer: _____

_____

_____

_____

Reference: _____

12. Does the *Code* permit threadless connectors to be used on the threaded end of RMC?

Answer: _____

_____

_____

_____

Reference: _____

13. Are flat cable assemblies Type FC cable permitted to be installed where subject to physical damage?

Answer: _____

_____

_____

_____

Reference: _____

14. Does the minimum wire bending space at terminals in Table 312.6(B) apply to aluminum conductors?

Answer: _____

_____

_____

_____

Reference: _____

15. Would the metal faceplate (cover) grounding requirements apply to fan speed controls or spring-wound timers?

Answer: _____

_____

_____

_____

Reference: _____

16. Are flash protection warning markings required for switchboards, panelboards, and motor control centers that are likely to require servicing while energized?

    Answer: _____

    _____

    _____

    _____

    Reference: _____

17. Is non–metallic-sheathed cable Type NM permitted as open runs in dropped or suspended ceilings in other than one- and two-family and multifamily dwellings?

    Answer: _____

    _____

    _____

    _____

    Reference: _____

18. A service point of attachment is anchored to a trade size 2 (53 designator) RMC riser 3 feet (900 m) above the roofline. There is a threaded coupling on the riser between the highest strap and the weather head. Is this a *Code* violation?

    Answer: _____

    _____

    _____

    _____

    Reference: _____

19. Could the required emergency disconnect for gasoline dispensers serve as the maintenance disconnect required?

    Answer: _____

    _____

    _____

    _____

    Reference: _____

20. Do the dedicated space requirements for switchboards and panelboards now apply to a fusible switch as well?

    Answer: _____

    _____

    _____

    _____

    Reference: _____

CODE QUIZ 7

1. Does the AFCI branch-circuit protection for most dwelling circuits require the use of AFCI circuit breakers?

   Answer: _____

   _____

   _____

   _____

   Reference: _____

2. Could an individual refrigerator branch circuit be considered in the small-appliance branch-circuit calculation, Or would this individual circuit be included as part of the general lighting load?

   Answer: _____

   _____

   _____

   _____

   Reference: _____

3. Type AC cable permitted to be used as a service entrance wiring method?

   Answer: _____

   _____

   _____

   _____

   Reference: _____

4. Would the bonding jumper connection to metal gas piping be required to be accessible after installation?

   Answer: _____

   _____

   _____

   _____

   Reference: _____

5. Could Type NM cable be run unsupported in lengths not exceeding 4½ feet (1.4 m) between luminaires in an accessible ceiling?

   Answer: _____

   _____

   _____

   _____

   Reference: _____

6. Is nonferrous IMC permitted or manufactured?

   Answer: _____

   _____

   _____

   _____

   Reference: _____

7. Is electrical nonmetallic tubing (ENT) permitted outside, exposed to the direct rays of the sun?

   Answer: _____

   _____

   _____

   _____

   Reference: _____

8. Does the *Code* allow non–metallic-sheathed cable (NM) to be installed in a complete system of IMC conduit?

   Answer: _____

   _____

   _____

   _____

   Reference: _____

9. Is there a maximum number of bends allowed between pull points for liquidtight flexible metal conduit?

   Answer: _____

   _____

   _____

   _____

   Reference: _____

10. Would a raceway used as physical protection for direct burial conductors of over 600 volts emerging from the ground, be required to be listed for this purpose?

    Answer: _____

    _____

    _____

    _____

    Reference: _____

11. Luminaires shall not be used as a raceway for circuit conductors unless _____ for use as a raceway.

    Answer: _____

    _____

    _____

    _____

    Reference: _____

12. What is Type NUCC? Is it required to be listed? Are the conductors inside NUCC required to be listed? Is the raceway required to be listed?

    Answer: _____

    _____

    _____

    _____

    Reference: _____

13. Is Type ENT permitted where subject to physical damage?

    Answer: _____

    _____

    _____

    _____

    Reference: _____

14. Cable trays shall be supported at intervals _____.

    Answer: _____

    _____

    _____

    _____

    Reference: _____

15. What is the maximum water level for a swimming pool?

    Answer: _____

    _____

    _____

    _____

    Reference: _____

16. Is EMT an equipment grounding conductor in accordance with the *Code* or is a separate equipment-grounding conductor required?

    Answer: _____

    _____

    _____

    _____

    Reference: _____

17. Do Tables 352.44(A) and (B) take into account heating of PVC conduit by direct sunlight or are the tables only based on ambient temperature?

    Answer: _____

    _____

    _____

    _____

    Reference: _____

18. Are nonmetallic wireways required to be listed?

    Answer: _____

    _____

    _____

    _____

    Reference: _____

19. A service mast contains a 6 AWG bare grounded conductor. Is this permitted by the *Code*?

    Answer: _____

    _____

    _____

    _____

    Reference: _____

20. Where in the *Code* does it require that a voltage drop for branch circuits not exceed 1.5%?

    Answer: _____

    _____

    _____

    _____

    Reference: _____

CODE QUIZ 8

1. Is liquidtight flexible nonmetallic conduit (LFNC) permitted to enclose conductors above 600 volts for other than electric signs?

   Answer: _____

   _____

   _____

   _____

   Reference: _____

2. An equipment bonding jumper for LFNC must be installed in accordance with _____.

   Answer: _____

   _____

   _____

   _____

   Reference: _____

3. Are metal wireways required to be listed?

   Answer: _____

   _____

   _____

   _____

   Reference: _____

4. The ampacity of the conductors from the generator terminals to the first distribution device(s) containing overcurrent protection shall not be less than _____ of the nameplate current rating of the generator.

   Answer: _____

   _____

   _____

   _____

   Reference: _____

5. Single-phase cord-and-plug-connected air conditioners shall be provided with factory-installed _____ protection. The protection shall be an integral part of the attachment plug or be located in the power supply cord within 12 inches(300 mm) of the attachment plug.

   Answer: _____

   _____

   _____

   _____

   Reference: _____

6. Is each motor circuit over 600 volts required to include coordinated protection to automatically interrupt overload and fault currents in the motor, the motor-circuit conductors, and the motor control apparatus?

   Answer: _____

   _____

   _____

   _____

   Reference: _____

7. Unless protected by GFCI protection for personnel, the secondary winding of the isolation transformer connected to the impedance heating elements shall not have an output voltage greater than_____ volts (alternating current).

   Answer: _____

   _____

   _____

   _____

   Reference: _____

8. Are snap switches and control switches required to be connected to an equipment grounding conductor?

   Answer: _____

   _____

   _____

   _____

   Reference: _____

9. If a motor were accessible from a raised platform, would the *Code* allow the disconnect to be mounted higher than 6 feet, 7 inches (2 m) above the platform?

   Answer: _____

   _____

   _____

   _____

   Reference: _____

10. Is it acceptable to install more than one neutral under a terminal in a panelboard?

    Answer: _____

    _____

    _____

    _____

    Reference: _____

11. Exposed vertical risers from industrial machinery or fixed equipment shall be permitted to be supported at intervals not exceeding 20 feet (6 m). If the conduit is made up with threaded couplings, the conduit is firmly supported at the top and bottom of the riser, and no other means of intermediate support is readily available. What types of conduit are permitted to be used?

Answer: _____

_____

_____

_____

Reference: _____

12. May a 25-ampere branch circuit supply two or more outlets for fixed electric space-heating equipment?

Answer: _____

_____

_____

_____

Reference: _____

13. The essential electrical distribution system for healthcare facilities, other than hospitals, nursing homes and limited care facilities shall be a _____ system.

Answer: _____

_____

_____

_____

Reference: _____

14. Where Class II, Group E dusts are present in hazardous quantities, there are only division _____ locations.

Answer: _____

_____

_____

_____

Reference: _____

15. Overcurrent protection shall not be required for conductors from a battery rated less than _____ if the battery provides power for starting, ignition, or control of prime movers.

Answer: _____

_____

_____

_____

Reference: _____

16. All circuits and circuit modifications shall be legibly identified as to its clear, evident, and specific _____ or _____ on a circuit directory located on the face or inside of the panel door in the case of a panelboard and at each switch on a switchboard.

    Answer: _____

    _____

    _____

    _____

    Reference: _____

17. Does 430.245(B) require a motor lead to be a solid conductor if larger than a 10 AWG?

    Answer: _____

    _____

    _____

    _____

    Reference: _____

18. A ballast in a fluorescent luminaire that is used for egress lighting and is energized only during an emergency shall not have _____.

    Answer: _____

    _____

    _____

    _____

    Reference: _____

19. Would the *Code* allow the specified disconnecting means to be mounted on a removable cover of air-conditioning or refrigeration equipment if the disconnect were supplied from a flexible wiring method?

    Answer: _____

    _____

    _____

    _____

    Reference: _____

20. A separable connector or a plug-and-receptacle combination in the supply line to an oven or cooking unit shall be approved for the _____ of the space in which it is located.

    Answer: _____

    _____

    _____

    _____

    Reference: _____

CODE QUIZ 9

1. Can the electrical "service" supplying a building or structure be supplied from a private generator?

   Answer: _____

   _____

   _____

   _____

   Reference: _____

2. Where are the requirements for tunnel installations over 600 volts covered?

   Answer: _____

   _____

   _____

   _____

   Reference: _____

3. A dedicated space equal to the depth and width of the equipment from the floor to a height of _____ or to the structural ceiling, whichever is _____.

   Answer: _____

   _____

   _____

   _____

   Reference: _____

4. Grounding and bonding requirements for tunnel installations over 600 volts are covered in *NEC*® _____.

   Answer: _____

   _____

   _____

   _____

   Reference: _____

5. Generally, grounded conductors 6 AWG or smaller shall be identified by a continuous white or gray finish or by _____ on other than green insulation along its entire length.

   Answer: _____

   _____

   _____

   _____

   Reference: _____

6. Is a receptacle installed in an accessory building, such as a storage building, at grade level associated with a dwelling required to have GFCI protection for personnel?

Answer: _____

_____

_____

_____

Reference: _____

7. What is an AFCI?

Answer: _____

_____

_____

_____

Reference: _____

8. May service conductors originate at a generator where there is no utility serving the premises?

Answer: _____

_____

_____

_____

Reference: _____

9. Electrical equipment associated with the electrical installation located above or below the equipment shall not be allowed to extend beyond the front of electrical equipment. True or False?

Answer: _____

_____

_____

_____

Reference: _____

10. The requirements for dedicated electrical space are located in Article _____.

Answer: _____

_____

_____

_____

Reference: _____

11. Equipment requirements for over 600 volts nominal are located in Article _____.

Answer: _____

_____

_____

_____

Reference: _____

12. An insulated grounded conductor of 6 AWG or smaller shall be identified by a continuous white or gray or

_____.

Answer: _____

_____

_____

_____

Reference: _____

13. Receptacles located in garages or accessory buildings having a floor located at or below grade and are limited to storage areas, work areas, or areas of similar use are required to be of the _____ type.

Answer: _____

_____

_____

_____

Reference: _____

14. All branch circuits supplying 125-volt, single-phase, 15- and 20-ampere branch circuits installed in dwelling unit bedrooms shall be protected by _____.

Answer: _____

_____

_____

_____

Reference: _____

15. Are the required two small-appliance circuits permitted to serve more than one kitchen?

Answer: _____

_____

_____

_____

Reference: _____

16. In dwelling units, at least one receptacle shall be located not more than _____ feet (_____ mm) of the outside edge of each basin in bathrooms.

    Answer: _____

    _____

    _____

    _____

    Reference: _____

17. In guest rooms of hotels, motels, and similar occupancies, at least _____ of the receptacles required by 210.52(A) shall be _____.

    Answer: _____

    _____

    _____

    _____

    Reference: _____

18. An additional service or feeder load of _____ shall be included for each 2 feet (600 mm) of track lighting or fraction thereof in all occupancies except dwelling units or guest rooms of hotels and motels.

    Answer: _____

    _____

    _____

    _____

    Reference: _____

19. Outside feeders and branch circuits not exceeding 300 volts-to-ground over residential property and drive-ways and commercial areas not subject to truck traffic shall have a minimum clearance from ground of _____ feet (_____ m).

    Answer: _____

    _____

    _____

    _____

    Reference: _____

20. For a one-family dwelling, the feeder disconnecting means shall have a rating of not less than _____, 3-wire.

    Answer: _____

    _____

    _____

    _____

    Reference: _____

CODE QUIZ 10

1. Which section of the *Code* permits one additional set of service entrance conductors to supply a separate meter for public or common area loads not permitted to be supplied from an individual dwelling unit?

   Answer: _____

   _____

   _____

   _____

   Reference: _____

2. For a one-family dwelling, the service-disconnecting means shall have a rating of not less than _____, 3-wire.

   Answer: _____

   _____

   _____

   _____

   Reference: _____

3. Do the overcurrent protection requirements in 240.4(D) for small conductors apply to motor or motor-operated appliances?

   Answer: _____

   _____

   _____

   _____

   Reference: _____

4. Tap conductor are defined in section _____.

   Answer: _____

   _____

   _____

   _____

   Reference: _____

5. Where a tap is located outdoors except for the point of termination, the conductors are protected from physical damage and terminate in a single overcurrent device, the disconnecting means is readily accessible either outside or inside the building nearest the point of entrance, and the overcurrent device is integral or immediately adjacent to the disconnect, the length shall not exceed _____ feet (_____ m).

   Answer: _____

   _____

   _____

   _____

   Reference: _____

6. Where in the *Code* is a "Supervised Industrial Installation" defined?

   Answer: _____

   _____

   _____

   _____

   Reference: _____

7. Is GFCI protection for personnel required for carnival and exhibition shows?

   Answer: _____

   _____

   _____

   _____

   Reference: _____

8. I have been told that a tap conductor could not be tapped from a conductor that had been tapped from another conductor. Where is this stated in the *NEC*®?

   Answer: _____

   _____

   _____

   _____

   Reference: _____

9. An individual 15-ampere branch circuit is installed for refrigeration equipment in the kitchen of a dwelling unit per 210.52(B)(1), Exception No. 2. Does *NEC*® 220.52(A) require additional 1500 volt-amperes for this 15-ampere circuit?

   Answer: _____

   _____

   _____

   _____

   Reference: _____

10. Do all the receptacle outlets in a hotel or motel room located conveniently for permanent furniture layout have to be accessible?

    Answer: _____

    _____

    _____

    _____

    Reference: _____

11. If outlets end up behind nonstationary furniture, such as a recliner chair, bed table, or desk, such that they would have to be moved in order to plug into a receptacle outlet, does this mean they are not considered accessible?

    Answer: _____

    _____

    _____

    _____

    Reference: _____

12. When should a transfer-switch switch the grounded conductor as well as the ungrounded conductors between a utility service and a generator service?

    Answer: _____

    _____

    _____

    _____

    Reference: _____

13. Can a receptacle be located below the rim of a pedestal sink in a bathroom, and if so, how far below before it would not be considered as the required receptacle outlet for the bathroom basin?

    Answer: _____

    _____

    _____

    _____

    Reference: _____

14. The *NEC*® permits a single 20-ampere circuit to supply all the receptacles in dwelling bathrooms. Does this *NEC*® section permit other loads to be connected to this circuit?

    Answer: _____

    _____

    _____

    _____

    Reference: _____

15. Many large assembly and exhibition halls utilize cable trays for distribution of temporary wiring for an event. Is it permissible to use SO or SJT flexible cords in these cable trays for such temporary installations?

    Answer: _____

    _____

    _____

    _____

    Reference: _____

16. The *Code* requires that electric motors used on fire pumps be _____ for fire pump service.

    Answer: _____

    _____

    _____

    _____

    Reference: _____

17. Do the equipotential plane requirements of agriculture buildings apply to chicken houses?

    Answer: _____

    _____

    _____

    _____

    Reference: _____

18. If an equipment grounding conductor is not run with a 4-wire, 3-phase feeder to a second building, can the grounded conductor of the feeder circuit be grounded to the same contiguous metal water pipe as the main electrical service in the first building where that water line runs to the second building?

    Answer: _____

    _____

    _____

    _____

    Reference: _____

19. Many new dwellings have a conduit or tubing sleeve installed from the top of a panelboard up into the attic space so that future circuits can be gained from the panelboard without having to fish cables or cut sheetrock. Are such installations permitted because future cables cannot be secured to the panelboard?

    Answer: _____

    _____

    _____

    _____

    Reference: _____

20. Are all 15- and 20-ampere receptacles for dwelling-unit unfinished accessory buildings required to be GFCI protected?

    Answer: _____

    _____

    _____

    _____

    Reference: _____

CODE QUIZ 11

1. Where multiple-conductor power cables are installed in parallel, is it necessary for the equipment grounding conductor in each cable to be sized per the feeder or branch-circuit overcurrent device protecting the circuit according to Table 250.122?

   Answer: _____

   _____

   _____

   _____

   Reference: _____

2. What is a panelboard?

   Answer: _____

   _____

   _____

   _____

   Reference: _____

3. Does the *Code* allow splicing of service entrance conductors?

   Answer: _____

   _____

   _____

   _____

   Reference: _____

4. Is it permissible to use NM cable for temporary wiring on a construction site?

   Answer: _____

   _____

   _____

   _____

   Reference: _____

5. Is it permissible to prewire ENT before installing it in a concrete slab?

   Answer: _____

   _____

   _____

   _____

   Reference: _____

6. Is there a limit on how many individual branch circuits are permitted to be run from one building to another on the same property under single management?

   Answer: _____

   _____

   _____

   _____

   Reference: _____

7. Are Type MC cable and AC cable allowed in assembly-type occupancies with over 100 occupants?

   Answer: _____

   _____

   _____

   _____

   Reference: _____

8. Do the requirements that state *receptacle outlets shall not be installed in a face-up position in the work surfaces or countertops* apply to wet dwelling unit bar sink locations?

   Answer: _____

   _____

   _____

   _____

   Reference: _____

9. What is the definition of a supervised industrial installation?

   Answer: _____

   _____

   _____

   _____

   Reference: _____

10. All 15- and 20-ampere, 125-volt receptacles in pediatric wards, rooms, or areas must be _____ or employ _____ covers.

    Answer: _____

    _____

    _____

    _____

    Reference: _____

11. Is it ever permissible to mount a small and lightweight luminaire to a device box if the device box is in a wall?

    Answer: _____

    _____

    _____

    _____

    Reference: _____

12. How many inches (mm) of individual conductor must be left from the front of a 2 × 4 inch (50 × 100 mm) outlet box for the connection of or future replacement of toggle switches and receptacles?

    Answer: _____

    _____

    _____

    _____

    Reference: _____

13. Does the *NEC*® permit NM cable to be used for lease space wiring in a 4-story building with 6000 square feet per floor that houses novelty shops? The Type of construction is type IIA

    Answer: _____

    _____

    _____

    _____

    Reference: _____

14. Where in the *Code* is a "curb" around the floor opening required for busways passing vertically through more than one floor of all buildings?

    Answer: _____

    _____

    _____

    _____

    Reference: _____

15. How many ground rods are required where the rod is to supplement a metallic water line electrode?

    Answer: _____

    _____

    _____

    _____

    Reference: _____

16. Does Article 830 cover high-power communication systems?

Answer: _____

_____

_____

_____

Reference: _____

17. Is at least one general-care patient-bed receptacle outlet always required to be served by the normal system power supply or can all general-care patient-bed receptacles be served through emergency system connected panelboards?

Answer: _____

_____

_____

_____

Reference: _____

18. How many "main power feeders" may a single-family dwelling have and do the loads they serve make any difference? For example, a large single-family dwelling has a 600-ampere-rated service with two 100-ampere fused service switches, one 200-ampere fused switch, and a 400-ampere fused switch each feeding interior located panelboards. Are all "main power feeders"?

Answer: _____

_____

_____

_____

Reference: _____

19. Is GFCI protection for 125-volt receptacle required on the third-floor outdoor balcony of a dwelling building?

Answer: _____

_____

_____

_____

Reference: _____

20. Where in the *Code* are "handholes" covered?

Answer: _____

_____

_____

_____

Reference: _____

PRACTICE EXAM 1

> Each lesson is purposely designed to require the student to apply the entire *NEC®* text and not specific chapters or articles. It has been found that when studying for a timed, open-book examination, the student must gain proficiency in the Table of Contents, the Index, and the ability to move quickly from cover to cover to find the correct answer to each question in a timely fashion.

1. Are all 125-volt, 15- and 20-ampere receptacles in the service area of a commercial garage required to be protected by a GFCI?

   Answer: _____

   _____

   _____

   _____

   Reference: _____

2. A luminaire is installed over a hydromassage bathtub. Is GFCI protection required for the luminaire?

   Answer: _____

   _____

   _____

   _____

   Reference: _____

3. A non–grounding-type receptacle is to be replaced with a GFCI receptacle because a grounding means does not exist within the box. Can you supply downstream receptacles from the GFCI? Are these downstream receptacles required to be 2-wire, nongrounding type?

   Answer: _____

   _____

   _____

   _____

   Reference: _____

4. A dwelling unit kitchen range has a built-in 125-volt receptacle and is within 6 feet (1.8 m) of the sink. Does the *Code* require such a receptacle to be protected with a GFCI?

   Answer: _____

   _____

   _____

   _____

   Reference: _____

5. Does the *NEC®* prohibit the use of nonmetallic outlet boxes with metal raceways?

   Answer: _____

   _____

   _____

   _____

   Reference: _____

6. Does the *NEC®* permit luminaires to be installed on trees?

   Answer: _____

   _____

   _____

   _____

   Reference: _____

7. Electrical nonmetallic tubing in buildings exceeding three floors can be used concealed within walls, ceilings, and floors where the walls, ceilings, and floors provide a thermal barrier of material that has at least _____.

   Answer: _____

   _____

   _____

   _____

   Reference: _____

8. Does the *Code* require any extra protection for ENT when it is installed through openings in metal studs?

   Answer: _____

   _____

   _____

   _____

   Reference: _____

9. What are the support requirements for ENT when installed in metal studs?

   Answer: _____

   _____

   _____

   _____

   Reference: _____

10. Does the *Code* require any extra protection for Type NM cable when it is installed through openings in metal studs?

Answer: _____

_____

_____

_____

Reference: _____

11. When UF cable is used for interior wiring, are the conductors required to be rated at 90°C?

Answer: _____

_____

_____

_____

Reference: _____

12. Must "hospital grade" receptacles be installed throughout hospitals?

Answer: _____

_____

_____

_____

Reference: _____

13. When railings or lattice work is used for room dividers, does the *Code* require receptacles to be installed as if they were solid walls?

Answer: _____

_____

_____

_____

Reference: _____

14. In an apartment project, multiwire branch circuits are routed through outlet boxes to supply receptacles. Can the neutral conductor continuity be assured by two connections to the screw terminals of the receptacle?

Answer: _____

_____

_____

_____

Reference: _____

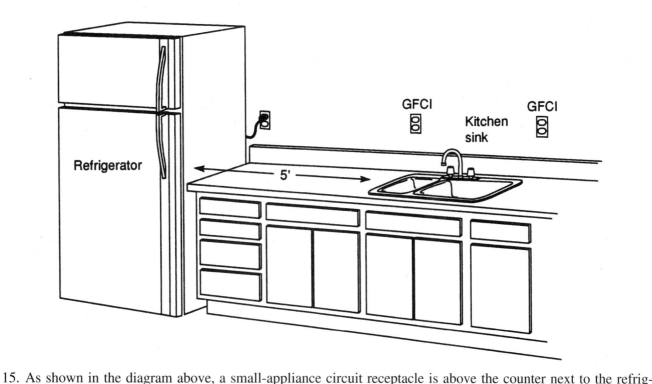

15. As shown in the diagram above, a small-appliance circuit receptacle is above the counter next to the refrigerator. The refrigerator is plugged into this receptacle, which is 5 feet (1.5 m) from the kitchen sink. Is the receptacle required to have GFCI protection?

Answer: _____

_____

_____

_____

Reference: _____

16. Does the *Code* require outdoor receptacles to be installed on balconies of high-rise apartment buildings?

Answer: _____

_____

_____

_____

Reference: _____

17. The *Code* did require that switches and circuit breakers used as switches be installed so that the center of the grip of the operating handle when in its highest position was no more than 6 feet (1.8 m) above the floor or working platform. Is this still the case?

Answer: _____

_____

_____

_____

Reference: _____

18. When can luminaires be used as raceways for circuit conductors?

    Answer: _____

    _____

    _____

    _____

    Reference: _____

19. In what areas of a hospital are hospital-grade receptacles required?

    Answer: _____

    _____

    _____

    _____

    Reference: _____

20. Is it a requirement that receptacles over a wet bar be spaced according to cabinet or floor line requirements?

    Answer: _____

    _____

    _____

    _____

    Reference: _____

21. A receptacle cover must be weatherproof when in use (attachment cap inserted) when installed in a _____ location.

    Answer: _____

    _____

    _____

    _____

    Reference: _____

22. Would ENT be an acceptable wiring method for a wet-niche luminaire or some portion of the circuit?

    Answer: _____

    _____

    _____

    _____

    Reference: _____

23. Can non–metallic-sheathed cable be used to supply a recessed fluorescent luminaire?

    Answer: _____

    _____

    _____

    _____

    Reference: _____

24. Do conductors for festoon lighting have to be rubber covered?

    Answer: _____

    _____

    _____

    _____

    Reference: _____

25. Can an electric discharge luminaire, such as a fluorescent strip, be used as a raceway for circuit conductors?

    Answer: _____

    _____

    _____

    _____

    Reference: _____

PRACTICE EXAM 2

1. Are all luminaires located in agriculture buildings to be listed as dusttight or watertight?

    Answer: _____

    _____

    _____

    _____

    Reference: _____

2. When installing Type AC cable horizontally through metal studs, are insulators needed?

    Answer: _____

    _____

    _____

    _____

    Reference: _____

3. What types of cable assemblies are acceptable for use as feeders to a floating building?

Answer: _____

_____

_____

_____

Reference: _____

4. Would non–metallic-sheathed cable be acceptable to wire a pool-associated motor in a one family dwelling?

Answer: _____

_____

_____

_____

Reference: _____

5. Would liquidtight flexible nonmetalic conduit be acceptable for enclosing conductors from an adjacent weatherproof box to an approved swimming pool junction box?

Answer: _____

_____

_____

_____

Reference: _____

6. Is the cord-and-plug connection of an electric water heater acceptable under the *NEC*®?

Answer: _____

_____

_____

_____

Reference: _____

7. Does the *Code* permit motor controllers and disconnects containing overcurrent devices to be used as junction boxes or wireways?

Answer: _____

_____

_____

_____

Reference: _____

8. In determining the locked-rotor kVA per horsepower of a dual voltage electric motor, which voltage must be considered?

Answer: _____

_____

_____

_____

Reference: _____

9. Would ENT be acceptable to enclose the insulated conductors for a panelboard (not part of service equipment) feeding pool-associated equipment inside of a building?

Answer: _____

_____

_____

_____

Reference: _____

10. Does the *Code* permit the installation of 15-ampere-rated receptacles on the 20-ampere small-appliance branch circuits required in a dwelling?

Answer: _____

_____

_____

_____

Reference: _____

11. Does the *Code* permit the installation of a 15-ampere-rated duplex receptacle on a 20-ampere separate circuit for a microwave oven with a nameplate reading of 13 amperes?

Answer: _____

_____

_____

_____

Reference: _____

12. What does the term "in free air" mean at the top of Table 310.15(B)(17)?

Answer: _____

_____

_____

_____

Reference: _____

13. Has the *NEC®* restricted PVC rigid conduit from being installed in cold weather areas?

    Answer: _____

    _____

    _____

    _____

    Reference: _____

14. All temporary wiring 15-ampere and 20-ampere, 125-volt receptacles not a part of the permanent building structure need to have GFCI protection. Does this mean that if a permanently installed receptacle is installed and energized, no GFCI protection is required even though this outlet is to be used by construction personnel?

    Answer: _____

    _____

    _____

    _____

    Reference: _____

15. In the *NEC®*, would a GFCI-protected receptacle be required on a wet bar in a dwelling?

    Answer: _____

    _____

    _____

    _____

    Reference: _____

16. Must the garbage disposal receptacle located under the sink of a dwelling unit kitchen be GFCI protected because it is within 6 feet (1.8 m) of the sink?

    Answer: _____

    _____

    _____

    _____

    Reference: _____

17. What is the maximum number of duplex receptacles permitted on a 2-wire, 15-ampere circuit in a dwelling?

    Answer: _____

    _____

    _____

    _____

    Reference: _____

18. How many receptacles are required at the patient's bed location in a critical care facility?

    Answer: _____

    _____

    _____

    _____

    Reference: _____

19. A ¼ horsepower circulating pump is to be located in a closet space housing a gas-fired water heater in a large single-family residence. Can the circulating pump be cord-and-plug connected?

    Answer: _____

    _____

    _____

    _____

    Reference: _____

20. In lieu of switching a ceiling luminaire, the *Code* permits switching a receptacle. Can the switched receptacle be counted as one of the receptacles required for the 12-foot (3.6 m) rule?

    Answer: _____

    _____

    _____

    _____

    Reference: _____

21. A laundry room in a single-family residence has a receptacle outlet on one wall and a second receptacle on the opposite wall. Can these two outlets be fed by the same laundry branch outlet?

    Answer: _____

    _____

    _____

    _____

    Reference: _____

22. Antique stores and others are selling hanging luminaires without an equipment of grounding conductor. Are these luminaires in violation of the *NEC*®?

    Answer: _____

    _____

    _____

    _____

    Reference: _____

23. Can 6 feet (1.8 m) of flexible metal conduit be installed between rigid nonmetallic raceways used for service entrance conductors?

Answer: _____

_____

_____

_____

Reference: _____

24. Many kitchen and bathroom sinks, tubs, and showers are provided with short sections of metal pipe (6 inches [150 mm] to 2 feet [600 mm]) for connection to nonmetallic pipe systems. Is it required to bond the several short sections of metal pipe to the service equipment?

Answer: _____

_____

_____

_____

Reference: _____

25. Is there a limit on the number of extension boxes that can be joined together?

Answer: _____

_____

_____

_____

Reference: _____

PRACTICE EXAM 3

1. A switch box was installed in a floor of a dwelling. A receptacle outlet with a standard plate was installed on the box. Does this standard outlet box and receptacle installed in the floor comply with the *Code*?

Answer: _____

_____

_____

_____

Reference: _____

2. A 4 inch × 2⅛ inch-square NM box is marked with the number of 14, 12, and 10 AWG conductors that can be installed in it. Does this mean that 8 AWG conductors are not permitted in the box?

   Answer: _____

   _____

   _____

   _____

   Reference: _____

3. Are outdoor receptacle outlets required at each dwelling unit of a multifamily dwelling located above grade?

   Answer: _____

   _____

   _____

   _____

   Reference: _____

4. Do portable lamps wired with flexible cord require a polarized attachment plug?

   Answer: _____

   _____

   _____

   _____

   Reference: _____

5. When figuring receptacles for a bedroom in a residence, must you include the space behind the door?

   Answer: _____

   _____

   _____

   _____

   Reference: _____

6. Are metal boxes required for splices in temporary wiring?

   Answer: _____

   _____

   _____

   _____

   Reference: _____

7. What is the maximum number of 12 AWG conductors permitted in a 4 inch × 1½ inch-deep octagon box?

Answer: _____

_____

_____

_____

Reference: _____

8. When a junction box contains a combination of conductors, such as 14 AWG and 12 AWG wires, what table is used to determine the size of the box?

Answer: _____

_____

_____

_____

Reference: _____

9. Can you use threaded IMC in a Class 1, Division 1 location?

Answer: _____

_____

_____

_____

Reference: _____

10. Does the *Code* consider EMT as conduit?

Answer: _____

_____

_____

_____

Reference: _____

11. Can suspended fluorescent luminaires be connected together with unsupported EMT between them?

Answer: _____

_____

_____

_____

Reference: _____

12. Must lay-in fluorescent luminaires be fastened to the grid T-bars of a suspended ceiling system?

    Answer: _____

    _____

    _____

    _____

    Reference: _____

13. Can luminaires marked "suitable for damp locations" be used in wet locations?

    Answer: _____

    _____

    _____

    _____

    Reference: _____

14. A listed fluorescent luminaire is equipped with a Class "P" ballast. Does the *Code* permit such a luminaire to be surface mounted on a combustible low-density cellulose fiberboard ceiling?

    Answer: _____

    _____

    _____

    _____

    Reference: _____

15. Are all recessed incandescent luminaires required to have thermal protection?

    Answer: _____

    _____

    _____

    _____

    Reference: _____

16. Is it necessary to identify the high leg of a 240/120-volt system at the motor disconnect of a 3-phase motor?

    Answer: _____

    _____

    _____

    _____

    Reference: _____

17. Is GFCI protection required for a receptacle located on a food preparation island in a dwelling unit kitchen that is less than 6 feet (1.8 m) from the countertop sink?

Answer: _____

_____

_____

_____

Reference: _____

18. Can conductors that pass through a panelboard also be spliced in the panelboard?

Answer: _____

_____

_____

_____

Reference: _____

19. Are the enclosures for mercury vapor luminaires installed 8 feet (2.4 m) above grade on poles outdoors required to be grounded?

Answer: _____

_____

_____

_____

Reference: _____

20. Would an exit luminaire with an emergency pack capable of supplying more than 1½ hours of power have to be connected to an emergency circuit ahead of the main disconnect?

Answer: _____

_____

_____

_____

Reference: _____

21. A grounded 125 volt 20-ampere receptacle is installed in a residential garage to serve an air compressor. Would this receptacle require GFCI protection?

Answer: _____

_____

_____

_____

Reference: _____

22. If additional receptacles are installed in a bathroom, not adjacent to the washbasin, are they required to be GFCI protected?

Answer: _____

_____

_____

_____

Reference: _____

23. Must a receptacle in a dwelling-unit kitchen installed for a luminaire be GFCI protected?

Answer: _____

_____

_____

_____

Reference: _____

24. When installing recessed luminaires (hi-hats) in the ceiling, does 314.20 apply to the luminaire housing?

Answer: _____

_____

_____

_____

Reference: _____

25. What are the support requirements for service entrance cables?

Answer: _____

_____

_____

_____

Reference: _____

PRACTICE EXAM 4

1. An underground cable is routed from the service equipment in the building underground and up a pole in a commercial parking lot to supply an overhead luminaire. What is the minimum burial depth for the cable?

Answer: _____

_____

_____

_____

Reference: _____

2. Does the *Code* permit 15-ampere duplex receptacles to be installed on a 20-ampere branch circuit?

   Answer: _____

   _____

   _____

   _____

   Reference: _____

3. We are designing a new hotel with 250 guest rooms. It is our intent to install permanent beds, desks, and dressers in each guest room, bolted to the wall. If receptacles are located in accordance with 210.52, the receptacles will be located behind the headboard of the bed and behind the dresser. Are the receptacles required behind these pieces of furniture? Would the number of required receptacles be less if the furniture was not bolted to the wall?

   Answer: _____

   _____

   _____

   _____

   Reference: _____

4. Can a paddle fan be supported only by an outlet box if it weighs less than 35 pounds (16 kg)?

   Answer: _____

   _____

   _____

   _____

   Reference: _____

5. Where can luminaires be used as raceways for circuit conductors?

   Answer: _____

   _____

   _____

   _____

   Reference: _____

6. Do recessed incandescent luminaires require thermal protection?

   Answer: _____

   _____

   _____

   _____

   Reference: _____

7. Are medium-base lampholders in industrial occupancies permitted on 277-volt luminaire circuits?

Answer: _____

_____

_____

_____

Reference: _____

8. Can a battery-powered (unit equipment) emergency light be supplied directly from a circuit in a panelboard that supplies normal lumination in the area that will be illuminated by a battery luminaire under emergency conditions?

Answer: _____

_____

_____

_____

Reference: _____

9. A single transfer switch is permitted to serve one or more branches of an essential electrical system in a small hospital. What is the maximum demand load on this system?

Answer: _____

_____

_____

_____

Reference: _____

10. Would a 3-phase, 4-wire, 1200-ampere, 480/277-volt service switch with 900-ampere fuses be required to have ground-fault protection?

Answer: _____

_____

_____

_____

Reference: _____

11. Are luminaires operating at less than 15 volts between conductors required to be protected by a GFCI over hot tubs or spas?

Answer: _____

_____

_____

_____

Reference: _____

12. Could you use EMT in the earth below a concrete slab?

Answer: _____

_____

_____

_____

Reference: _____

13. An industrial building will be served by several services of different voltage ratings. Is it required that each system be connected to separate grounding electrodes?

Answer: _____

_____

_____

_____

Reference: _____

14. Can a service panel be located over a counter or an appliance (such as a washing machine or a dryer) that extends away from the wall?

Answer: _____

_____

_____

_____

Reference: _____

15. A surface-mounted fluorescent luminaire with an integral receptacle is installed above the countertop in the kitchen of a dwelling and is not more than 5½ feet (1.7 m) above the floor. Is this luminaire required to be supplied by a 20-ampere circuit as are other kitchen receptacles? Is the luminaire receptacle required to be protected by GFCI?

Answer: _____

_____

_____

_____

Reference: _____

16. What are the requirements for grounding an agricultural building fed from central distribution point where livestock are housed?

Answer: _____

_____

_____

_____

Reference: _____

17. Can a fused disconnect switch or circuit breaker located on the meter pole of a farm installation be used as the service equipment for the residence?

    Answer: _____

    _____

    _____

    _____

    Reference: _____

18. A Type UF cable feeds a 120-volt yard luminaire location on residential property. The circuit is protected by a 15-ampere overcurrent protection device. It is not GFCI protected. What is the minimum burial depth permitted for the cable?

    Answer: _____

    _____

    _____

    _____

    Reference: _____

19. Can a fixed storage-type water heater having a capacity of 120 gallons or less be cord-and-plug connected?

    Answer: _____

    _____

    _____

    _____

    Reference: _____

20. A TV dish antenna is located in the side yard of a single-family dwelling. (a) What is the minimum depth required by the *NEC*® for the coaxial signal cable? (b) How are these antenna units required to be grounded?

    Answer: _____

    _____

    _____

    _____

    Reference: _____

21. Does the *Code* permit low-voltage control cables to be supported from the conduit containing the power circuit conductors feeding an air-conditioning unit?

    Answer: _____

    _____

    _____

    _____

    Reference: _____

22. Is it true that if the underground metallic water piping is not at least 10 feet (3 m) long, the underground piping system is not adequate as a grounding electrode?

Answer: _____

_____

_____

_____

Reference: _____

23. Does the luminaire outlet in the attic have to be switch controlled?

Answer: _____

_____

_____

_____

Reference: _____

24. With regard to recessed luminaires installed over showers, is it permissible to use a gasketed vapor-proof trim only to comply with the *Code* or must the entire luminaire be rated for a wet location?

Answer: _____

_____

_____

_____

Reference: _____

25. May overhead outside branch-circuit conductors of 10 AWG copper be used in an unsupported open span up to 50 feet (15 m)? Is 10 AWG listed copper multiconductor UF cable suitable for such an application?

Answer: _____

_____

_____

_____

Reference: _____

PRACTICE EXAM 5

1. Is a nonmetallic cable tray permitted by the *NEC*®?

Answer: _____

_____

_____

_____

Reference: _____

2. The *Code* requires clearance for luminaires installed in clothes closets. (a) What are the clearance requirements for a surface-mounted luminaire mounted on the wall above the door? (b) What are clearance requirements for a surface-mounted luminaire mounted on the ceiling? (c) What are the clearance requirements for a recessed incandescent or fluorescent luminaire ceiling mounted?

Answer: _____

_____

_____

_____

Reference: _____

3. Are four 16 AWG luminaire wires in a domed canopy counted for the number of conductors in outlet boxes to determine the box fill?

Answer: _____

_____

_____

_____

Reference: _____

4. Are switchboards and control panels rated 1200 amperes or more, 600 volts, and over 6 feet (1.8 m) wide required to have one entrance at each end of a room?

Answer: _____

_____

_____

_____

Reference: _____

5. What size grounding-electrode conductor is required where a grounding electrode conductor is routed to a driven rod and from the rod to a concrete-encased electrode?

Answer: _____

_____

_____

_____

Reference: _____

6. Can handle locks on circuit breakers be locked so that the power to loads, such as emergency lighting, sump pumps, alarm-warning circuits, and other types of equipment, cannot be cut off by mistake? What section of the *NEC*® applies?

Answer: _____

_____

_____

_____

Reference: _____

7. Can a service be mounted on a mobile home if the home is designed to be installed on a foundation?

   Answer: _____

   _____

   _____

   _____

   Reference: _____

8. Is an insulated equipment grounding conductor required for a pool panelboard feeder fed from the service equipment?

   Answer: _____

   _____

   _____

   _____

   Reference: _____

9. Would the *Code* allow the use of stainless steel ground rods?

   Answer: _____

   _____

   _____

   _____

   Reference: _____

10. The *Code* allows flexible metal conduit for services. Does the *Code* permit liquidtight flexible conduit, metallic or nonmetallic?

    Answer: _____

    _____

    _____

    _____

    Reference: _____

11. The grounding electrode conductor can be connected to the grounding electrode by listed connectors, clamps, or other listed means and _____.

    Answer: _____

    _____

    _____

    _____

    Reference: _____

12. Does the *Code* require an equipment grounding conductor of the wire type in all flexible raceways?

    Answer: _____

    _____

    _____

    _____

    Reference: _____

13. Is AC cable a permitted wiring method in places of assembly?

    Answer: _____

    _____

    _____

    _____

    Reference: _____

14. Where are AFCI protection devices required by the *NEC*®?

    Answer: _____

    _____

    _____

    _____

    Reference: _____

15. Can LFNC be used on residential work?

    Answer: _____

    _____

    _____

    _____

    Reference: _____

16. Are there any *Code* requirements against installing a bare neutral wire inside the service conduit?

    Answer: _____

    _____

    _____

    _____

    Reference: _____

17. Can the branch-circuit and Class 2 control wires for two 3-phase motors be installed in a common raceway of the proper size?

    Answer: _____

    _____

    _____

    _____

    Reference: _____

18. The *Code* explicitly requires bonding around concentric knockouts on service equipment, but does the *Code* require that equipment downstream from the service equipment be bonded in the same manner?

    Answer: _____

    _____

    _____

    _____

    Reference: _____

19. Is non–metallic-sheathed cable permitted to run through cold air returns if it is sleeved with EMT or flexible metal conduit in short lengths?

    Answer: _____

    _____

    _____

    _____

    Reference: _____

20. Is it necessary to pigtail the neutral when connecting receptacles to a multiwire circuit?

    Answer: _____

    _____

    _____

    _____

    Reference: _____

21. When a motor-control circuit extends beyond the control enclosure, what is the maximum overcurrent protection acceptable for 12 AWG copper control wires?

    Answer: _____

    _____

    _____

    _____

    Reference: _____

22. Is an equipment bonding conductor required to bond a cover-mounted receptacle to a surface-mounted outlet box that has cover screws on a raised portion of the cover?

    Answer: _____

    _____

    _____

    _____

    Reference: _____

23. When calculating branch-circuit and feeder loads, what are the voltages that shall be used?

    Answer: _____

    _____

    _____

    _____

    Reference: _____

24. Is it permissible to run four 12 AWG THW copper conductors in a cable or raceway where they encounter 100°F ambient temperatures to supply cord-and-plug-connected loads protected at 20 amperes?

    Answer: _____

    _____

    _____

    _____

    Reference: _____

25. When flat conductor cable, Type FCC, is installed under carpet, what is the maximum size carpet squares permitted?

    Answer: _____

    _____

    _____

    _____

    Reference: _____

PRACTICE EXAM 6

1. Are bonding jumpers required on metal feeder and branch-circuit raceways containing circuits of more than 250 volts-to-ground where oversize concentric or eccentric knockouts are encountered?

   Answer: _____

   _____

   _____

   _____

   Reference: _____

2. What is the maximum number of service disconnects allowed for a set of service entrance conductors installed in an apartment per the *NEC*®?

   Answer: _____

   _____

   _____

   _____

   Reference: _____

3. (a) Is a GFCI required on or within an outdoor portable sign? (b) Would a GFCI-protected branch circuit supplying the sign be acceptable?

   Answer: _____

   _____

   _____

   _____

   Reference: _____

4. Is a luminaire outlet required in the crawl space of a manufactured building if equipment requiring service is installed in the space?

   Answer: _____

   _____

   _____

   _____

   Reference: _____

5. Can an office partition assembly be cord-and-plug connected?

   Answer: _____

   _____

   _____

   _____

   Reference: _____

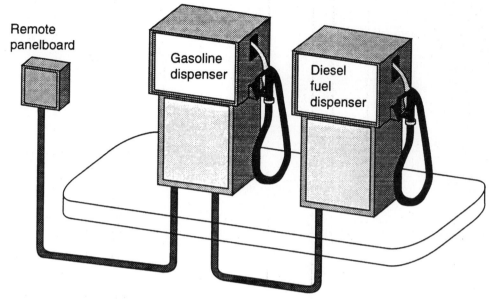

6. In the diagram above, a diesel fuel dispenser is mounted adjacent to a gasoline dispenser on a service station island. A sealing fitting is installed in the conduit entering the gasoline dispenser. Is a sealing fitting required in the conduit entering the diesel fuel dispenser?

Answer: _____

_____

_____

_____

Reference: _____

7. In residential-type occupancies, does the wall space behind doors where the door opens against the wall count when spacing the receptacles in a room?

Answer: _____

_____

_____

_____

Reference: _____

8. A four-family apartment building is supplied by an overhead drop to a 300-kcmil THW aluminum riser. Four electric meters are grouped and have 100-ampere breakers below the meters to supply a panel in each apartment. 3 AWG THW copper feeders are run to each apartment unit. An individual water meter is provided for each apartment. Each water meter is supplied with nonmetallic pipe. The water system for each apartment is copper above grade. (a) What size bonding jumper is required? (b) How would this conductor be attached to the metallic water piping in each apartment? (c) Is a supplementary driven electrode required?

Answer: _____

_____

_____

_____

Reference: _____

9. Is a 14 AWG copper conductor allowed in a kitchen of a dwelling unit if it serves a single fixed appliance, such as a dishwasher?

Answer: _____

_____

_____

_____

Reference: _____

10. Can Type UF cable be used to supply a swimming pool motor?

Answer: _____

_____

_____

_____

Reference: _____

11. Are 16 AWG and 18 AWG luminaire wires counted for the number of conductors in outlet boxes for box fill?

Answer: _____

_____

_____

_____

Reference: _____

12. Is a receptacle outlet required for a wet bar and the outlet is installed within 6 feet (1.8 m) of the wet bar sink, would it be required to be GFCI protected?

Answer: _____

_____

_____

_____

Reference: _____

13. When calculating the demand for a 3-kW and a 6-kW range in a dwelling, would you use Table 220.55, Column B, for the 3-kW and Column C for the 6-kW range or would you combine them and use either B or C for the two ranges?

Answer: _____

_____

_____

_____

Reference: _____

14. When sizing the branch-circuit conductors to a 4.5-kW dryer, would you use 4.5 kW or 5 kW in making the calculations?

Answer: _____

_____

_____

_____

Reference: _____

15. What is the maximum weight of a ceiling fan that can be supported by an outlet box?

Answer: _____

_____

_____

_____

Reference: _____

16. Where does the *NEC*® list or calculate the number of conductors allowed in flexible metal conduit?

Answer: _____

_____

_____

_____

Reference: _____

17. Are ceiling grid support wires that rigidly support a lay-in ceiling acceptable for the sole support of junction boxes above the lay-in ceiling?

Answer: _____

_____

_____

_____

Reference: _____

18. Can electrical equipment approved for a Class I location be used in a Class II location?

Answer: _____

_____

_____

_____

Reference: _____

19. Can THHN insulated conductors be used outside for service entrance conductors if exposed to the elements?

Answer: _____

_____

_____

_____

Reference: _____

20. Who has the authority to determine acceptability of electrical equipment and materials?

Answer: _____

_____

_____

_____

Reference: _____

21. Can more than one receptacle in a laundry room be supplied by the laundry branch circuit?

Answer: _____

_____

_____

_____

Reference: _____

22. Many air-conditioning units are being installed on roof tops of apartment buildings for different reasons. Is a receptacle adjacent to such equipment required for servicing this equipment?

Answer: _____

_____

_____

_____

Reference: _____

23. *NEC®* 430.72 applies to motor-control circuits that are tapped from the load side of motor branch-circuit, short-circuit, and ground-fault protective devices. Because Column B of Table 430.72(B) covers control circuits that do not extend beyond the motor control equipment enclosure and Column C covers those that do extend beyond, what does Column A cover?

Answer: _____

_____

_____

_____

Reference: _____

24. A 240-volt, single-phase circuit with a rated load of 1.8 kW draws 9 amperes of current. The power factor is _____ %.

    Answer: _____

    _____

    _____

    _____

    Reference: _____

25. Can you run multioutlet assembly from one room to another room through a sheetrock wall?

    Answer: _____

    _____

    _____

    _____

    Reference: _____

PRACTICE EXAM 7

1. Illuminated exit signs are required by local building codes, and battery packs will be used. Do these signs have to be installed on a luminaire circuit in the area of the sign or could they be on a separate branch-circuit?

    Answer: _____

    _____

    _____

    _____

    Reference: _____

2. Can a metal cable tray be used as an equipment grounding conductor?

    Answer: _____

    _____

    _____

    _____

    Reference: _____

3. Can surface nonmetallic raceway be used outside in a wet location?

    Answer: _____

    _____

    _____

    _____

    Reference: _____

4. Switchboards and control panels rated 1200 amperes or more, 600 volts or less, and over 6 feet (1.8 m) wide are required to have one entrance at each end of a room. What conditions would permit only one entrance to this room?

Answer: _____

_____

_____

_____

Reference: _____

5. Is Type MC cable required to have a bushing installed like the bushings used on Type AC armored cable?

Answer: _____

_____

_____

_____

Reference: _____

6. Can ENT be surface mounted in a warehouse building?

Answer: _____

_____

_____

_____

Reference: _____

7. How deep must a residential branch circuit be buried if it passes under the driveway?

Answer: _____

_____

_____

_____

Reference: _____

8. Where more than one building or structure is on the same property and under single management, the *Code* requires each building to have a disconnecting means at each such building or structure of 600 volts or less. Under what conditions can the disconnecting means be located elsewhere on the premises?

Answer: _____

_____

_____

_____

Reference: _____

9. Are cable trays permitted in a grain elevator?

   Answer: _____

   _____

   _____

   _____

   Reference: _____

10. If a luminaire is listed for use as a raceway, can this raceway be used for conductors to supply a circuit or circuits beyond the luminaire?

    Answer: _____

    _____

    _____

    _____

    Reference: _____

11. What *Code* section prohibits bare neutral conductors?

    Answer: _____

    _____

    _____

    _____

    Reference: _____

12. Would a large mercantile store be considered a place of assembly if it can hold more than 100 people?

    Answer: _____

    _____

    _____

    _____

    Reference: _____

13. Can the neutral be reduced by 70% for data-processing equipment over 200 amperes?

    Answer: _____

    _____

    _____

    _____

    Reference: _____

14. What ampacity correction factor would you use for 12 AWG THW copper conductors used in an ambient temperature of 78°F (not more than three conductors in raceway, in free air, 240-volt circuit)?

Answer: _____

_____

_____

_____

Reference: _____

15. How are box volumes calculated when the box contains different-size conductors and a combination of clamps, devices, and studs?

Answer: _____

_____

_____

_____

Reference: _____

16. What is the minimum overhead service conductor clearance required over a residential yard where the conductors do not exceed 150 volts-to-ground?

Answer: _____

_____

_____

_____

Reference: _____

17. An equipment room containing electrical, telephone, and air-handling equipment uses the space in the room for air-handling purposes. Is this room considered a plenum used for environmental air?

Answer: _____

_____

_____

_____

Reference: _____

18. As applied to illumination in a clothes closet, is the entire 12-inch (300 mm) or width-of-shelf area considered as storage space?

Answer: _____

_____

_____

_____

Reference: _____

19. Would an outdoor luminaire be considered as grounded by the physical connection to a grounded metal support pole or would an equipment grounding conductor connection directly to the luminaire be required?

    Answer: _____

    _____

    _____

    _____

    Reference: _____

20. Would luminaires that are installed under roof overhangs, porch roofs, or canopies be required to be suitable for damp locations?

    Answer: _____

    _____

    _____

    _____

    Reference: _____

21. Table 430.147 through Table 430.150 list the horsepower and the full-load currents of motors. How could you determine the full-load current of a motor not listed, such as a 35-horsepower, 460-volt motor?

    Answer: _____

    _____

    _____

    _____

    Reference: _____

22. Where a single Type NM cable is pulled into a raceway, what % may the cross-section of the conduit be filled?

    Answer: _____

    _____

    _____

    _____

    Reference: _____

23. Can you use Table 310.15(B)(7) for a feeder to each apartment in a building regardless of whether the mains in the apartment units or on the outside of the building?

    Answer: _____

    _____

    _____

    _____

    Reference: _____

24. Can swimming pool equipment such as a pump motor be calculated as a fixed appliance when applying 220.14(A) to a dwelling?

Answer: _____

_____

_____

_____

Reference: _____

25. Is a GFCI protected receptacle required in a detached garage of a dwelling?

Answer: _____

_____

_____

_____

Reference: _____

PRACTICE EXAM 8

1. Are all recessed luminaires that are to be installed in a suspended ceiling required to be thermally protected?

Answer: _____

_____

_____

_____

Reference: _____

2. How can you ground an agricultural building and comply with the *Code*?

Answer: _____

_____

_____

_____

Reference: _____

3. Is Table 220.55 applicable to microwave ovens and convection cooking ovens?

Answer: _____

_____

_____

_____

Reference: _____

4. Is an autotransformer recognized for use in a motor-control circuit?

   Answer: _____

   _____

   _____

   _____

   Reference: _____

5. Can service equipment be mounted on a floating dwelling unit that is moved frequently?

   Answer: _____

   _____

   _____

   _____

   Reference: _____

6. Is it permissible to use neon lighting on or in a dwelling unit?

   Answer: _____

   _____

   _____

   _____

   Reference: _____

7. In a large industrial plant, does the *Code* allow a branch circuit over 50 amperes to feed several outlets such as to supply power to portable welders that move from place to place?

   Answer: _____

   _____

   _____

   _____

   Reference: _____

8. Can nonmetallic raceways be used in healthcare facilities?

   Answer: _____

   _____

   _____

   _____

   Reference: _____

9. Can a conductor be protected by an overcurrent device sized for its full ampacity even though the allowable load current is only 50% of the ampacity after ampacity adjustment?

Answer: _____

_____

_____

_____

Reference: _____

10. Does the *Code* permit single-pole circuit breakers in panelboards to be used as a switch for controlling a lighting circuit as an off and on switch?

Answer: _____

_____

_____

_____

Reference: _____

11. When determining the required size box, items such as fittings, cable clamps, hickeys, and where combinations of different size conductors are used, which conductor size do you deduct for the device and fittings?

Answer: _____

_____

_____

_____

Reference: _____

12. Where RMC is used as service raceways, are double locknuts permitted for the service equipment required continuity?

Answer: _____

_____

_____

_____

Reference: _____

13. A receptacle outlet is to be added to an existing installation, and Type NM cable is to be "fished in" the partition wall. Does the *Code* allow a nonmetallic box without cable clamps to be used for such an installation?

Answer: _____

_____

_____

_____

Reference: _____

14. What size general-use switch is required on an air-conditioning compressor, sealed (hermetic-type) when the nameplate states FLA of 100 amperes, 3-phase, 230 volts, LRA 580?

Answer: _____

_____

_____

_____

Reference: _____

15. A fluorescent luminaire is supplied with trade size ⅜ (12) flexible metal conduit 5 feet (1.5 m) long. Is an equipment grounding required to ground the luminaire?

Answer: _____

_____

_____

_____

Reference: _____

16. What is the full-load current of an electric furnace rated 240 volts, single-phase, 10 kW when connected to a 208-volt circuit? (Show calculations.)

Answer: _____

_____

_____

_____

Reference: _____

17. Can four 400-watt wet-niche luminaires be connected to one GFCI for a swimming pool?

Answer: _____

_____

_____

_____

Reference: _____

18. How does one determine how many conductors conduit bodies (condulets) are approved for?

Answer: _____

_____

_____

_____

Reference: _____

19. How can a recessed luminaire approved for direct contact with thermal insulation be identified?

   Answer: _____

   _____

   _____

   _____

   Reference: _____

20. Does the 10-foot (3 m) tap rule apply only to conductors, such as wires, and not panelboard bus?

   Answer: _____

   _____

   _____

   _____

   Reference: _____

21. How are working clearances measured between enclosed electrical equipment facing each other across an aisle?

   Answer: _____

   _____

   _____

   _____

   Reference: _____

22. Can either the disposal, dishwasher, or trash compactor be installed on a 20-ampere kitchen small-appliance circuit?

   Answer: _____

   _____

   _____

   _____

   Reference: _____

23. What is the required horizontal distance between overhead service conductors and swimming pools?

   Answer: _____

   _____

   _____

   _____

   Reference: _____

24. Can a bare 4 AWG copper grounding electrode conductor be buried below a concrete floor slab, and is there any depth requirement?

    Answer: _____

    _____

    _____

    _____

    Reference: _____

25. Where the grounding electrode conductor connects to the water system on the house side of the water meter, is bonding required around valves with sweated connections? Also, what other equipment is required to be bonded around?

    Answer: _____

    _____

    _____

    _____

    Reference: _____

PRACTICE EXAM 9

1. Can a floodlight and/or receptacle be on the GFCI circuit protecting an underwater luminaire?

   Answer: _____

   _____

   _____

   _____

   Reference: _____

2. For a 600-ampere service where 350 kcmil copper conductors are parallel in two conduits and only 3-phase, 3-wire is needed, but the transformer is Y-connected and grounded: (a) Must the grounded conductor be brought into the service? (b) Can the grounded conductor be brought in through a single raceway or is it required to be installed in each of the parallel raceways? (c) What is the minimum size of the grounded conductor?

   Answer: _____

   _____

   _____

   _____

   Reference: _____

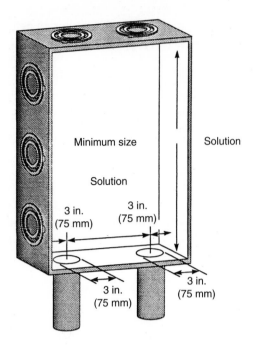

3. What are the minimum dimensions of a junction box illustrated above with only a U-pull of two trade size 3 (78) conduits in one side?

Answer: _____

_____

_____

_____

Reference: _____

4. With two 42-circuit panelboards bolted together, can the 42 circuits from one panelboard feed through the gutter of the second panelboard if the 40% fill in that gutter is not exceeded?

Answer: _____

_____

_____

_____

Reference: _____

5. Can suspended fluorescent luminaires be connected together with unsupported EMT between them if the length of the EMT is less than 3 feet (900 mm)?

Answer: _____

_____

_____

_____

Reference: _____

6. (a) Are 3-phase generators permitted to be connected to a three-pole transfer switch with a solid neutral or is a four-pole transfer switch that breaks the neutral required? (b) If either is OK, must a grounding electrode be provided at the generator in both cases?

Answer: _____

_____

_____

_____

Reference: _____

7. A 600-ampere underground service with four 4/0 AWG conductors per phase paralleled is installed in two PVC conduits. Must the phase conductors and neutral be grouped in a single raceway if the conduits are nonmetallic?

Answer: _____

_____

_____

_____

Reference: _____

8. The raceway for a 240-volt motor circuit includes both the power circuit and Class 2 control circuit conductors. The power circuit conductors use 600-volt insulation. Must the control circuit conductors also have 600-volt insulation or would 300-volt insulation be acceptable?

Answer: _____

_____

_____

_____

Reference: _____

9. Does the *Code* permit or prohibit a sprinkler head over a switchboard?

Answer: _____

_____

_____

_____

Reference: _____

10. A 2000-ampere, 3-phase, 480-volt ungrounded service is located 300 feet (91.4 m) from the waterline within a building. Can the grounding electrode conductor from this service be connected to the building steel in the vicinity of the service and then 300 feet (91.4 m) away bond the steel to the waterline such that the building steel is being used as the grounding electrode conductor?

Answer: _____

_____

_____

_____

Reference: _____

11. Four 12 AWG THHN conductors in one conduit supply a continuous lighting load. What is the maximum permissible circuit-breaker size and permissible load allowed on each conductor?

Answer: _____

_____

_____

_____

Reference: _____

12. A receptacle outlet is located in a soffit or roof overhang for connection of Christmas decorations or roof heating cable at a dwelling. When this receptacle is located out of reach from the grade, approximately 9 feet (2.75 m) above grade, is this receptacle required to be GFCI protected?

Answer: _____

_____

_____

_____

Reference: _____

13. Does the *NEC*® permit more than one Type NM cable under one device box cable clamp?

Answer: _____

_____

_____

_____

Reference: _____

14. A 400-ampere two 200-ampere panelboards, each with a main breaker. Additional circuit are load is added so that one more 100-ampere service disconnect is required. Must the service entrance conductors be increased in size? (The calculated load is not increased.)

    Answer: _____

    _____

    _____

    _____

    Reference: _____

15. Can an attachment plug be used as a disconnecting means for a 3-horsepower motor?

    Answer: _____

    _____

    _____

    _____

    Reference: _____

16. Where in the *Code* does it permit using 4 AWG copper THW conductors for service conductors supplying a 100-ampere service?

    Answer: _____

    _____

    _____

    _____

    Reference: _____

17. In three-way and four-way switch circuitry in a metal raceway, is the neutral required to be in the same raceway with the travelers (switch legs)?

    Answer: _____

    _____

    _____

    _____

    Reference: _____

18. Single-conductor UF cable comes underground from a submersible pump through a building wall to the pump controller. Are these conductors permitted to be run inside the building?

    Answer: _____

    _____

    _____

    _____

    Reference: _____

19. In commercial garages, can regular receptacles and plates be used over 18 inches (450 mm) above the floor? What are the wiring restrictions in a garage that has a minimum of four air changes per hour?

Answer: _____

_____

_____

_____

Reference: _____

20. A single-family dwelling has two electric dryer receptacle outlets wired with two 10 AWG black conductors to each and one 10 AWG white conductor neutral. Each outlet has a separate 30-ampere circuit breaker. Is one 10 AWG neutral conductor permitted to be used with this installation?

Answer: _____

_____

_____

_____

Reference: _____

21. Can trade size ⅜ (12) flexible metal conduit be used on machinery and boilers for limit switches, flow switches, can switches, and solenoids?

Answer: _____

_____

_____

_____

Reference: _____

22. What NEC section tamper resistant receptacles in dwelling units? Which areas in the dwelling unit are they required? _____ (do not consider exceptions)

Answer: _____

_____

_____

_____

Reference: _____

23. Can unit equipment used for emergency lighting be directly connected on the branch circuit rather than use a plug-in receptacle?

Answer: _____

_____

_____

_____

Reference: _____

24. Are receptacles required to be GFCI protected when outdoors at healthcare facilities?

    Answer: _____

    _____

    _____

    _____

    Reference: _____

25. Does a gas furnace in a dwelling require a disconnect switch? May the furnace be directly connected to a basement lighting circuit?

    Answer: _____

    _____

    _____

    _____

    Reference: _____

PRACTICE EXAM 10

1. Are open tube fluorescent luminaires acceptable in a basement garage under an apartment building?

   Answer: _____

   _____

   _____

   _____

   Reference: _____

2. Are there size and depth requirements to classify hot tubs as a pool or are they classified as a pool regardless of size for wiring of circulation pumps, heaters, and so forth?

   Answer: _____

   _____

   _____

   _____

   Reference: _____

3. Rigid metal conduit runs in a commercial garage floor and extends up through an 18-inch (450 mm) hazardous area with no couplings or fittings in that 18-inch (450 mm) area. Is a seal-off fitting required in this conduit?

   Answer: _____

   _____

   _____

   _____

   Reference: _____

4. Three single-phase, 480-volt transformers rated at 104 amperes each are connected in a wye bank for a 3-phase system. What is the maximum overcurrent protection permitted for the primary when the secondary is also protected? Would the primary protection be different if connected in a delta bank?

Answer: _____

_____

_____

_____

Reference: _____

5. How is the load for a section of multioutlet assembly determined?

Answer: _____

_____

_____

_____

Reference: _____

6. A motor is connected to a branch-circuit breaker in a panel that is out of sight of the motor location. The breaker is used as the controller and also the disconnecting means for the motor. If this breaker is of the "lock off" type, would this meet *Code* requirements?

Answer: _____

_____

_____

_____

Reference: _____

7. A 12 AWG remote-control circuit is protected by 80-ampere fuses in the motor disconnect (magnetic contact). Is this control circuit properly protected?

Answer: _____

_____

_____

_____

Reference: _____

8. Bus duct is reduced in size from 1000 amperes to 400-ampere bus duct and runs 40 feet (12 m) to where it terminates in a distribution panel. Is overcurrent protection required?

Answer: _____

_____

_____

_____

Reference: _____

9. Can UF cable be used from the meter, down the pole, then underground 20 feet (6 m) to the disconnect for a pasture pump?

   Answer: _____

   _____

   _____

   _____

   Reference: _____

10. On a 480/277-volt, 3-phase, 4-wire system where the neutral is not used, must it be grounded at the service panel?

    Answer: _____

    _____

    _____

    _____

    Reference: _____

11. If present on the premises, the *Code* requires the bonding of all grounding electrodes together to form the grounding electrode system. Are these requirements the same for a separately derived system?

    Answer: _____

    _____

    _____

    _____

    Reference: _____

12. Define tap conductors. Is it permissible to make a tap from a tap?

    Answer: _____

    _____

    _____

    _____

    Reference: _____

13. Can tandem circuit breakers mounted adjacent to each other in a panelboard with handles tied together be used in place of two double-pole breakers to supply two 290-volt electric baseboard heaters?

    Answer: _____

    _____

    _____

    _____

    Reference: _____

14. What minimum size copper grounding electrode conductor is required a 400-ampere service when two 3/0 AWG copper conductors in parallel are used for the phase conductors?

Answer: _____

_____

_____

_____

Reference: _____

15. Can *aluminum* RMC be used in hazardous locations, such as in sewer plants?

Answer: _____

_____

_____

_____

Reference: _____

16. A two-lamp exit luminaire is supplied from both a battery-operated unit of equipment and a normal source. Must the normal supply to the exit luminaire be from a recognized emergency source, such as ahead of the service disconnect?

Answer: _____

_____

_____

_____

Reference: _____

17. A service is changed and moved to a different location. The old electric range run is Type SE cable that is too short to reach the new service, so a metal junction box is installed and a short piece of new Type SE cable is installed to reach the new service. This is in a residential basement on a 7-foot (2 m) ceiling. Does this metal junction box require grounding, and if so, does it require a separate equipment grounding conductor to the service equipment (not the neutral of the 3-wire Type SE cable)?

Answer: _____

_____

_____

_____

Reference: _____

18. Can service conductors be run directly to a fire pump controller?

    Answer: _____

    _____

    _____

    _____

    Reference: _____

19. A trucking firm has added a second service. They have 240-volt service to the old part of the building and 208-volt service to the new part. Common computer equipment is connected to both services, with shielded TWINAX (signal cable) between devices. The two services are not tied to the same ground. This causes a ground loop on signal shield. What is the best way per the *NEC*® to achieve equal potential bonding on network?

    Answer: _____

    _____

    _____

    _____

    Reference: _____

20. A luminaire is installed over a hydromassage bathtub. Is GFCI protection required for the luminaire?

    Answer: _____

    _____

    _____

    _____

    Reference: _____

21. Frequently, the *NEC*® requires GFCI protection of receptacles. Is there any requirement that limits the number of receptacles that can be protected by a single GFCI circuit breaker or a feed-through receptacle?

    Answer: _____

    _____

    _____

    _____

    Reference: _____

22. Can an electrical discharge luminaire, such as a fluorescent strip, be used as a branch-circuit junction box or termination point for circuit conductors?

    Answer: _____

    _____

    _____

    _____

    Reference: _____

23. Does the *Code* allow ENT to be installed above a suspended ceiling provided that the thermal barrier of material has at least a 15-minute rating of fire-rated assemblies?

    Answer: _____

    _____

    _____

    _____

    Reference: _____

24. Does the *Code* permit 8 AWG conductors to be installed in a 4 × 2⅛ inch-square nonmetallic box that is marked with the number of 14 AWG, 12 AWG, and 10 AWG conductors that can be installed in it?

    Answer: _____

    _____

    _____

    _____

    Reference: _____

25. Must lay-in fluorescent luminaires be fastened or can they lay in grid T-bars of a suspended ceiling system?

    Answer: _____

    _____

    _____

    _____

    Reference: _____

PRACTICE EXAM 11

1. When Type UF cable is used for interior wiring as permitted, are the conductors required to be rated at 90°C?

   Answer: _____

   _____

   _____

   _____

   Reference: _____

2. Are swimming pool underwater luminaires operating at less than 15 volts between conductors required to be protected by a GFCI device or circuit breaker?

   Answer: _____

   _____

   _____

   _____

   Reference: _____

3. Can handle locks on circuit breakers be locked so that the power to loads, such as emergency lighting, sump pumps, alarm-warning circuits, and other types of equipment, cannot be cut off by mistake? If so, what sections of the *NEC®* would apply?

   Answer: _____

   _____

   _____

   _____

   Reference: _____

4. Can non–metallic-sheathed cable be run parallel to framing members through cold air returns?

   Answer: _____

   _____

   _____

   _____

   Reference: _____

5. Is a GFCI properly protected circuit supplying an outdoor portable sign an acceptable method for installation in accordance with the *NEC®*?

   Answer: _____

   _____

   _____

   _____

   Reference: _____

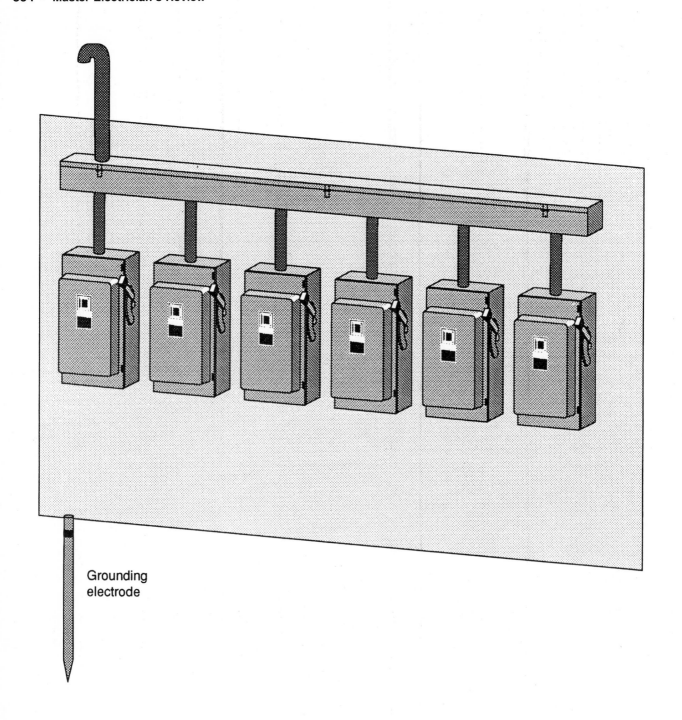

Grounding
electrode

6. *Refer to the diagram above.* Where should the grounding electrode conductor terminate in a multiple disconnect service?

Answer: _____

_____

_____

_____

Reference: _____

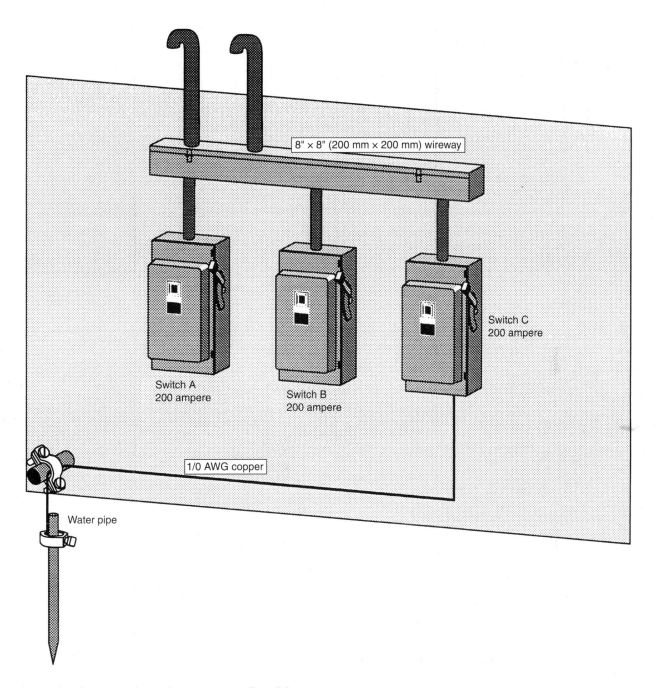

Refer to the diagram above for Questions 7 and 8.

7. A 600-ampere service is built with parallel sets of 350-kcmil CU THWN service conductors. Three 200-ampere fused service disconnects are supplied with 4/0 AWG THWN service taps within an 8″ × 8″ (200 mm × 200 mm) wireway. All service bonding is continuous from the CT cabinet through all service enclosures. Does the *Code* permit only one grounding electrode conductor connection at switch "C" if sized for 700-kcmil service conductors?

Answer: _____

_____

_____

_____

Reference: _____

8. With reference to Question 7, how should the trough above the three service disconnects be sized? Is the trough to be sized as a wireway or a junction box?

Answer: _____

_____

_____

_____

Reference: _____

9. An EMT and steel box system was installed throughout an office/warehouse building. An additional equipment grounding conductor was pulled with all receptacle branch circuits and terminated at all receptacle grounding screws. Must the equipment grounding conductor also be attached to each metal box?

Answer: _____

_____

_____

_____

Reference: _____

10. Is an insulated equipment grounding conductor required in non–metallic-sheathed cable that is supplying a swimming pool pump motor installed in the garage of a single-family dwelling?

Answer: _____

_____

_____

_____

Reference: _____

11. A 225-ampere rated panel has main lugs only and powers a calculated load of 180 amperes. It is supplied by a 500-kcmil CU feeder with 400-ampere overcurrent protection. Does this installation comply with the *Code*?

Answer: _____

_____

_____

_____

Reference: _____

12. If a small dental office uses limited amounts of inhalation anesthetics, does *NEC*® 517.31 require standby emergency power to supply emergency lighting and the anesthetic equipment?

Answer: _____

_____

_____

_____

Reference: _____

13. Does the *Code* allow any demand factors in a commercial laundry? The normal load of a washing machine is 4 amperes; the spin cycle draws 9 amperes. Because it is unlikely that all machines will be spinning at the same time, what load figure should be used for load calculations?

    Answer: _____

    _____

    _____

    _____

    Reference: _____

14. Is the examining room in a dental office considered to be a general-care patient care area? Is Type NM cable a permitted wiring method in this area?

    Answer: _____

    _____

    _____

    _____

    Reference: _____

15. Do the metal enclosures for individual channel-type neon letters (either front-lit or back-lit) require connection to an equipment grounding conductor?

    Answer: _____

    _____

    _____

    _____

    Reference: _____

16. Is it permissible to ground the secondary of a separately derived system back to the neutral terminal of a service switchboard instead of going directly to the waterline when there is no effectively grounded structural steel nearby?

    Answer: _____

    _____

    _____

    _____

    Reference: _____

17. When conductors are paralleled, does the ampacity double (assuming the same type, size, and length)?

    Answer: _____

    _____

    _____

    _____

    Reference: _____

18. Suppose you use three 300-kcmil conductors in a raceway r with 75°C temperature rating of 285 amperes for each conductor. Instead of this, can you use six 1/0 AWG conductors with 75°C rating and two conductors in parallel per phase, all in the same conduit at 300 amperes ampacity, or must you apply an ampacity adjustment because you have more than three conductors? (Assume that you are using Type THWN conductors in a trade size 2½ (63) conduit with device terminations rated for 75°C.)

Answer: _____

_____

_____

_____

Reference: _____

19. Are medium-base HID lampholders permitted to be installed on a 277-volt circuit?

Answer: _____

_____

_____

_____

Reference: _____

20. Is GFCI protection required for 125V, 15A or 20A Ampere receptacles located on decks where the receptacles are located less than 6 feet, 6 inches (2 m) above grade and access to the deck is by steps or stepping onto a low deck? (The receptacle cannot be reached while standing on the ground.)

Answer: _____

_____

_____

_____

Reference: _____

21. Is the bathroom GFCI-protected circuit permitted to supply luminaire outlets in addition to the bathroom receptacles?

Answer: _____

_____

_____

_____

Reference: _____

22. Is it acceptable to connect a garbage disposal and dishwasher on the same branch circuit?

Answer: _____

_____

_____

_____

Reference: _____

23. Are receptacles required in a four-car garage at an apartment building if the garage is detached and provided with electric power and lights?

    Answer: ———

    ——

    ——

    ——

    Reference: ————————————————————————

24. Does an outdoor entrance luminaire with integral photocell satisfy the *Code* without a wall switch control? What about a motion detector used to activate the luminaire?

    Answer: ———

    ——

    ——

    ——

    Reference: ————————————————————————

25. Does the *Code* require motor feeder conductors to have an ampacity of 125% of the continuous load plus the noncontinuous load or is it only necessary for the feeder overcurrent device to have this rating?

    Answer: ———

    ——

    ——

    ——

    Reference: ————————————————————————

PRACTICE EXAM 12

1. In calculating feeder loads for electric space heating, is it required to figure 125% of space heating loads or is the 125% required only for branch-circuit loads?

    Answer: ———

    ——

    ——

    ——

    Reference: ————————————————————————

2. For household range load calculations, is it acceptable to combine ovens and countertops and treat them as one appliance for feeder calculations?

    Answer: _____

    _____

    _____

    _____

    Reference: _____

3. Because the disconnecting means for separate buildings on one property are required to be suitable for use as service equipment, are the number of disconnects at each building's entrance limited to six?

    Answer: _____

    _____

    _____

    _____

    Reference: _____

4. The *Code* requires multiple services to be grounded to the same electrode. Must the grounding connections be made to the same point on the electrode or can they be grounded to the same electrode (waterline or structural steel) at different locations?

    Answer: _____

    _____

    _____

    _____

    Reference: _____

5. Does Article 514 apply to underground wiring (such as to a sign) that is at least 30 feet (9 m) from any hazardous location in a service station?

    Answer: _____

    _____

    _____

    _____

    Reference: _____

6. What is the required minimum clearance between a thermally protected Type IC recessed incandescent luminaire and wood framing?

    Answer: _____

    _____

    _____

    _____

    Reference: _____

7. Does the *Code* permit the metal coaxial sheath of CATV cable to be grounded to a separate driven rod or is it required to be bonded to the building's power system ground or service equipment enclosure?

   Answer: _____

   _____

   _____

   _____

   Reference: _____

8. An automotive repair garage is classified as Class I, Division 2 up 18 inches (450 mm) above the floor. An attached sales and office area is not classified. Is Type NM-B cable permitted to be used in the sales and office area?

   Answer: _____

   _____

   _____

   _____

   Reference: _____

9. Do Type AC and MC cables installed on 277-volt circuits require bonding if connected into concentric knockouts?

   Answer: _____

   _____

   _____

   _____

   Reference: _____

10. A swimming pool is installed at a single-family dwelling. Can 2-wire Type NM-B cable with a bare equipment grounding conductor be used to connect the pump motor located in the basement?

    Answer: _____

    _____

    _____

    _____

    Reference: _____

11. Ceiling lighting consists of lay-in fluorescent luminaires. Is it acceptable to use trade size ⅜(12) flexible metal conduit from luminaire to luminaire in lengths less than 6 feet (1.8 m) provided an equipment grounding conductor is installed in the conduit?

    Answer: _____

    _____

    _____

    _____

    Reference: _____

12. Can a 12-volt dry-niche luminaire mounted on the outside of a permanent aboveground swimming pool be supplied by flexible cord connected to a receptacle located 11 feet (3.35 m) from the pool?

    Answer: _____

    _____

    _____

    _____

    Reference: _____

13. Can a timer switch for a spa be located within 5 feet (1.5 m) of the spa? What if it is located on the unit and not on the wall?

    Answer: _____

    _____

    _____

    _____

    Reference: _____

14. Is there any requirement relative to the location of the wall switch for an outdoor entrance luminaire? As an example, must the switch for an entrance luminaire at a sunroom door be located at the entrance door from outside or can it be located 10 feet (3 m) away at another door leading to the sunroom?

    Answer: _____

    _____

    _____

    _____

    Reference: _____

15. When calculating the demand for a 3 kW and a 6 kW range, would you use Table 220.55, Column B, for the 3-kW and Column C for the 6 kW or would you combine them and use either B or C for the two ranges?

    Answer: _____

    _____

    _____

    _____

    Reference: _____

16. Can more than one receptacle in a laundry room be supplied by the laundry branch circuit?

    Answer: _____

    _____

    _____

    _____

    Reference: _____

17. For a large bathroom with provisions for a washing machine, is GFCI protection required for an inaccessible receptacle located behind and dedicated to the washing machine?

    Answer: _____

    _____

    _____

    _____

    Reference: _____

18. Is there an acceptable method whereby control wiring can be run in the same conduit with power conductors to a central air-conditioning unit?

    Answer: _____

    _____

    _____

    _____

    Reference: _____

19. Does the *Code* require grounding the metal sheath of CATV cable entering a dwelling?

    Answer: _____

    _____

    _____

    _____

    Reference: _____

20. The *Code* requires bonding a metallic conduit enclosing a grounding electrode conductor. How is the bonding jumper sized when the conduit enters a concentric knockout?

    Answer: _____

    _____

    _____

    _____

    Reference: _____

21. Does the *Code* require a GFCI-protected receptacle when installed within 6 feet (1.8 m) of a dwelling wet bar or laundry sink?

    Answer: _____

    _____

    _____

    _____

    Reference: _____

22. A 500-kcmil Type THW CU feeder protected with 400-ampere overcurrent device is routed through a junction box. What is the minimum size tap conductor required to supply a 30-ampere load using the 25-foot (7.5 m) tap rule? What is the minimum size using the 10-foot (3 m) tap rule?

    Answer: _____

    _____

    _____

    _____

    Reference: _____

23. A large recreation room in a dwelling has three sets of sliding doors opening out to a covered porch. Is a switch required at each door to control porch lumination (lighting)?

    Answer: _____

    _____

    _____

    _____

    Reference: _____

24. Existing feeder conductors are being installed using conductors with 90°C Type THHN insulation. What ampacity values are allowed for these conductors that are connected to a circuit breaker or panel main lug terminals?

    Answer: _____

    _____

    _____

    _____

    Reference: _____

25. Branch circuits serving patient care areas must be installed in metal raceways or other specified methods. Which areas in nursing homes and residential custodial care facilities are designated as patient care areas?

    Answer: _____

    _____

    _____

    _____

    Reference: _____

FINAL EXAMINATION

Examination Instructions:

For a positive evaluation of your knowledge and preparation awareness, you must:

1. locate yourself in a quiet atmosphere (room by yourself).
2. have with you at least two sharp No. 2 pencils, the 2011 *NEC®*, and a handheld calculator.
3. time yourself (three hours) with no interruptions. Do not spend more than three minutes per question.
4. grade yourself honestly and concentrate your studies on the sections of the *NEC®* in which you missed the questions.

Caution: Do not just refer to the answer key for the correct answers because the questions in this examination are only an exercise and not actual test questions. Therefore, it is important that you be able to quickly find answers from throughout the *NEC®*.

1. According to the *NEC®*, open conductors for communication equipment on a building shall be separated at least _____ feet (_____ m) from *lightning* conductors.
   A. 2 (600 mm)
   B. 4 (1.2)
   C. 6 (1.8)
   D. 8 (2.4)

   Answer: _____   Reference: _____

2. A megohmmeter is an instrument used for
   A. polarizing a circuit.
   B. measuring high resistances.
   C. shunting a generation system.
   D. determining amperes.

   Answer: _____   Reference: _____

3. The total opposition to alternating current in a circuit that includes resistance, inductance, and capacitance is called
   A. reactance.
   B. resistance.
   C. reluctance.
   D. impedance.

   Answer: _____   Reference: _____

4. A 240-volt, single-phase circuit has a resistive load of 8500 watts. The net calculated current to supply this load is _____ amperes.
   A. 35
   B. 39
   C. 44
   D. 71

   Answer: _____   Reference: _____

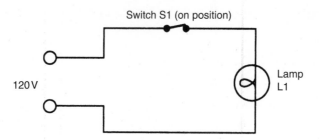

Refer to the diagram above for Questions 5–6.

5. In the diagram above, switch S1 is in the "on" position, but luminaire L1 does not come on. Voltage across L1 is measured to be 120 volts. Voltage across S1 is measured to be 0 volt. The luminaire does not come on because
A. the luminaire is open (burned out).
B. the luminaire and switch are shorted.
C. the luminaire is good, but the switch does not make contact.
D. there is a break in the wire of the circuit.

Answer: _____   Reference: _____

6. In the diagram above, with a 9-ampere current in the circuit, the power factor is _____ %.
A. 71
B. 83
C. 93
D. 108

Answer: _____   Reference: _____

480V 3-phase

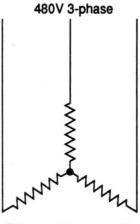

Balanced 3-phase load

7. In the diagram above, three balanced resistive 100-ampere loads are connected to a 480-volt, 3-phase, 3-wire circuit. The total power of this circuit is _____ kilowatts.
A. 48
B. 72
C. 83
D. 144

Answer: _____   Reference: _____

8. A 100-horsepower induction motor is loaded to 30 horsepower. To improve the power factor of this motor, the amount of load should be
   A. increased.
   B. left unchanged because load has no effect.
   C. decreased.
   D. removed to give 0 load.

   Answer: _____   Reference: _____

9. An electrical installation requires a total of 200 feet (61 m) of conductor with a maximum line resistance of 0.5 ohm. The minimum size conductor must have a resistance of no more than _____ ohm(s) per 1000 feet (305 m).
   A. 0.5
   B. 1.0
   C. 2.5
   D. 5.0

   Answer: _____   Reference: _____

10. To get the maximum total resistance using three resistors,
    A. all three resistors should be connected in series.
    B. all three resistors should be connected in parallel.
    C. two resistors should be connected in parallel, then one connected in series.
    D. two resistors should be connected in series, then one connected in parallel.

    Answer: _____   Reference: _____

11. Four resistance heaters are connected in parallel. Their resistances are heater 1, 20 ohms; heater 2, 30 ohms; heater 3, 60 ohms; and heater 4, 10 ohms. The total resistance of the parallel circuit is _____ ohms.
    A. 5
    B. 12
    C. 50
    D. 120

    Answer: _____   Reference: _____

12. For circuits supplying luminaire units having ballasts, transformers, or autotransformers, the calculated load shall be based on the _____ of the luminaire units.
    A. total ampere rating
    B. size of the conductors
    C. total wattage of the lamps
    D. voltage rating

    Answer: _____   Reference: _____

13. A building on a blueprint is 16 inches × 10 inches. If the drawing scale is ¼ inch = 1 foot, what is the area of the building in square feet?
    A. 160 square feet
    B. 640 square feet
    C. 2560 square feet
    D. 5120 square feet

    Answer: _____   Reference: _____

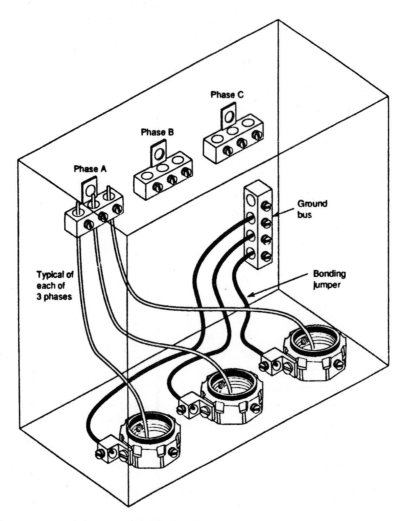

14. In the diagram above, each of the supply-side EMT conduits contains three 500-kcmil copper THW service conductors in parallel. As shown, a separate bonding jumper is installed from each conduit to the grounded bus terminal. Each bonding jumper must be at least _____ AWG copper.
    A. 1/0
    B. 2/0
    C. 3/0
    D. 4/0

    Answer: _____   Reference: _____

15. A group of conveyors is operated by individual motors in a manufacturing facility. The total ampacity required to run the group is 58 amperes at 240 volts, single-phase. There are 250 feet of conduit between the main service panel and the subpanel. The voltage drop in the feeder conductors can be a maximum of 3%.

    Each conductor in the feeder circuit to supply these conveyors must be at least _____ AWG copper THW. (Use resistance values in the *NEC*®, Chapter 9, Table 8.)
    A. 2
    B. 3
    C. 4
    D. 6

    Answer: _____   Reference: _____

16. A single-phase, 3-wire service has two ungrounded conductors of 2/0 AWG copper THWN. The neutral conductor is 1 AWG copper THWN. The conduit size needed for the service entrance conductors is at least trade size _____ (_____).
    A. 1½ (41)
    B. 2 (53)
    C. 2½ (63)
    D. 3½ (91)

    Answer: _____   Reference: _____

17. Two 1 AWG, one 1/0 AWG, and one 2/0 AWG copper conductors, Type THHN, are in a 10-foot (3.05-m) conduit. The size of IMC conduit must be at least _____ inches (_____ mm).
    A. 1¼ (35)
    B. 1½ (41)
    C. 2 (53)
    D. 2½ (63)

    Answer: _____   Reference: _____

18. A one-family dwelling contains the following:
    - 1200 square feet of floor space
    - Service: 120/240 volt, single-phase
    - Heat (separate control): two 500-watt, 240-volt baseboard heaters; three 1000-watt, 240-volt baseboard heaters; one 2000-watt, 240-volt baseboard heater
    - Range: one 11,000-watt, 240-volt
    - Dryer: one 6000-watt, 240-volt
    - Water: one 4000-watt, 240-volt water heater

    Using the optional calculation method, each ungrounded conductor in the service for this dwelling has a total net calculated load of _____ amperes.
    A. 146
    B. 120
    C. 99
    D. 84

    Answer: _____   Reference: _____

19. A metallic cold water piping system is present within the structure being supplied by an electrical service with 1/0 AWG THW copper service entrance conductors. The copper grounding electrode conductor run to the metal water pipe shall have a minimum size of _____ AWG.
    A. 8
    B. 6
    C. 4
    D. 2

    Answer: _____   Reference: _____

20. Ground-fault protection is required to be installed on a 2000-ampere, solidly grounded wye service. What is the maximum setting of the ground-fault protection?
    A. 800 amperes
    B. 1000 amperes
    C. 1200 amperes
    D. 1600 amperes

    Answer: _____ Reference: _____

21. Overhead service conductors shall be installed so that the minimum clearance from a window opening is _____ feet (_____ mm/m).
    A. 2 (600 mm)
    B. 3 (900 mm)
    C. 6 (1.8 m)
    D. 7 (2.4 m)

    Answer: _____ Reference: _____

22. The 240-volt, single-phase feeder shown in the diagram above supplies a branch-circuit panel that has an improperly balanced load of 57 amperes on one ungrounded conductor and 45 amperes on the other ungrounded conductor. What is the load on the grounded (neutral) conductor?
    A. no load
    B. 12 amperes
    C. 57 amperes
    D. 102 amperes

    Answer: _____ Reference: _____

23. The maximum number of power or lighting conductors that can be installed in a raceway before the ampacity adjustment factors must be applied is
    A. 1.
    B. 3.
    C. 5.
    D. 7.

    Answer: _____ Reference: _____

24. A set of blueprints for an electrical installation calls for a trade size ¾ (21) conduit to be installed as a raceway to enclose a 3-wire circuit consisting of 8 AWG, Type TW conductors. The load to be served is 36 amperes. After the conduit is installed, it is discovered that 10 feet (3 m) of the conduit pass through an area where the ambient temperature is 115° Fahrenheit (115°F). Which of the following actions is an acceptable action that can be taken to correct the condition? The total circuit length is 100 feet.
    I. 8 AWG THW conductors can be substituted for the 8 AWG TW conductors.
    II. 6 AWG TW conductors can be substituted for the 8 AWG TW conductors.
    A. I only
    B. II only
    C. Both I and II
    D. Neither I nor II

    Answer: _____ Reference: _____

25. A circuit consisting of three 6 AWG THW insulated copper conductors is run through an area of a building where the temperature is normally 120° Fahrenheit (120°F). Which of the following is the maximum allowable load current for each conductor?
    A. 48.75 amperes
    B. 56.25 amperes
    C. 60.00 amperes
    D. 65.50 amperes

    Answer: _____ Reference: _____

26. A 240-volt, single-phase, 100-ampere circuit is installed in a nonmetallic raceway. Which of the following conditions must apply to the equipment grounding conductor installed with the circuit?
    I. It must be counted when determining conductor fill of the raceway.
    II. It must be the same size as that of the circuit conductor.
    A. I only
    B. II only
    C. both I and II
    D. neither I nor II

    Answer: _____ Reference: _____

27. Three 4 AWG THWN and four 1/0 AWG THW conductors are to be installed in a single run of IMC conduit. The minimum trade size of conduit permitted is
    A. 1 (27).
    B. 2 (53).
    C. 3 (78).
    D. 4 (103).

    Answer: _____ Reference: _____

28. Eight 4 AWG THWN conductors are to be installed in a single run of rigid PVC nonmetallic conduit. The minimum size of conduit permitted is
    A. 1 (27).
    B. 1½ (41).
    C. 2 (53).
    D. 2½ (63).

    Answer: _____   Reference: _____

29. A municipality has adopted the *NEC®* without amendments. A conflict occurs about the interpretation of a section of the adopted electrical code. Which of the following is responsible for making the interpretation of the *Code*?
    A. The engineer overseeing the construction
    B. The electrical contractor performing the work
    C. The chief electrical inspector (authority having jurisdiction)
    D. The International Association of Electrical Inspectors

    Answer: _____   Reference: _____

30. A service-disconnecting means can be installed at which of the following locations?
    I. Outside a building at a readily accessible point nearest the point of entrance of the service entrance conductors.
    II. Inside a building at a readily accessible point nearest the point of entrance of the service entrance conductors.
    A. I only
    B. II only
    C. either I or II
    D. neither I nor II

    Answer: _____   Reference: _____

Conductors crossing and connecting

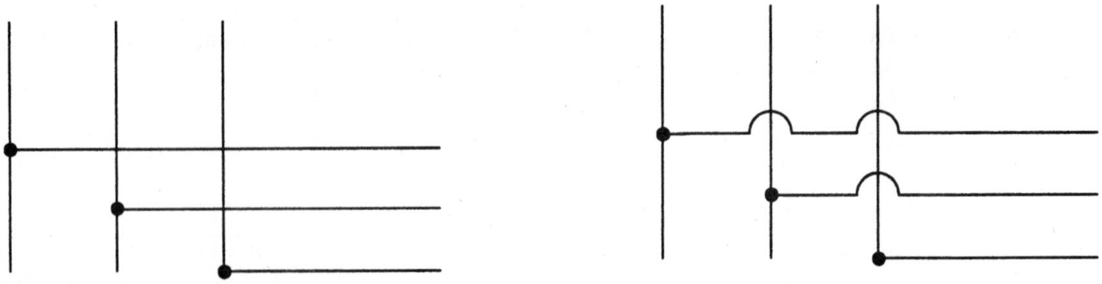

Figure 1   **Figure 2**

31. Figure 1 and Figure 2 shown above are standard electrical symbols used on blueprints representing crossing conductors. Which symbol represents two conductors crossing but not connecting?
    A. Figure 1 only
    B. Figure 2 only
    C. Figure 1 and Figure 2
    D. Neither Figure 1 nor Figure 2

    Answer: _____   Reference: _____

32. The following figure is a standard symbol used on blueprints. The symbol represents which of the following?
    A. an air circuit breaker
    B. a lightning arrester
    C. a fuse
    D. a thermal element

    Answer: _____   Reference: _____

33. An electrical contractor installs the wiring in a new 1400-square-foot single-family dwelling unit before a permit is obtained from the city electrical inspector. The electrical inspector stops the contractor from continuing the job. The city where the structure is located has adopted an electrical code that contains the following provisions:

Permits for New Construction

1. One- and two-family dwelling: On all new single- and two-family dwelling construction, the electrical fee shall be as follows: $.02/square feet under roof.

2. All other

Penalty for Failure to Obtain Permit

In case it shall be discovered that any electrical work has been installed or put into use for which no permit has been issued, the violator shall pay a fee equal to three times the permit fee that shall have been paid for work done in violation thereof and no additional permits shall be granted until all fees have been paid.

Before the electrical contractor can complete the work, a permit fee must be paid. Which of the following is the minimum permit fee required?
    A. $2.80
    B. $28.00
    C. $56.00
    D. $84.00

    Answer: _____   Reference: _____

• **Master Electrician's Review**

34. A service disconnect is supplied by conductors having 38,500 amperes RMS fault current available at the supply terminals of the disconnect. Which of the following statements is correct?
    A. Only the overcurrent device has to have an interrupting rating at or above the 38,500 amperes.
    B. The overcurrent device and the panel must be rated for the maximum fault current available.
    C. The overcurrent device must be rated at 200,000 amperes RMS.
    D. Neither the panel nor the overcurrent device has to be rated with fault-current interrupting rating.

    Answer: _____   Reference: _____

35. A motor is protected against short circuit and ground fault by an adjustable instantaneous trip circuit breaker that is part of a combination controller having motor overload and short-circuit and ground-fault protection in each conductor. The setting of the instantaneous trip breaker shall be permitted to exceed the AC 3-phase motor full-load current by not more than _____ percent.
    A. 250
    B. 800
    C. 1100
    D. 1300

    Answer: _____   Reference: _____

36. Thermal overload relays are used for the protection of polyphase induction motors. Their primary purpose is to protect the motor in case of
    A. reversal of phases in the supply.
    B. low-line voltage.
    C. short circuit between phases.
    D. sustained overload.

    Answer: _____   Reference: _____

37. A motor controller and motor branch-circuit disconnecting means a 2300-volt motor shall have a continuous ampere rating of not less than the
    A. trip setting of the short-circuit protective device rating.
    B. trip setting of the overload protection device.
    C. trip setting of the fault-current protection device.
    D. locked-rotor rating of the motor.

    Answer: _____   Reference: _____

38. The disconnecting means for both a 2300-volt motor and the controller shall be
    A. permitted within the same enclosure as the controller.
    B. located separately from the controller enclosure.
    C. located at the service equipment.
    D. permitted within the watt-hour meter enclosure.

    Answer: _____   Reference: _____

39. Speed-limiting devices shall be provided with which of the following?
    A. polyphase squirrel-cage motors
    B. synchronous motors
    C. compound motors
    D. series motors

    Answer: _____ Reference: _____

40. Which of the following appliances can be grounded to the grounded (neutral) conductor in an existing dwelling?
    A. electric water heater
    B. kitchen disposal
    C. dishwasher
    D. electric dryer

    Answer: _____ Reference: _____

41. A metal luminaire shall be grounded if located
    A. 10 feet (3 m) vertically or 6 feet (1.8 m) horizontally from a kitchen sink.
    B. 8 feet (2.4 m) vertically or 5 feet (1.5 m) horizontally from a kitchen sink.
    C. 6 feet (1.8 m) vertically or 3 feet (900 mm) horizontally from a kitchen sink.
    D. 8 feet (2.4 m) vertically or 3 feet (900 mm) horizontally from a kitchen sink.

    Answer: _____ Reference: _____

42. When used, a driven ground rod shall be installed so that the soil will be in contact with a length of the rod not less than
    A. 4 feet (1.2 m).
    B. 6 feet (1.8 m).
    C. 8 feet (2.4 m).
    D. 10 feet (3 m).

    Answer: _____ Reference: _____

43. Portable and stationary electrically heated appliances having metal frames are connected to an electrical system operating at 277 volts to ground. Which of the following statements about the metal frames of each appliance is (are) correct?
    A. It must be grounded.
    B. It must be grounded only if supplied by a portable cord.
    C. It can be permanently insulated from ground by special permission.
    D. It is connected by the manufacturer to the grounded conductor thermal.

    Answer: _____ Reference: _____

44. Which of the following most accurately describes the condition of a motor known as "locked-rotor"?
    A. When the electrician places a lock on the motor controller to keep the motor from being energized
    B. Mechanical brakes used on the motor shaft to stop the motor during shutdown
    C. An electronic control device used to lock the speed of the motor at that specified by the manufacturer
    D. When the circuits of a motor are energized but the rotor is not turning.

    Answer: _____ Reference: _____

45. The following diagram represents overhead conductors between two buildings on an industrial site. The voltage is 240/480 volts (alternating current). The conductors pass over a driveway leading to a loading dock at one of the buildings. What is the minimum vertical clearance permitted between the overhead conductors and the driveway?

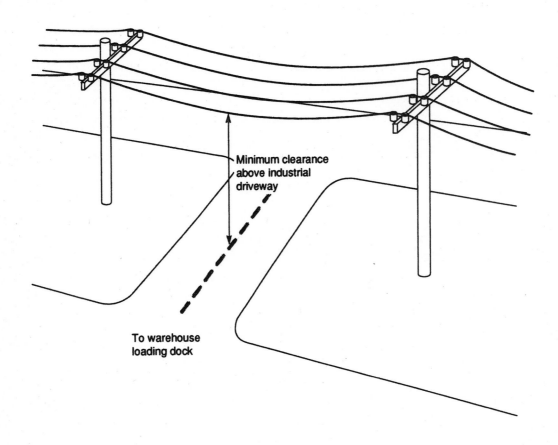

Minimum clearance above industrial driveway

To warehouse loading dock

A. 10 feet (3 m)
B. 12 feet (3.7 m)
C. 15 feet (4.5 m)
D. 18 feet (5.5 m)

Answer: _____   Reference: _____

46. When a ground-fault protection for equipment is installed within service equipment, it shall be performance tested
    A. at the factory before shipment.
    B. before being installed on-site.
    C. when first installed on-site.
    D. after the electrical system has been used for one week.

Answer: _____   Reference: _____

47. Three continuous-duty motors with full-load current ratings of 5.6 amperes, 4.5 amperes, and 4.5 amperes, respectively, are to be installed on a single branch circuit. The circuit conductors shall have a minimum ampacity of _____ amperes.
    A. 16
    B. 20
    C. 24
    D. 30

    Answer: _____   Reference: _____

48. Which of the following statements about the number of electrical services to a building is (are) correct?
    I. A service supplied by a wind-powered generator can be installed on a building in addition to the service supplied by the local electrical utility.
    II. A service supplied by a solar photovoltaic system can be installed on a building in addition to the service supplied by the local electrical utility.
    A. I only
    B. II only
    C. Both I and II
    D. Neither I nor II

    Answer: _____   Reference: _____

49. Which of the following statements about the termination and bonding of conductors at service equipment is (are) true?
    I. The grounded service conductor and the equipment grounding conductor shall be bonded together within the service equipment.
    II. The grounding electrode conductor and the system grounded conductor shall be bonded together within the service equipment.
    A. I only
    B. II only
    C. Both I and II
    D. Neither I nor II

    Answer: _____   Reference: _____

50. When a universal series ac motor is caused to run without being connected to a load, the motor will
    A. run at a constant speed for an indefinite time.
    B. vary in running speed from about 80% to 125% of the normal rated speed.
    C. decrease in running speed until a locked-rotor condition occurs.
    D. increase in running speed to a dangerous level that can damage the motor.

    Answer: _____   Reference: _____

51. The following diagram represents an electrical service with a fused switch as the main disconnect. What is the maximum height the center of the grip of the operating handle of the switch is permitted to be located above the ground?

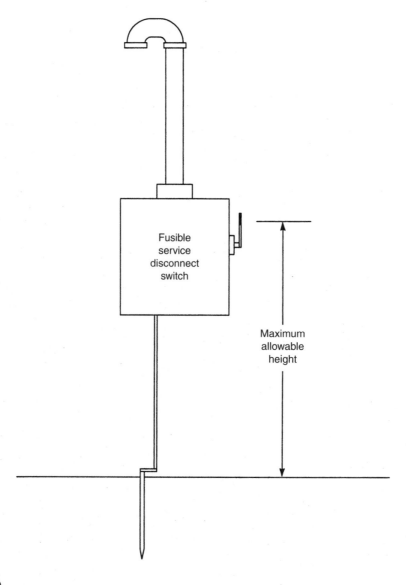

Fusible
service
disconnect
switch

Maximum
allowable
height

   A. 5 feet (1.5 m)
   B. 6 feet, 7 inches (2 m)
   C. 7 feet (2.1 m)
   D. ½ foot (150 mm)

Answer: _____   Reference: _____

52. When fuses are used for motor overload protection for a 3-wire, 3-phase motor, the fuse shall protect
   A. each ungrounded conductor.
   B. only two ungrounded conductors.
   C. only one ungrounded conductor.
   D. the equipment-grounding conductor.

Answer: _____   Reference: _____

53. How is the size of an electrical conductor supplying a circuit determined?
    A. voltage
    B. amperage
    C. the length
    D. all of the above

    Answer: _____ Reference: _____

54. Ohm's law is
    A. the measurement of the I2R losses.
    B. the relationship between voltage, current, and power.
    C. an equation for determining power.
    D. a relationship between voltage, current, and resistance.

    Answer: _____ Reference: _____

55. Electrical pressure is measured in
    A. volts.
    B. amperes.
    C. coulombs.
    D. watts.

    Answer: _____ Reference: _____

56. The resistance of a 1500-watt, 120-volt resistance heater element is
    A. 14.4 ohms.
    B. 9.6 ohms.
    C. 11.2 ohms.
    D. 12.5 ohms.

    Answer: _____ Reference: _____

57. Total ampacity of stranded or solid conductors is the same if they have the same
    A. diameter.
    B. circumference.
    C. cross-sectional area.
    D. insulation.

    Answer: _____ Reference: _____

58. Two 6-volt batteries connected in parallel will give
    A. longer service than one battery.
    B. 12 volts.
    C. higher currents.
    D. lower currents.

    Answer: _____ Reference: _____

59. A branch circuit that supplies a number of outlets for lighting and appliances is called a _____ branch circuit.
    A. general-purpose
    B. utility
    C. multipurpose
    D. none of the above

    Answer: _____   Reference: _____

60. An electric heater will produce less heat on low voltage because
    A. its total watt output decreases.
    B. the current will decrease.
    C. the resistance does not change.
    D. all of the above

    Answer: _____   Reference: _____

61. A 10-ohm resistance carrying 10 amperes of current uses _____ watts of power.
    A. 100
    B. 200
    C. 500
    D. 1000

    Answer: _____   Reference: _____

62. A _____ stores energy in much the same manner as a spring stores mechanical energy.
    A. coil
    B. capacitor
    C. resistor
    D. none of the above

    Answer: _____   Reference: _____

63. More heat is created when current flows through which of the following?
    A. 10-ohm capacitor
    B. 10-ohm inductance coil
    C. 10-ohm resistor
    D. All of the above are equal.

    Answer: _____   Reference: _____

64. _____ means that it is constructed or protected so that exposure to the weather will not interfere with its successful operation.
    A. Weatherproof
    B. Weathertight
    C. Weather-resistant
    D. all of the above

    Answer: _____   Reference: _____

65. The service conductor clearance from windows can be less than 3 feet if it is
    A. run along the left side of the window.
    B. run above the top level of the window.
    C. run below the bottom level of the window.
    D. none of the above

    Answer: _____   Reference: _____

66. The *NEC*® states that electrical equipment shall be installed
    A. not exceeding the provisions of the *Code*.
    B. not less than the *Code* permits.
    C. according to the *Code* and local code amendments.
    D. none of the above

    Answer: _____   Reference: _____

67. The equivalent resistance of three resistors of 8 ohms, 8 ohms, and 4 ohms that are connected in parallel is
    A. 2 ohms.
    B. 4 ohms.
    C. 6 ohms.
    D. 8 ohms.

    Answer: _____   Reference: _____

68. Because fuses are rated by both amperage and voltage, a fuse will operate correctly on
    A. ac only.
    B. ac or dc.
    C. dc only.
    D. any voltage.

    Answer: _____   Reference: _____

69. A 10-horsepower, 480-volt, 3-phase motor operating at 80% efficiency has an input current of _____ amperes. (1 hp = 746 watts)
    A. 11.2
    B. 21.5
    C. 33.3
    D. 42

    Answer: _____   Reference: _____

70. A branch circuit supplying a detached garage requires a total of 380 feet (116 m) of circuit conductor with a maximum conductor resistance of 0.8 ohm. The minimum size conductor must have a resistance of no more than _____ ohms per 1000 feet (305 m).
    A. 1.6
    B. 2.1
    C. 3.4
    D. 4.8

    Answer: _____   Reference: _____

71. An electric resistance heater is rated for 2400 watts at 240 volts. What power is consumed when the heater is operated at 120 volts?
    A. 600 watts
    B. 1200 watts
    C. 2400 watts
    D. 4800 watts

    Answer: _____   Reference: _____

72. A 230-volt, single-phase circuit has 10 kilowatts of load and 50 amperes of current. The power factor is _____ %.
    A. 78
    B. 87
    C. 94
    D. 115

    Answer: _____   Reference: _____

73. A 1½-horsepower single-phase motor that has an efficiency of 80% operates at 230 volts and has an input current of _____ amperes. (1 hp = 746 watts)
    A. 1.3
    B. 3.2
    C. 6.1
    D. 6.8

    Answer: _____   Reference: _____

74. A 30-horsepower, wound-rotor induction motor with no code letter is to be installed with 460-volt, 3-phase alternating current. Disregarding all the exceptions, the non–time-delay fuse for short-circuit protection of the motor branch circuit must be rated at a maximum _____ amperes.
    A. 40
    B. 50
    C. 60
    D. 80

    Answer: _____   Reference: _____

75. The following 480-volt, 3-phase, 3-wire, intermittent-use equipment is in a commercial kitchen: two 5000-watt water heaters; four 3000-watt fryers; and two 6000-watt ovens. Each ungrounded conductor in the feeder circuit for this kitchen equipment must be sized to carry a minimum calculated load of _____ amperes.

    A. 15
    B. 27
    C. 41
    D. 46

    Answer: _____   Reference: _____

Appendix 1: Symbols

Caution: These are standard symbols. Some variation is common from region to region.

(Courtesy of American Iron and Steel Institute)

Electrical Wiring Symbols

Selected from American National Standard Graphic for
Electrical Wiring and Layout Diagrams Used in Architecture and Building Construction

ANSI Y32.9-1972

1. Lighting Outlets

Ceiling   *Wall*

1.1 Surface or pendant incandescent, mercury-vapor, or similar lamp fixture

1.2 Recessed incandescent, mercury-vapor, or similar lamp fixture

1.3 Surface of pendant individual fluorescent fixture

1.4 Recessed individual fluorescent fixture

1.5 Surface or pendant continuous-row fluorescent fixture

1.6 Recessed continuous-row fluorescent fixture

1.8 Surface or pendant exit light

1.9 Recessed exit light

1.10 Blanked outlet

1.11 Junction box

1.12 Outlet controlled by low-voltage switching when relay is installed in outlet box

2. Receptacle Outlets

Grounded   *Ungrounded*

2.1 Single receptacle outlet

2.2 Duplex receptacle outlet

2.3 Triplex receptacle outlet

2.4 Quadrex receptacle outlet

2.5 Duplex receptacle outlet – split wired

2.6 Triplex receptacle outlet – split wired

2.7 Single special-purpose receptacle outlet – split wired

NOTE 2.7A: Use numeral or letter as a subscript alongside the symbol, keyed to explanation in the drawing list of symbols, to indicate type of receptacle or use

2.8 Duplex special-purpose receptacle outlet
See note 2.7A

2.9 Range outlet (typical)
See note 2.7A

3. Switch Outlets

2.10 Special-purpose connection or provision for connection
Use subscript letters to indicate function
(SW – dishwasher; CD – clothes dryer, etc).

DW   UNG
            DW

3.1 Single-pole switch

S

3.2 Double-pole switch

S2

2.12 Clock hanger receptacle

C   C UNG

3.3 Three-way switch

S3

2.13 Fan hanger receptacle

F   F UNG

3.4 Four-way switch

S4

2.14 Floor single receptacle outlet

UNG

3.5 Key-operated switch

SK

2.15 Floor duplex receptacle outlet

UNG

3.6 Switch and pilot lamp

SP

2.16 Floor special-purpose outlet
See note 2.7A

UNG

3.7 Switch for low-voltage switching system

SL

3.8 Master switch for low-voltage switching system

SLM

2.17 Floor telephone outlet – public

3.9 Switch and single receptacle

S

2.18 Floor telephone outlet – private

3.10 Switch and double receptacle

S

2.19 Underfloor duct and junction box for triple, double, or single duct system (as indicated by the number of parallel lines)

3.11 Door switch

SD

3.12 Time switch

ST

2.20 Cellular floor header duct

3.13 Circuit breaker switch

SCB

3.14 Momentary contact switch or pushbutton for other than signaling system

SMC

3.15 Ceiling pull switch

Ⓢ

5.13 Radio outlet

R

5.14 Television outlet

TV

6. Panelboards, switchboards, and related equipment

6.1 Flush-mounted panel board and cabinet
NOTE 6.1A: Identify by notation or schedule

6.2 Surface-mounted panel board and cabinet
See note 6.1A

6.3 Switchboard, power control center, unit substations (should be drawn to scale)
See note 6.1A

6.4 Flush-mounted terminal cabinet
See note 6.1A

NOTE 6.4A: In small-scale drawings the TC may be indicated alongside the symbol

TC

6.5 Surface-mounted terminal cabinet
See note 6.1A and 6.4A

TC

6.6 Pull box
Identify in relation to wiring system section and size

6.7 Motor or other power controller
See note 6.1A

MC

6.8 Externally operated disconnection switch
See note 6.1A

6.9 Combination controller and disconnection m
See note 6.1A

7. Bus Ducts and Wireways

7.1 Trolley duct
See note 6.1A

| | T | | T | | T | |

7.2 Busway (service, feeder, or plug-in)
See note 6.1A

| | B | | B | | B | |

7.3 Cable through, ladder, or channel
See note 6.1A

| | BP | | BP | | BP | |

7.4 Wireway
See note 6.1A

| | W | | W | | W | |

9. Circuiting

Wiring method indentification by notation on drawing or in specifications.

9.1 Wiring concealed in ceiling or wall

NOTE 9.1A: Use heavy weight line to identify service and feed runs.

9.2 Wiring concealed in floor
See note 9.1A

Appendix 2: Basic Electrical Formulas

Basic Electrical Formulas
(Courtesy of American Iron and Steel Institute)

DC Circuit Characteristics

Ohm's Law:

$$E = IRI = \frac{E}{R}R = \frac{E}{I}$$

E = voltage impressed on circuit (volts)

I = current flowing in circuit (amperes)

R = circuit resistance (ohms)

Resistances in Series:

$$R_t = R_1 + R_2 + R_3 + \ldots$$

R_T = total resistance (ohms)

R_1, R_2 etc. = individual resistances (ohms)

Resistances in Parallel:

$$R_t = \frac{1}{\frac{1}{R_1} + \frac{1}{R_2} + \frac{1}{R_3} + \ldots}$$

Formulas for the conversion of electrical and mechanical power:

$$HP = \frac{watts}{746} (watts \times .00134)$$

$$HP = \frac{kilowatts}{.746} (kilowatts \times 1.34)$$

Kilowatts = HP × .746

Watts = HP × 746

HP = Horsepower

In direct-current circuits, electrical power is equal to the product of the voltage and current:

$$P = EI = I_2 R = \frac{E_2}{R}$$

P = power (watts)

E = voltage (volts)

I = current (amperes)

R = resistance (ohms)

Solving the basic formula for I, E, and R gives

$$I = \frac{P}{E} = \sqrt{\frac{P}{R}}; E = \frac{P}{I} = \sqrt{RP}; R = \frac{E_2}{P} = \frac{P}{T^2}$$

Energy

Energy is the capacity for doing work. Electrical energy is expressed in kilowatt-hours (kWhr), one kilowatt-hour representing the energy expended by a power source of 1 kW over a period of 1 hour.

Efficiency

Efficiency of a machine, motor or other device is the ratio of the energy output (useful energy delivered by the machine) to the energy input (energy delivered to the machine), usually expressed as a percentage:

$$Efficiency = \frac{output}{input} \times 100\%$$

$$or\ Output = Input \times \frac{efficiency}{100\%}$$

Torque

Torque may be described as a force tending to cause a body to rotate. It is expressed in pound-feet or pounds of force acting at a certain radius:

Torque (pound-feet) = force tending to produce rotation (pounds) × distance from center of rotation to point at which force is applied (feet).

Relations between torque and horsepower:

$$Torque = \frac{33,000 \times HP}{6.28 \times rpm}$$

$$HP = \frac{6.28 \times rpm\ time\ torque}{33,000}$$

rpm = speed of rotating part (revolutions per minute)

AC Circuit Characteristics

The instantaneous values of an alternating current or voltage vary from zero to maximum value each half cycle. In the practical formula that follows, the "effective value" of current and voltage is used, defined as follows:

Effective value = 0.707 × maximum instantaneous value

Inductances in Series and Parallel:

The resulting circuit inductance of several inductances in series or parallel is determined exactly as the sum of resistances in series or parallel as described under dc circuit characteristics.

Impedance:

Impedance is the total opposition to the flow of alternating current. It is a function of resistance, capacitive reactance and inductive reactance. The following formulae relate these circuit properties:

$$X_L = 2\pi HzLXc = \frac{1}{2\pi HzC}Z = \sqrt{R^2 + (X_L - X_C^2)}$$

X_L = inductive reactance (ohms)

X_c = capacitive reactance (ohms)

Z = impedance (ohms)

Hz = (Hertz) cycles per second

C = capacitance (farads)

L = inductance (henrys)

R = resistance (ohms)

$\pi = 3.14$

In circuits where one or more of the properties L, C, or R is absent, the impedance formula is simplified as follows:

Resistance only: Inductance only: Capacitance only:

$Z = R$ $Z = X_L$ $Z = XC$

Reistance and Resistance and Inductance and
Inductance only: Capacitance only: Capacitance only:

$Z = \sqrt{R^2 + X_L^2}$ $Z = \sqrt{R^2 + X_C^2}$ $Z = \sqrt{X_L - X_C^2)}$

Ohm's law for AC circuits:

$E = 1 \times Z$ $I = \dfrac{E}{Z}$ $Z = \dfrac{E}{I}$

Capacitances in Parallel:

$C_t = C_1 + C_2\, C_3 + ...$

C_t = total capacitance (farads)

$C_1 C_2 C_3 ...$ = individual cpacitances (farads)

Capacitances in series:

$C_t = \dfrac{1}{\dfrac{1}{C_1} + \dfrac{1}{C_2} + \dfrac{1}{C_3} + ...}$

Phase Angle

An alternating current through an inductance lags the voltage across the inductance by an angle computed as follows:

Tangent of angle of lag = $\dfrac{X_L}{R}$

An alternating current through a capacitance leads the voltage across the capacitance by an angle computed as follows:

Tangent of angle of lead = $\dfrac{X_C}{R}$

The resultant angle by which a current leads or lags the voltage in an entire circuit is called the phase angle and is computed as follows:

Cosine of phase angle = $\dfrac{R \text{ of circuit}}{Z \text{ of circuit}}$

Power Factor

Power factor of a circuit or system is the ratio of actual power (watts) to apparent power (volt-amperes), and is equal to the cosine of the phase angle of the circuit:

$PF = \dfrac{\text{actual power}}{\text{apparent power}} = \dfrac{\text{watts}}{\text{volts} \times \text{amperes}} = \dfrac{kW}{kVA} = \dfrac{R}{Z}$

KW = kilowatts

kVA = kilowatt-amperes = volt-amperes + 1,000

PF = power factor (expressed as decimal or percent)

Single-Phase Circuits

$kVA = \dfrac{EI}{1,000} = \dfrac{kW}{PF} \quad kW = kVA \times PF$

$I = \dfrac{P}{E \times PF} \quad E = \dfrac{P}{I \times PF} \quad PF = \dfrac{P}{E \times I}$

$P = E \times I \times PF$

P = power (watts)

Two-Phase Circuits

$I = \dfrac{P}{2 \times E \times PF} \quad E = \dfrac{P}{2 \times I \times PF} \quad PF = \dfrac{P}{E \times I}$

$kVA = \dfrac{2 \times E \times I}{1000} = \dfrac{kW}{PF} \quad kW = kVA \times PF$

$P = 2 \times E \times 1 \times PF$

E = phase voltage (volts

Three-Phase Circuits, Balanced Star or Wye

$I_N = 0 \quad I = I_P \quad E = \sqrt{3}E_P = 1.73E_P$

$E_P = \dfrac{E}{\sqrt{3}} = \dfrac{E}{1.73} = 0.577E$

I_N = current in neutral (amperes)

I = line current per phase (amperes)

I_P = current in each phase winding (amperes)

E = voltage, phase to phase (volts)

E_P = voltage, phase to neutral (volts)

Three-Phase Circuits, Balanced Delta

$I = 1.732 \times I_P \quad I_P = \dfrac{1}{\sqrt{3}} = 0.577 \times I$

$E = E_P$

Power:

Balanced 3-Wire, 3-Phase Circuit, Delta or Wye

For unit power factor (PF = 1.0):

$P = 1.732 \times E \times I$

$I = \dfrac{P}{\sqrt{3}} \quad E = 0.577\dfrac{P}{E} \quad\quad E = \dfrac{P}{\sqrt{3}} \times I = 0.577\dfrac{P}{I}$

P = total power (watts)

For any load:

$P = 1.732 \times E \times I \times PF \quad VA = 1.732 \times E \times I$

$E = \dfrac{P}{PF \times 1.73 \times I} = 0.577 \times \dfrac{P}{PF \times I}$

$I = \dfrac{P}{PF \times 1.73 \times E} = 0.577 \times \dfrac{P}{I} \times E$

$PF = \dfrac{P}{1.73 \times I \times E} = \dfrac{0.577 \times P}{I \times E}$

VA = apparent power (volt-amperes)

P = actual power (watts)

E = line voltage (volts)

I = line current (amperes)

Power Loss:

Any AC or DC Circuit

$P = I_2 R I = \sqrt{\dfrac{P}{R}} \quad R = \dfrac{P}{I^2}$

P = power heat loss in circuit (watts)

I = effective current in conductor (amperes)

R = conductor resistance (ohms)

Load Calculations

Branch Circuits—Lighting & Appliance 2-Wire:

$$I = \frac{\text{total connected load (watts)}}{\text{line voltage (volts)}}$$

I = current load on conductor (amperes)

3-Wire:

Apply same formula as for 2–wire branch circuit, considering each line to neutral separately. Use line-to-neutral voltage; result gives current in line conductors.

USEFUL FORMULAS

TO FIND	SINGLE PHASE	THREE PHASE	DIRECT CURRENT
AMPERES when kVA is known	$\frac{\text{kVA} \times 1000}{E}$	$\frac{\text{kVA} \times 1000}{E \times 1.73}$	not applicable
AMPERES when horsepower is known	$\frac{HP \times 746}{E \times \% \text{ eff.} \times pf}$	$\frac{HP \times 746}{E \times 1.73 \times \% \text{ eff.} \times pf}$	$\frac{HP \times 746}{E \times \% \text{ eff.}}$
AMPERES when kilowatts are known	$\frac{kW \times 1000}{E \times pf}$	$\frac{kW \times 1000}{E \times 1.73 \times pf}$	$\frac{kW \times 1000}{E}$
HORSEPOWER	$\frac{I \times E \times \% \text{ eff.} \times pf}{746}$	$\frac{I \times E\ 1.73 \times \% \text{ eff.} \times pf}{746}$	$\frac{I \times E \times \% \text{ eff.}}{746}$
KILOVOLT AMPERES	$\frac{I \times E}{1000}$	$\frac{I \times E \times 1.73}{1000}$	not applicable
KILOWATTS	$\frac{I \times E \times pf}{1000}$	$\frac{I \times E \times 1.73 \times pf}{1000}$	$\frac{I \times E}{1000}$
WATTS	$E \times I \times pf$	$E \times I \times 1.73 \times pf$	$E \times I$

$$\text{ENERGY EFFICIENCY} = \frac{\text{Load Horsepower} \times 746}{\text{Load Input kVA} \times 1000}$$

$$\text{POWER FACTOR (pf)} = \frac{\text{Power Consumed}}{\text{Apparent Power}} = \frac{W}{VA} = \frac{kW}{kVA} = \cos\varnothing$$

I = Amperes	E = Volts	kW = Kilowatts	kVA = Kilovolt-amperes
HP = Horsepower	% eff. = Percent Efficiency		pf = Power Factor
	e.g., 90% eff. is 0.90		e.g., 95% pf is 0.95

EQUATIONS BASED ON OHM'S LAW:

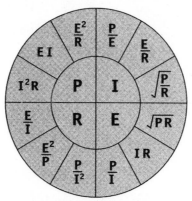

P = POWER, IN WATTS
I = CURRENT, IN AMPERES
R = RESISTANCE, IN OHMS
E = ELECTROMOTIVE FORCE, IN VOLTS

Appendix 3: Answer Key

SAMPLE EXAM QUESTION REVIEW

1. A. Ref.—350.10
2. B. Ref.—352.26
3. C. Ref.—360.20(B)
4. C. Ref.—314.28(A)(2)
5. B. Ref.—300.6(D)
6. B. Ref.—368.30
7. C. Ref.—Chapter 9, Table 1
8. B. Ref.—400.13 and 240.5(B)(1)
9. D. Ref.—310.106.(B)
10. D. Ref.—300.4(G)
11. D. Ref.—334.30
12. A. Ref.—Table 310.15(B)(7)
13. D. Ref.—300.5(D)(1)
14. C. Ref.—110.12 378.22 and 378.12
15. D. Ref.—Table 310.15(B)(6)
16. D. Ref.—and 285.28
17. D. Ref.—408.54
18. B. Ref.—230.54(C) Exception
19. A. Ref.—230.31(B) Exception
20. B. Ref.—230.9(A) Exception
21. C. Ref.—230.24(A)
22. D. Ref.—240.83(C)
23. D. Ref.—110.26(D)
24. B. Ref.—230.95
25. B. Ref.—Knowledge
26. D. Ref.—General knowledge; Appendix 2
27. B. Ref.—Table 430.37
28. D. Ref.—Knowledge
29. D. Ref.—430.24 and Table 430.250, Table 310.15 (B)(16)
30. B. Ref.—430.22 and Table 430.250, Tables 310.15(e)(16), 110.14(c) (1), 430.53(D)(2)
31. C. Ref.—430.22, and Table 430.250, Tables 310.15(B)(16), 240.4(E)(6), 430.53(D)(2)
32. A. Ref.—430.81(B)
33. B. Ref.—404.8(B)
34. C. Ref.—450.43(A)
35. A. Ref.—300.7(A)
36. B. Ref.—410.116(A)(1)
37. C. Ref.—410.136(B)
38. A. Ref.—700.12(B)(2)
39. C. Ref.—520.25(A) and (C)
40. D. Ref.—501.15(A) 514.9
41. A. Ref.—500.5(C)
42. D. Ref.—511.7(A)(1)
43. B. Ref.—517.18(B)
44. C. Ref.—675.15, 250.106 (FPN) 2
45. A. Ref.—440.61, 440.3(A), and 250.4(A)(5)(3)
46. B. Ref.—250.66(A)
47. D. Ref.—250.52(A)
48. C. Ref.—Knowledge; Appendix 2
49. A. Ref.—210.52 and 210.11(C)
50. B. Ref.—422.13, 422.10, and 310.15(B)(7) 422.11(E) 240.4(D)
51. C. Ref.—422.11(E)
52. C. Ref.—Chapter 3, P = EI (Appendix 2)
53. A. Ref.—Chapter 3; Appendix 2, P = I $\times$ R
54. B. Ref.—230.95(A)
55. C. Ref.—Chapter 9, Tables 1, 4, 5
56. B. Ref.—430.24
57. A. Ref.—430.62(A) Table 430.52
58. C. Ref.—424.3(B)
59. B. Ref.—Chapter 3, this text
60. C. Ref.—424.3(B)
61. D. Ref.—90.2(A) 90.2(B)
62. C. Ref.—General knowledge
63. C. Ref.—Chapter 2 text
64. B. Ref.—392.46
65. B. Ref.—312.2(A)
66. C. Ref.—344.30(A)
67. B. Ref.—Table 314.16(A)
68. D. Ref.—Table 110.28
69. D. Ref.—410.130(F)(1) and 410.130(F)(4)
70. A. Ref.—450.47
71. B. Ref.—450.13(A)

72. C. Ref.—460.2
73. C. Ref.—440.32, Table 310.15(B)16, 310.15(e)(3)(6)
74. D. Ref.—Table 430.52
75. D. Ref.—450.21(B)

QUESTION REVIEW CHAPTER 2

1. No. However, see 356.12(4) for electrical signs. Ref.—356.12(4)
2. Governmental bodies exercising legal jurisdiction over the electrical installation, insurance inspectors, or others with the authority and responsibility for governing the electrical installation. Ref.—Art 100
3. Ships, railway rolling stock, aircraft, or automotive vehicles. Ref.—90.2(B)
4. Informational notes (FPN). Ref.—90.5(C)
5. 110.3(B)
6. When adopted by the regulatory authority over the intended use. Ref.—Page 70-i, which states that the *Code* is advisory as far as NFDA and ANSI are concerned but is offered for use in law for regulatory purposes.
7. No. It is the intent that all premise wiring or other wiring of the utility-owned meter read equipment on the load side of service point of buildings, structures, or other premises not owned or leased by the utility are covered. It is also the intent that buildings used by the utility for purposes other than listed in 90.2(B)(5), such as office buildings, warehouses, machine shops, or recreational buildings, which are not an integral part of the generating plant, substation, or control center, are also covered. Ref.—90.2(B)(5)
8. There has been a change in this edition of the *NEC®* from the wording of the previous edition.
9. (A) 2011, (B) September 2010.
10. 1881, The National Fire Engineers met in Richmond, Virginia, which resulted in the first document that led to the *NEC®*. Various meetings were held after that; then in 1897, some various allied interests of the church, electrical, and architectural concerns met, and in 1911, the NFPA was set up as the sponsor.
11. Yes. Ref.—590.4(C)
12. Chapter 8, Ref.—90.3
13. No, it is not a training manual for untrained persons. Ref.—90.1(C)

14. No, only underground mines and rolling stock for surface mines. All surface mining other than the rolling stock are covered by the *NEC®*. Ref.—90.2(B)(2)
15. Nowhere in the *Code* does it state that it is the minimum allowed. Only when adopted by a local jurisdiction would it be considered the minimum of that jurisdiction. When an installation is installed for the *NEC®*, it provides a safeguarding of persons and property, and the provisions considered necessary for safety, and compliance with the *NEC®* will result in an installation essentially free from hazards. However, it may not be efficient or convenient or adequate in some installations. Ref.—90.1(A) and (B)
16. Definitions, Article 100, to continue for three hours or more
17. 2005. The *NEC®* is adopted every three years; there being a 2005, a 2008, and a 2011. Therefore, not adopting the two editions, they would be using the 2005 *NEC®*.
18. Class 2 or Class 3 cable. 725.130
19. Article 517—Health Care Facilities, 517.2, Definitions
20. Present (250.50). The water pipe must be used if available, and supplemented with another grounding electrode of one of the types listed in 250.52 (A)(1) through 250.52(A)(7), per 250.53(A)(2).
21. Article 422 lists specific requirements for specific appliances. Ref.—422.16(B)
22. Yes. Ref.—Article 770
23. Phase converters are covered in *NEC®* Article 455. Both static and rotary types are covered.
24. Article 426 covers fixed outdoor electrical deicing- and snow-melting equipment.
25. Article 680 for swimming pools, fountains, and similar installations. 680.1 clearly states decorative pools are covered.

CHAPTER 3—PRACTICE EXAM

1. C
2. D
3. D
4. C
5. C
6. A

7. D
8. B
9. C
10. B
11. C
12. A
13. B
14. C
15. A

QUESTION REVIEW CHAPTER 3

(Reduce fractions to their lowest terms.)

1. $\frac{1}{4}$
2. $\frac{1}{10}$
3. $\frac{2}{5}$
4. $\frac{3}{4}$
5. $\frac{3}{4}$
6. $\frac{1}{2}$
7. $\frac{3}{5}$
8. $\frac{3}{5}$
9. $\frac{1}{5}$
10. $\frac{1}{4}$
11. $\frac{1}{4}$

(Change mixed numbers to improper fractions.)

1. $\frac{3}{2}$
2. $\frac{13}{4}$
3. $\frac{43}{8}$
4. $\frac{17}{4}$
5. $\frac{57}{8}$
6. $\frac{13}{6}$
7. $\frac{23}{4}$
8. $\frac{25}{3}$
9. $\frac{46}{7}$
10. $\frac{29}{8}$

(Change improper fractions to mixed numbers.)

1. $4\frac{1}{2}$
2. $2\frac{2}{5}$
3. $12\frac{4}{5}$
4. $3\frac{1}{4}$
5. $9\frac{2}{3}$
6. $15\frac{1}{2}$
7. $1\frac{2}{3}$
8. $4\frac{1}{3}$

(Multiply whole numbers and fractions. Answer in proper reduced form.)

1. 4
2. $\frac{3}{8}$
3. $\frac{1}{40}$
4. $\frac{9}{2} = 4\frac{1}{2}$
5. $2\frac{1}{4} = 5\frac{1}{4}$
6. $\frac{5}{2} = 2\frac{1}{2}$
7. 9
8. $\frac{15}{7} = 2\frac{1}{7}$

(Convert fractions to decimal equivalents.)

1. 0.25
2. 0.625
3. 0.75
4. 0.866
5. 0.375

(Resolve to whole numbers.)

1. 16
2. 81
3. 2500
4. 1
5. 1000

(Find the square root.)

1. 5
2. 9
3. 7
4. 56
5. 101
6. 29
7. 78

(Solve the following problems.)

1. 0.05 or 5%. Ref.—Ohm's law; Appendix 2
2. 100%. Ref.—Chapter 3
3. 5 amperes. Ref.—Chapter 3; Ohm's law; Appendix 2
4. 1.33 ohms. Ref.—Chapter 3
5. 14.4 ohms. Ref.—Chapter 3

6. 6 volts. Ref.—Chapter 3

7. 33.34 ohms

8. 83.34 ohms

9. 23.81 ohms

10. 107.15 ohms

11. 37 volts

12. 56 volts

13. 27 volts

14. 1.12 amperes

QUESTION REVIEW CHAPTER 4

1. Yes. Ref.—90.2(A). A floating restaurant must be wired as a typical commercial restaurant.

2. Chapter 8. Article 250 does not apply in Chapter 8 unless specifically referenced as in 820, Part IV.

3. Article 406

4. False. The rules for calculating branch circuits are referenced in many articles in the *NEC®*.

5. Article 516.

6. Article 422 for appliances, specifically 422.2. Electrical heating is covered in Article 424 and heat pumps in Article 440.

7. Chapter 9, Tables 1 through 8 and all Notes

8. Lighting and power in mine shafts are not covered by the *NEC®*, Article 90, 90.2(B)

9. 300.50(A) and 300, Part II

10. Article 225, which covers outside branch circuits and feeders, 225.31

11. Table 352.44. Expansion fittings are required by 300.7(B), 352.44.

12. Places of assembly, Article 518.

13. 250.34

14. 680.26(B)(4)

15. 90.9(C)(2)

16. These requirements appear in the building codes or the NFPA Life Safety *Code*; they are not found in the *NEC®*.

17. Article 760

18. No. Ref.—590.6(A). No cordsets with GFCI permitted.

19. Article 502—(a) Class 2, (b) Division 1. Article 500 for the definitions, and 502.10(A) and (B) for the specific requirements.

20. It is a violation to put over 1000-volt wiring in a residence. Neon signs requirements are found in Article 600. However, Article 410 requirements limit installation under 1000 volts in dwellings. Ref.—410.140(B)

21. Article 500 defines this area. Article 501—Volatile Liquids would be the article required in which to make the installation. Articles 511 would not apply, see 511.1 scope, and 514.1 scope

22. Yes, Article 220, 220.88

23. Article 300, 300.4(D)

24. Article 702

25. Article 705, which covers interconnected electrical power production sources, otherwise known as co-generation

QUESTION REVIEW CHAPTER 5

1. 4 feet (1.2 m). Ref.—110.26, Table 110.26(A).

2. Two. Ref.—110.26(C)(2)

3. Yes. Ref.—110.26(A)(3)

4. 6½ feet (2 m). Ref.—110.26(A)(3)

5. One. Where the required working space is doubled, only one door is required. Ref.—110.26(C)(2)(b)

6. No. The required disconnecting means and supplementary overcurrent device must have a working clearance in accordance with 110.26. Clear working clearance of 30 inches (750 mm) wide and 36 inches (1 m) deep is required from the ground level up 6½ feet (2 m). Table 110.26(A)(1)

7. 30 inches (750 mm). Ref.—110.26(A)(2)

8. Ungrounded. Ref.—240.15(A)

9. 60%. Ref.—Chapter 9, Table Note (4), 310.15(B)(3)(a)

10. Branch-circuit overcurrent devices are not permitted to be located where subject to physical damage. They may not be placed in the vicinity of easily ignitible material, such as in clothes closets, and may not be located in bathrooms, in dwelling units and guest rooms, or guest suites of hotels and motels. Ref.—240.24(C), (D), and (E).

11. 1¼ inches (32 mm). Ref.—300.4(A)(1)

12. 600 volts. Ref.—362.12(5)

13. Yes. Ref.—230.43(16)

14. Yes. It shall be bonded to the electrode at each end and all intervening raceway boxes and enclosures

between the service equipment and the grounding-electrode conductor. Ref.—250.64(E)

15. The rating of the overcurrent device. Ref.—250.122

16. Two. Ref.—210.52(E)(1)

17. Yes. All metal piping systems and all grounded metal parts in contact with the circulating water shall be bonded together using a solid copper bonding jumper, insulated, covered, or bare, not smaller than 8 AWG. The bonding jumper shall be connected to the terminal on the circulating pump motor that is intended for this purpose. The bonding jumper shall not be required to be connected to a double-insulated circulating pump motor. Ref.—680.74

18. Yes. Ref.—210.52(A)(2)(3)

19. No. Ref.—210.52(A) for site-built dwellings; 550.13(D)

20. 210.63. Ref.—Article 210

21. Article 312

22. Article 410, 410.10(D) 680.72

23. No. 310.10(H)(1) states that parallel conductors shall not be permitted unless they are of size 1/0 AWG or larger.

24. 2/0 AWG. Table 310.15(B)(7)

25. Chapter 9, Table 8

QUESTION REVIEW CHAPTER 6

1. 18.04 amperes—P = E × 1.73/I or I = P/E × 1.73

2. 10, 80% loading. Ref.—220.19(A)(1) and 220.20(A)

3. 1120 square feet. Ref.—Knowledge

4. 2. Ref.—Tables 220.12 and 210.11(C)(2)

5. 5—2 kitchen appliance circuits, 1 laundry circuit, 1 heating circuit (Ref.—422.12), and 1 bathroom receptacle (210.11(C)). 422.12 requires an individual branch circuit for the furnace. This would likely be a 15-ampere circuit depending on the calculated load.

6. 2. Ref.—210.11(C)(1)

7. Yes. Ref.—210.11(C)(2)

8. 30 amperes 150% nameplate. Ref.—422.13 and 422.11(E)(3)

9. None. The *NEC*® does not require power to the structure. Ref.—210.52(G)

10. 2. Ref.—210.52(E)

11. Five minimum, one for each of the two small appliance branch circuits, one for the outdoor receptacles, one for the bathroom receptacles, and one in the crawl space or exterior. Ref.—210.8(A), also see 210.52 and all subsections 210.63

12. Yes. Illumination is required at these locations. Ref.—210.70 (A)(2)(b)

13. No. Ref.—210.70

14. Yes. Ref.—210.63

15. 20 amperes. Ref.—210.11(C)(2)

16. No. Ref.—422.12, Definition of Individual Circuit (Article 10)

17. No. Ref.—410.16 and 210.70. Lighting is permitted but not required

18. Yes. Ref.—210.52(B)(2) Exception 2

19. Yes. Ref.—Definition of direct grade access, 210.8(3)

20. Yes, two receptacles accessible while standing at grade level are required. Ref.—210.52(E)

21. All the 120-volt, 15-, and 20-ampere branch circuits in all areas except the kitchen, the bathroom, and the garage are required to be protected by a combination type arc-fault circuit interrupter type. 210.12(A) and (B)

22. 100 amperes Ref: 230.79(C)

    (a) 220.12(J) and Table 220.12; 28 × 40 = 1120 sq. ft.; 3 VA × 1120 = 3360 VA

    (b) 220.52; 3000 volt-amperes

    (c) 220.14(A); 4500 volt-amperes. Ref.—422.13

    (d) 220.52(B); 1500 volt-amperes

    (e) 220.54; 5000 volt-amperes

    (f) 0. Electricity not required; however, if supplied, the load would have to be included in calculation. Ref.—210.52(G)

    (g) 220.14(A); size not stated, therefore, a 20-ampere branch-circuit is recommended.

    (h) 0. 220.12; 210.52(B) Example 2 could be general lighting or on appliance load

    (i) 0. None. Read question again.

    (j) 0. Gas range igniter can be fed from appliance circuit, 210.52(B)(2) Exception 2

    (k) 0. None. This load is included in general lighting calculations, 220.3.

23. 125%. Ref.—430.22

24. 1 VA per square feet or $100 \times 200 = 20,000$ VA Ref.—Table 220.12

25. Two small appliance circuits minimum required. Ref.—210.11(C)(1)

26. Size; Formulas in Appendix 2

QUESTION REVIEW CHAPTER 7

1. 6. Ref.—230.208

2. 94%. Ref.—Chapter 3

3. A disconnecting means is required at each building. Ref.—Article 225, Part II, 225.31

4. Yes, you are limited to not more than six disconnects on this secondary side. Ref.—450.3(B)—conductors must be protected in accordance with 240.4(F), Table 430.3(B) Note 2

5. No; however, 230.54(E) requires that the conductors be separated where of a different potential.

6. No, 240.24(E) prohibits branch-circuit overcurrent devices from being located within a residential bathroom. Ref.—240.24(E) and 230.70(A)(2)

7. Yes. SE cable can be used in the same manner as Type NM cable. In this application, however, it must comply with Articles 300 and 334, Part II. Ref.—338.10(B)(4)(a)

8. Usually no. Yes, if the window is not designed to be opened, or where the conductors run above the top level of the window. Where windows are designed for opening and closing, then a minimum of 3 feet (900 mm) from the windows, doors, porches, fire escapes, or similar locations is required. Ref.—230.9(A)

9. (a) Yes. Ref.—250.64(D)(1)

 (b) They shall be sized in accordance with 250.66 for the largest conductor entering each respective enclosure.

10. The requirement to open all of the conductors is a disconnect requirement and not an overcurrent requirement. Ref.—240.15(B)(1) gives the requirements for circuit breakers, 240.40 gives the requirements for fuses. 210.4 is the requirement for multiwire branch circuits. Disconnecting, not overcurrent, is the purpose.

11. A separately derived system is defined in Article 100, and the grounding requirements are found in 250.30. Ref.—Article 100 and 250.30

12. 450.3—Table 450.3(B) Note 2—not more than six disconnects are permitted on the secondary side of the transformer.

13. No. Identification of the hi-leg is only required where the neutral is present. Ref.—110.15

14. No. The grounding-electrode conductor must be continuous, and the connection shall be made solidly and used for no other purpose. Ref.—250.64(C)

15. Less than 0.25 (¼) inch (6.4 mm). Ref.—352.44

16. 90.5—mandatory rules of the *Code* are characterized by the word "shall." Explanatory material is in the form of informational note.

17. 250.52(A)(2). It is only permitted if it meets one of the conditions listed.

18. Article 334, nonmetallic-sheathed cable. Romex is a registered trade of the South wire company.

19. Article 360 covers the requirements of flexible nonmetallic tubing. It is manufactured in trade sizes ½ (16) and ¾ (21) only.

 Note: This is a flexible metallic tubing that is liquid tight without a nonmetallic jacket. Rarely used in construction installation.

20. Article 404

21. Chapter 9, Table 1, Footnote 9 states that multiconductor cable of two or more conductors shall be treated as a single conductor cable, and cables that have an elliptical cross-section, the cross-section layering calculation shall be based on using the major diameter for the ellipse as the circle diameter.

22. Article 511 does not apply. 511.1 and 511.3(A)(1) clearly exempt these parking structures. Therefore, Chapters 1–4 apply, and the wiring method would be taken from those four chapters.

23. Article 378. This is a limited-use product and cannot be used in any of the five conditions in 378.12

24. 180 volt-amperes each. Article 220, 220.14(I), and 220.14(K) covers the requirements.

25. It is not permitted. 240.24(E) prohibits the installation of overcurrent devices in bathrooms of residences.

QUESTION REVIEW CHAPTER 8

1. Yes. Ref.—240.15(B)(2). However, handle ties are required.

2. No maximum length, provided they are outdoors. Ref.—240.21(B)(5)

3. No. Ref.—240.60(A). Voltage to ground cannot exceed 300 volts on a wye connection and, therefore, would not be permitted.

4. Ref.—230.90(A) Exception 3

5. 100%. Ref.—220.51

6. No. Ref.—501.30(B)

7. Yes. Both the overcurrent device and conductors would have to meet these requirements. Ref.—215.2(A)(1)

8. There is no limit. Ref.—220.14(I).

   Note: Many local jurisdictions set limits on 15- and 20-ampere branch-circuits in all installations. However, the *NEC*® does not specify a limit on dwelling branch-circuits for general-use receptacles.

9. No. Ref.—220.14(I), 16 × 180 VA = 2880 ÷ 120 volt = 24 amperes.

10. No, you must use 310.15, see Note at the top of Annex B. The Annex is not a part of the *NEC*® requirements; it serves only for the purpose of reference.

11. (A) No. Ref.—220.19(A)(1) and 220.60

    (B) Yes. Ref.—220.19(A)(1) and 424.3(B)

12. No, only if installed in cable with 12 inches (300 mm) of cover. Otherwise, can be installed at the greater depth or in raceways. Ref.—300.5 and Table 300.5

13. Yes. Ref—240.15(B)(2)

14. No. Ref.—422.10. Branch circuits are required to be calculated at 125% for water heaters.

15. No, No, No. Ref.—240.21(C)(1)

16. Yes. Ref.—210.19, 210.21, 210.23

17. No. Ref.—210.6(A)—12 amperes or less must be 120 volts; over that can be 240 volts. 1400 VA ÷ 240 volts = 5.833 amperes

18. 180 VA. Ref.—220.14(I)

19. Yes, it complies with the tap rules in 240.21(B)(1) for the 10-foot tap rules.

20. Yes, it complies with the 25-foot tap rule. Ref.—240.21(B)(2)

21. No, it does not comply with any of the tap rules in 240.21 because the switch does not contain overcurrent protection.

22. True. Ref.—368.17(B)

23. B. Ref.—240.21(B)(1)

24. Yes. Ref.—250.122(B)

25. A fuse must be installed. Ref.—110.3(B). The manufacturer's instructions must be followed.

QUESTION REVIEW CHAPTER 9

1. Soldered. Ref.—250.8(B)

2. 14. Ref.—600.7(B)(7)

3. Not required to be accessible. Ref.—250.68(A) Exception 1

4. White or gray. Ref.—200.6(A)

5. 1/0. Ref.—250.102(C)(2)

6. 4. Ref.—250.102(D) and Table 250.122

7. Yes, bonding is not required. Ref.—250.92(B)(3)

8. They are required to be run in each nonmetallic raceway. Ref.—215.6, 250.122(F) and Table 250.122

9. Yes. Ref.—250.112(L)

10. No. Ref.—250.30(A)(7)

11. Yes, it must be bonded. No, it is not sufficient to bond at one end. It must be bonded at both ends. Ref.—250.64(E)

12. No, a driven rod is not acceptable. An equipment grounding conductor must be installed with the circuit. Ref.—551.76(A)

13. Yes, the requirements for separately derived transformers are found in 250.30. These requirements do not amend the requirements of 250.50, but they are the specific requirements. Therefore, a supplementary grounding electrode is required.

14. 8 AWG copper or equivalent. Ref.—551.56(C)

15. No. Ref.—250.104(B).

    Warning: The underground gas piping system may not be used as an electrode grounded. Electrolysis may occur, creating dangerous underground gas leaks.

16. 1 AWG copper or a 2/0 AWG aluminum or copper-clad aluminum wire is required in each conduit. Ref.—Table 250.122 and 250.122(F), which state that where run in parallel in multiple raceways or cables as permitted, the equipment grounding conductors also shall be run in parallel, each equipment grounding conductor shall be sized in accordance with Table 250.122.

17. No. Ref.—110.14(B)

18. 8 AWG. Ref.—Table 250.122. This is a minimum size conductor and may have to be larger if the fault current available exceeds that for which the 8 AWG conductor will carry.

19. Yes. The flexible metal conduit cannot exceed 6 feet (1.8 m) for the combined two pieces of flexible metal conduit. Ref.—250.118(5)(C), and 348.60

20. The maximum rating of the grounding impedance, no smaller than 8 AWG copper or 6 AWG aluminum. Ref.—250.36(B)

21. By a ground-fault circuit interrupter. Ref.—Article 680.71

22. Yes, backfed circuit breakers field installed must have a retainer kit. Ref.—408.36(D)

23. Yes 250.64(F)

24. 250.80 Exception for Services and 250.86 Exception 1 for Feeders or Other Conductors permit these metal 90s where maintained at a burial depth in accordance with Table 300.5 need not be grounded. Otherwise, they must be bonded and at most practical places where they emerge from the ground, such as turning up into a concrete slab or turning up under a panelboard, or turning to go up the side of a pole or other structure. Bonding is required and provides a greater degree of safety.

25. The common grounding-electrode conductor would be based on the four 750-kcmil service conductors, and they must be sized at 2/0 AWG copper or 4/0 AWG aluminum. The tap conductor from the grounding-electrode conductor to Switch A, would be sized only on the conductor feeding that switch, 3 AWG, and the size of the conductor, based on Table 250.66, would be 8 AWG copper or 6 AWG aluminum. Switch B would be based on the Table 250.66, and would be 4 AWG copper or 2 AWG aluminum. Switch C would be 8 AWG copper or 6 AWG aluminum, and Switch D would be the 2/0 AWG. Ref.—250.64(0) and Table 250.66

QUESTION REVIEW CHAPTER 10

1. Within 12 inches (300 mm) of the outlet or box, and every 4½ feet (1.4 m). Ref.—338.10(B)(4) and 334.30

2. Maximum number of conductors shall be permitted using the provisions of 314.16(B). Conduit bodies shall be supported in a rigid and secure manner. Ref.—314.16(C)(2)

3. Yes. Ref.—348.20(A)(1) and 350.20(A) Exception

4. Yes. Ref.—342.30(A)(2)

5. Yes. Ref.—362.30(B)—Metal studs are the required support when they do not exceed 3 feet (900 mm) apart.

6. Type MC cable must be supported at intervals not exceeding 6 feet (1.8 m), and it shall be secured within 12 inches (300 mm) from every outlet box, junction box, cabinet, or fitting, except where fished. Ref.—330.30; 4 or fewer 10 AWG conductors

7. NM cable shall be supported at intervals not exceeding 4½ feet (1.4 m) and within 12 inches (300 mm) of every cabinet, box, or fitting. Two conductor cables shall not be stapled on edge. Where type NM cable is run through holes in wood studs or rafters, it shall be considered supported. Ref.—334.30

8. No, unless specifically permitted by other Articles in the Code. 318.12(2)

9. Aboveground conductors must be installed in a rigid metal conduit, intermediate metal conduit, electrical metallic tubing, RTRC, PVC conduit cable tray, auxiliary Gutters or busway or cable bus or in other identified raceways, or as open runs of metal clad cable suitable for the use. Open runs of Type MV cable, bare conductors, or bare bus bars also are permitted as they apply. Ref.—300.37

10. Yes, provided that carpet squares not exceeding 36 inches square (914 mm2) are used in a wiring system in accordance with Article 324 Type FCC cable and associated accessories are used. Ref.—Article 324.10 and 324.41

11. For metal wireways, the number of conductors and their ampacity shall comply with **376.22(A)** and **(B)**. The sum of the cross-sectional areas of all contained conductors at any cross section of a wireway shall not exceed 20% of the interior cross-sectional area of the wireway. The adjustment factors in **310.15(B)(3)(a)** shall be applied only where the number of current-carrying conductors, including neutral conductors classified as current-carrying under the provisions of **310.15(B)(5)**, exceeds 30.

For nonmetallic wireways, the adjustment factors specified in **310.15(B)(3)(a)** shall be applicable to all current-carrying conductors, and the sum of cross-sectional areas of all contained conductors at any cross section of the nonmetallic wireway shall not exceed 20% of the interior cross-sectional area of the nonmetallic wireway.

Conductors for signaling circuits or controller conductors between a motor and its starter and used only for starting duty shall not be considered as current-carrying conductors.

up to and including the 20% fill specified above.

Note: This is a critical difference. Caution should be used when reading questions as they apply to a specific wiring method, metallic or nonmetallic, because the conditions and rules are very different.

12. 1¼ inches (32 mm) from the nearest edge of the stud. Ref.—300.4(D)

13. No, the electrical metallic tubing must be securely fastened in place at least every 10 feet (3 m) and within 3 feet (900 mm) of each outlet box, junction box, device box, cabinet, conduit body, or other tubing termination. Support within 3 feet (900 mm) of each coupling is not required. Ref.—358.30

14. There is no difference in the permitted uses or installation requirements in the two products. Article 342 and Article 344 are essentially the same. Ref.—Article 342 and Article 344

15. 1/0 AWG. Ref.—310.10(H)

16. Yes, in any building exceeding three floors, nonmetallic tubing must be concealed within a wall, floor, or ceiling where the material has at least a 15-minute finished rating as identified in the listings of the fire-rated assemblies. *Exception: Where a fire sprinkler system(s) is installed in accordance with NFPA 13-2010,* Standard for the Installation of Sprinkler Systems, *on all floors, ENT is permitted to be used within walls, floors, and ceilings, exposed or concealed, in buildings exceeding three floors above grade* [reprinted with permission from NFPA 70-2011]. Ref.—362.10(2)

17. No, they must be removed. They are not permitted for a period to exceed 90 days. Ref.—590.3(B)

18. Lengths not exceeding 4 feet (1.2 m) of flexible metal conduit can be used to connect physically adjustable equipment and devices that are permitted in the duct. Ref.—300.22(B)

19. ¼ inch (6 mm). Ref.—352.44

20. Yes, even though 300.3 requires that all conductors of the same circuit, including the neutral and equipment grounding conductors, must be run within the same raceway, cable trench, cable, or cord; Exception 2 to that section permits that with column-type panelboards; the auxiliary gutter and pull box may contain the neutral terminations. Ref.—300.3(B)(4)

21. No. Generally, each cable shall be secured to the panelboard individually. Ref.—312.5(C). The exception would allow this under conditions.

22. Yes. Ref.—408.36 Exception 1 and 230.71(8)

23. It is prohibited by the *Code.* The reference you are looking for is 404.8(B). The switches must be arranged so that the voltage does not exceed 300 volts, or permanently installed barriers between the adjacent switches must be installed. You are correct; it is a dangerous situation.

24. Yes, knob-and-tube wiring is still an acceptable wiring method and is covered in Article 394. Extensions are permitted in hollow spaces, walls, and ceilings. However, they are not permitted in the hollow spaces of walls, ceilings, and attics where the spaces are insulated by loose, rolled, or formed-in-place insulated material that covers the conductors. You can make the installation, but you will have to route the conductors along a running board above the insulation. Ref.—394.12(5)

25. 1/0 AWG or larger. Ref.—392.10(B)(1)(a)

QUESTION REVIEW CHAPTER 11

1. 430.6(A). Ref.—430.6(A)
2. 100 ampere. Ref.—430.6, Tables 430.250 and 430.52 (40 × 250% = 100)
3. 175%. Ref.—Tables 430.250 and 430.52
4. 8 AWG THW aluminum. Ref.—Table 430.248, 430.24, and 430.25
5. Same. Ref.—240.21
6. 20. Ref.—240.4.(f)
7. 40. Ref.—240.4, and Table 310.15(B)(16)
8. 100. Ref.—240.4, and Table 310.15(B)(16)
9. 200. Ref.—240.4, and Table 310.15(B)(16)
10. 125%. Ref.—450.3(B)
11. 175%. Ref.—Table 430.52
12. $V_D = 22 \times 2 \times 150 \times 1.24/1000 = 8.184$ volts
13. 3.5%
14. 8 AWG

15. 125%. Ref.—Article 430.32(A)(1)

16. 6. Ref.—Table 110.20

17. Indoors. Ref.—Table 410.130(F)(1), (4)

18. Fire protection or equipment cooling. Ref.—450.47

19. Readily accessible. Ref.—450.13(A)

20. 3. Ref.—460.2

21. 8. Ref.—440.32 and Table 310.15(B)(16)

22. 300%. Ref.—Table 430.52

23. 112.5 kVA. Ref.—450.21(B)

24. 600 watts. Ref.—Ohm's law (*Clue:* Resistance does not change); Appendix 2

25. 87%. Ref.—Chapter 3; Appendix 2

QUESTION REVIEW CHAPTER 12

1. Nonconductive, conductive, composite. Ref.—770.2

2. True. Ref.—511.7

3. Yes, where the raceway emerges from the ground, both at the office and at the dispenser. Ref.—514.9(A) and (B)

4. False. Ref.—700.1

5. Class 1, Division 2. Ref.—511.3(B)(2)

6. Inadvertent interconnection. Ref.—700.6

7. ⅝ (18 mm). Ref.—501.15(C)(3)

8. 12 inches (305 mm). Ref.—Table 490.24

9. Four receptacles. Ref.—517.18(B)

10. 6. Ref.—220.14, 430.24, Tables 430.248 and 310.15(B)(16)

11. 15 feet. (4.5 m) Ref.—600.10(D)(2)

12. Length and design. Ref.—600.41(A)

13. True. Ref.—240.54(B)

14. False. Ref.—240.53

15. True. Ref.—240.53

16. False. Ref.—240.53

17. False. Ref.—240.53

18. False. Ref.—240.53

19. True. Ref.—240.53

20. True. Ref.—240.53

21. (a) Field marking is required to state "Caution—Series combination system Rated, _____ Amperes. Identified Replacement Components Required." Ref.—110.22

    (b) Manufacturer

22. 6 amperes. Ref.— I = $I = \dfrac{HP \times 746}{E \times PF \times eff\%}$

    Note: When power factor is not noted, it must be calculated at 100%.

23. 60. Ref.—Tables 430.250 and 430.52

24. 27. Ref.—Ohm's law, Chapter 3, Table 220.56; Appendix 2

25. 18. Ref.—Ohm's law; Appendix 2

QUESTION REVIEW CHAPTER 13

1. Yes. Ref.—810.20(C) and 810.21

2. No. Chapter 8 stands alone. There are no burial depth requirements for satellite dishes installed anywhere there is no safety hazard, and, therefore, the only consideration is for the customer's protection of his or her own equipment, *NEC*® 800.10. Ref: 90.3

3. Yes. Ref.—800.93

4. 6 feet (1.8 m). Ref.—820.44(E)(3)

5. It is an intrinsically safe circuit covered in Article 504 of the *NEC*®.

6. Yes, motor control centers are covered in Part VIII of Article 430.

7. Definition of Article 100 defines a separately derived system. 250.30 covers the grounding requirements for this separately derived transformer.

8. 406.3(D) covers these isolated ground receptacles intended for the reduction of electrical noise.

9. Article 530 covers film vaults.

    Note: Scope 530.1 covers these areas in a portion of a building.

10. When a term is used only once in the *NEC*®, the definition appears where it is being used. The definition of a bathroom appears in Article 100—Definitions.

11. Yes. Ref.—Article 650

12. Yes, Article 110, 110.26 for installations of less than 600 volts, 110.32 for installations over 600 volts.

13. Yes. 110.12(B) clearly states that paint, plaster, and other abrasives may damage the bus bars, wiring terminals, insulators, and other surfaces in panelboards. Therefore, the internal parts of this equipment must be covered and protected.

14. 550.32 requires the service equipment to be located outside and not more than 30 feet (9 m)

from the exterior wall of the mobile home unless permitted by 550.32(B).

15. Optional standby system is covered in Article 702 and is defined in 702.2.

16. Color coding is not specifically required; however, 210.5(C) requires the identification of ungrounded conductors. This identification can be by color coding, marking tape tagging, or other equally effective means.

17. No. Ref.—240(4)(D)

18. 225.7(B) covers outdoor branch-circuits for lighting equipment and there is no maximum number. See 215.4(A) for feeders.

19. Yes. Article 547—Agriculture Buildings covers this type of installation.

20. Yes. Article 324.41 covers this type of installation.

21. Ref.—Articles 280 and 285

22. Article 590 permits a class less than would be required for permanent installation. However, there are time constraints.

   Note: 590.4(A) requires the service on that temporary pole to comply with Article 230.

23. 86%. Ref.—Chapter 3, Duff and Herman, *Alternating Current Fundamentals, 4E* by Delmar Cengage Learning; Appendix 2

24. 60 hp. Formula is $EFF = \dfrac{W}{\sqrt{3} \times E \times I}$

   Ref.—Loper, *Direct Current Fundamentals, 4E* by Delmar Cengage Learning; Appendix 2

25. 0.105. Ref.—Loper, *Direct Current Fundamentals, 4E* by Delmar Cengage Learning; Appendix 2

PRACTICE EXAMINATION

1. C. -Ref.—Article 100—Definition
2. A. -Ref.—310.120(C)(1)
3. E. -Ref.—724.43
4. D. -Ref.—312.8
5. B. -Ref.—240.51(A)
6. D. -Ref.—760.176(A)
7. C. -Ref.—Chapter 9, Note 4 to Table 1
8. D. -Ref.—240.80
9. C. -Ref.—720.5
10. D. -Ref.—Table 402.5
11. A. -Ref.—Chapter 9, Table 8, Col 6

12. B. -Ref.—760.136(A)
13. B. -Ref.—90.5(C)
14. D. -Ref.—344.30(B) and Table 344.30(B)(2)
15. D. -Ref.—Article 100, see definition "switch"
16. D. -Ref.—725.31(B)
17. B. -Ref.—250.12
18. C. -Ref.—Article 100—Definition
19. D. -Ref.—110.27(A)
20. B. -Ref.—110.6
21. B. -Ref.—90.9
22. B. -Ref.—Article 100—Definition
23. B. -Ref.—725.136
24. B. -Ref.—Article 100—Definition
25. A. -Ref.—200.6
26. B. -Ref.—110.19
27. A. -Ref.—424.39
28. C. -Ref.—660.5, 517.72(A)
29. A. -Ref.—225.14(C)
30. C or the height of the equipment, whichever is greater. Ref.—110.26(A)(3)
31. A. Ref.—220.14(H)
32. A. Ref.—810.11 Exception
33. B. Ref.—230.79(C)
34. A. Ref.—430.89
35. D. Ref.—225.4
36. D. Ref.—110.53
37. D. Ref.—225.36 Exception
38. A. Ref.—210.23(A)(2)
39. D. Ref.—366.23, copper is 1000 amperes per square inch. 4 × ½ = 2 square inches, 1000 × 2 = 2000 amperes maximum
40. D. Ref.—590.7
41. D. -Ref.—430.32(D)(2)
42. D. -Ref.—550.16(A)
43. B. -Ref.—600.9(A)
44. D. -Ref.—820.93 and 820.100
45. E. -Ref.—110.32
46. D. -Ref.—398.30(E)
47. A. -Ref.—Article 100—Definition
48. A. Ref.—225.6(A)(1)
49. B. Ref.—210.52(B)(1)
50. B. Ref.—210.52(B)(3)

CODE QUIZ 1

1. Article 682—Natural and Artificially Made Bodies of Water. Ref.—Table of Contents 680.1

2. -Article 353—High Density Polyethylene Conduit: Type HDPE Conduit. Ref.—Table of Contents

3. -Article 590—Temporary Installations. Ref.—Index

4. Bathroom. Ref.—Article 100—Definitions

5. Unclassified locations. Ref.—500.2

6. No. Annex G contains administrative and enforcement rules.

7. The system bonding jumper is used to bond the grounded conductor, supply side bonding jumper and equipment grounding conductors together Ref.—Article100

8. Grounding electrode conductor. Ref.—Article 100—Definitions

9. Handhole enclosure. Ref.—Article 100—Definitions

10. Other than dwelling kitchens. Ref.—210.8(B)(2)

11. Ground fault protection Ref.—210.8(B)(4)

12. Chapter 9, Table 2. Ref.—Same

13. 90.2(B)(4). Ref.—830.1 Informational note No. 2

14. The fill restrictions do not apply; therefore, there are no maximum requirements. Ref.—820.110(B)

15. False. Ref.—513.12. The GFCI requirements only apply to receptacles used for diagnostic equipment, electrical hand tools, or portable lighting equipment.

CODE QUIZ 2

1. Raceways. Ref.—645.5(G)

2. Enclosed in raceways. Ref.—610.11(A) and (B)

3. Rigid metal conduit; intermediate metal conduit. Ref.—550.15(H)

4. 230; 525.10(A) and (B). Ref.—Same

5. 506.15(A) (1) through (5) Ref.—Same

6. Resistors and reactors that are part of other apparatus are not covered by Article 470 but are generally covered by the listing for the apparatus. Ref.—470.1 Exception

7. 115. Ref.—430.110

8. Bathtubs; shower stalls. Ref.—406.9(C).

9. 410.30(B)(6); 300.19. Ref.—Same

10. 400.4 and Table 400.4. Ref.—Same

11. 600 volts. Ref.—398.12 and 398.10

12. Yes. Ref.—Section 374.17

13. False. Ref.—392.10(D).

14. Corrosive vapors; physical damage. Ref.—368.12(A)

15. 100 feet (30.5 m) of trade size 1 (27) intermediate metal conduit (IMC). Ref.—300.1(C), and 342.130

CODE QUIZ 3

1. Interior wiring. Ref.—326.12

2. Tamper resistant type. Ref.—517–18(c).

3. Earth. Ref.—250.4(A)(5)

4. NFPA 496. Ref.—500.2 (Information Note) under the definition of "purged and pressurized."

5. 225.6. Ref.—300.3(A)

6. Where the conductors of different systems originate. Ref.—200.6(D). They must be marked with one of the prescribed methods in 200.7.

7. True. Ref.—400.7 (B)

8. Bathtub; shower space; face-up; above. Ref.—550.13(F)

9. ½ (16); one. Ref.—600.32(A)(2) and (3)

10. 630. Ref.—630.1

11. The disconnecting switch must be in-sight. Ref.—430.102; also see Exceptions

12. 250.102

13. No, they must be approved. Ref.—300.5(K); Definition "approved" in Article 100

14. The *Code* requires listed or marked and listed. Ref.—310.10(D)(1) and (2) by adding sleeve to I.D. listed for "sunlight resistant"

15. No, the *Code* limits are 120 volts between conductors. Ref.—210.6(A)(1)

16. Yes. Ref.—210.8(A)(3)

17. 103 is the metric designator for trade size 4 conduit and tubing. Ref.—300.1(C)

18. No, EMT is an acceptable equipment grounding conductor. Ref.—250.118

19. Article 409, Part II. Ref.—Same

20. Ungrounded. Ref.—285.4

CODE QUIZ 4

1. Yes. Ref.—517.13(B) Exception No 2

2. Luminaire. See Definition, Article 100; luminaire is an international term

3. Yes. Ref.—*NEC*® Annex G

4. Yes. 90.1(D) and the term "International" appears on the cover.

5. Yes.

6. Yes. *NEC*® Annex A and referenced in 90.7, informational note 3

7. No, the *Code* now uses a dual measuring system; either can be used as applicable. Ref.—90.9(B)

8. Article 406 contains these requirements.

9. Yes, Ref.—406.1

10. No. Each disconnect must be marked to indicate its purpose, unless the purpose is evident.

11. Sometimes. 110.26(C)(2) requires panic hardware on doors to electrical equipment spaces that contain large equipment rated 1200 amperes or more and over 6 feet wide.

12. 910.24

13. Article 100, Definition of luminaire

14. Yes. Ref.—406.5(E)

15. Yes. Ref.—210.12(A)

16. No. Ref.—210.52(C)(5)

17. Yes. 424.44(G) requires GFCI protection in these locations.

18. Yes. Ref.—200.6

19. No. 210.6(E) only permits where conditions of maintenance and supervision ensure that only qualified persons service the installation.

20. Yes. 230.22 Exception permits the grounded conductor of a multiconductor cable to be bare.

CODE QUIZ 5

1. Yes. Ref.—230.33 in accordance with 110.14, 300.5(E), 300.13, and 300.15

2. No. 230.44 permits cable tray as a support system. Ref.—392.2 and 392.10

3. Yes. Ref.—110.14(B)

4. No. 230.70(A)(3) permits a shunt trip device, but the disconnecting means is still required Ref—230.66, 230.70, 230.70(A)

5. Yes. Ref.—550.16(C)(3)

6. Yes. Ref.—430.83(A)(3)

7. No. The service disconnecting means shall be installed at a readily accessible location either outside of a building or structure or inside nearest the point of entrance of the service conductors. Ref.—230.70(A)(1)

8. Article 647

9. 230.82(6).

10. The overcurrent protective device(s) shall be selected or set to carry indefinitely the sum of the locked-rotor current of the fire pump motor(s) and the pressure maintenance pump motor(s) and the full-load current of the associated fire pump accessory equipment when connected to this power supply. Ref.—695.4(B)(2)

11. Yes. Ref.—210.8(B)(2)

12. 240.24(B). However, 240.24(B)(1) and (2) permits them to be accessible to only authorized management personnel.

13. No circuit breakers used as switches in 120-volt and 277-volt fluorescent lighting circuits shall be listed and shall be marked SWD or HID. Circuit breakers used as switches in high-intensity discharge lighting circuits shall be listed and shall be marked as HID. Ref.—240.81 and 240.83, 404.11

14. Yes, unless they are in raceways. Ref.— 645.5(g), 725.25, 760.25, 770.25,800.25,820.25,830.25,840.25.

15. 590.3(B)

16. Yes. Ref.—250.53(D)(2)

17. Yes. 250.54 but the earth shall not be the sole equipment grounding conductor(250.4(A)(5).

18. Article 280 covers surge arresters over 1 kV. Article 285 covers surge protective devices under 1 kV referred to as SPDs [surge arresters and transient voltage surge suppressors (TVSSs)]. They are defined in Article 100.

19. No. Ref.— 314.20.

20. 50 pounds (23 kg). Ref.—314.27(A)(2)

CODE QUIZ 6

1. 6 feet (1.8 m). Ref.—320.30(D)(3)

2. Yes. Ref.—300.7(A)

3. No. Ref.—300.11(C)

4. No. Ref.—300.22 (B), and (C)

5. Article 590

6. It is a device intended to provide protection from the effects of arc faults by recognizing characteristics unique to arcing and by functioning to de-energize the circuit when an arc fault is detected. Ref.—Art 100

7. No. If the runs comply exactly with 310.15(B)(3)(a)(4), then they would not apply.

8. No. Ref.—396.12

9. Yes. Ref.—320.30(D)(3)

10. Yes. 511.1 would cover the repair. Article 625 does not cover the vehicle, only the supply related to vehicle charging. Ref.—625.1

11. Yes. 342.30(B)(3), where threaded and threadless couplings are used.

12. Yes, if the threadless fittings are listed for the purpose. Ref.—344.42(A)

13. No. Ref.—322.10(3)

14. Yes.Ref.—312.6

15. Yes. Ref.—404.9(B)

16. Yes. Ref.—110.16

17. No. Ref.—334.12(A)(1)

18. No, fittings such as threadless couplings are required to be identified for service masts. Ref.—230.28

19. Yes. 514.11(A), (B) and (C), and 514.13.

20. No. Ref.—110.26(E)

CODE QUIZ 7

1. The areas shall be protected by an arc-fault circuit interrupter listed to provide protection of the *entire branch circuit*. There are exceptions to this rule. Ref.—210.12(A)

2. It would be considered as part of the lighting load. Ref.—220.52(A) Exception

3. No. Ref.—230.43

4. Yes. Ref.—250.104(B), last sentence.

5. Yes. Ref.—334.30(B)(2) except as prohibited by 334.12(A)(1)

6. No, only steel IMC is permitted by Article 342. Ref.—342.2

7. No, unless identified as sunlight resistant. Ref.—362.12(8)

8. Yes, all raceway articles permit cables unless specifically prohibited by the respective cable article. Ref.—342.22

9. Yes, 4 quarter bends (360 degrees total). Ref.—350.26

10. No. However, the raceway is required to be listed. Ref.— 300.50(C)

11. Listed and marked. Ref.—410.64(A)

12. NUCC is nonmetallic underground conduit with conductors as covered by Article 354. Yes, it is required to be listed (Ref.—354.6). Yes, the conductors and raceway are required to be listed (Ref.—354.100(B) and 354.100(C)).

13. No. Ref.—362.12(9)

14. In accordance with the manufacturers' installation instructions Ref.—392.30

15. The highest level that water can reach before it spills out. Ref.—680.2

16. EMT is a permitted equipment grounding conductor. Ref.—250.118(4)

17. These tables are provided to assist the user. The amount of temperature change would be provided by the user. Ref.—352.44 and the Tables 352.44 (A) and (B)

18. Yes. Ref.—378.6

19. Yes. Ref.—230.41 Exception

20. 647.4(D).

> *Note:* Informational notes are not enforceable there. This and 695.7 are the only location in the *Code* where voltage drop is mandated.

CODE QUIZ 8

1. No. Ref.—356.12(4)

2. 250.102, 356.60

3. -No. Article 376.

4. 115%. Ref.—445.13

5. LCDI or AFCI. Ref.—440.65

6. No. Ref.—430.225(A) Exception permits alarm devices where the operation of the motor is vital to the safeguard of persons.

7. 30. Ref.—426.32

8. Yes. Ref.—404.9(B)

9. Yes. Ref.—404.8(A) Exception 2

10. No. Ref.—408.41

11. Intermediate metal conduit (Type IMC) and rigid metal conduit (Type RMC) Ref.—392.30(B)(3) and 399.30(B)(3)

12. Yes. Ref.—424.3(A)

13. Battery or generator. Ref.—517.45(A)

14. Division 1. Ref.—500.5(C)(1)(3)

15. 50 volts. Ref.—480.4

16. Purpose; use. Ref.—408.4

17. No. Ref.—430.245(B); 310.106(C)

18. Thermal protection. Ref.—410.130(E)(4)

19. No. Ref.—440.14

20. Temperatures. Ref.—422.16(B)(3)

CODE QUIZ 9

1. No. Ref.—Article 100, Definition of service; it must be supplied from the serving utility.

2. Article 110, Part IV. Ref.—Table of Contents

3. 6 feet (1.8 m); lower. Ref.—110.26(E)

4. 110.54. Ref.—Same

5. Three continuous white stripes. Ref.—200.6

6. Yes. Ref.—210.8(A)(2)

7. An arc-fault circuit interrupter *is a device intended to provide protection from the effects of arc faults by recognizing characteristics unique to arcing and by functioning to de-energize the circuit when an arc-fault is detected.* [Reprinted with permission from NFPA 70-2011.] Ref.— Article 100

8. No, that would be a feeder by definition. Ref.—Article 100—Definitions

9. False. Ref.—110.26(A)(3). Equipment located above or below other electrical equipment is permitted to extend up to 6 inches (150 mm) beyond the front of other electrical equipment.

10. 110. Ref.—110.26(F)(1)(A)

11. 110, Part III. Ref.—Same

12. Three continuous white stripes on other than green insulation along its entire length. Ref.—200.6

13. GFCI (ground-fault circuit-interrupter protection for personnel). Ref.—210.8(A)(2)

14. Arc-fault circuit interrupters (AFCI). Ref.—210.12(A)

15. No. Ref.—210.52(B)(3). Some dwellings have more than one kitchen, and this new requirement was added to reduce overloading.

16. 3; 900. Ref.—210.52(D)

17. Two; readily accessible. Ref.—210.60(B)

18. 150 volt-amperes. Ref.—220.43(B)

19. 12; (3.7 m). Ref.—225.18(2)

20. 100 amperes. Ref.—225.39(C)

CODE QUIZ 10

1. 230.40 Exception 4. Ref.—Same

2. 100 amperes. Ref.—230.79(C)

3. No, 240.4(D). Ref.—240.4(G)

4. 240.2. Ref.—240.2 Definitive

5. Unlimited length. Ref.—240.21(B)(5)

6. 240.2. Ref.—Same

7. They are required to be GFCI protected. Ref.—525.23(A)

8. First paragraph of 240.21, Ref.—240.21, second sentence, first paragraph

9. No, the exception to 220.52 allows this load to be excluded from the calculation. Ref.—210.52(B)(1) Exception 2 and 220.52(A) Exception

10. Yes, they would have to be accessible; however, only two would have to be readily accessible. Ref.—210.60(A) and (B)

11. No, they would still be accessible as defined by the *Code*. Ref.—Article 100, Definition of accessible and readily accessible

12. Yes where the generator is a separately derived system. Ref.—Article 100

13. 12 inches and it must be within 36 inches (900 mm) of the outside edge of each basin. Ref.—210.52(D)

14. Yes, an exception states that if the circuit supplies only a single bathroom, then it is permitted to supply other equipment within that single bathroom. Ref.—210.11(C)(3) and Exception

15. Yes, in places of assembly Article 518. Ref.—518.3(B) Exception

16. Listed. Ref.—695.10

17. No. The term livestock does not include poultry. Ref.—547.10

18. No, an equipment grounding conductor in accordance with 250.118 is required, and the grounded (neutral) conductor cannot be bonded to the grounding-electrode system where there are common metallic paths between the two buildings. Ref.—250.32(B) Exception applies only to existing premises

19. No, if the raceway must not penetrate the ceiling. Ref.— 312,5(C)Exception.

20. Yes, if they have a floor located at or below grade and are not intended as habitable, and are limited to storage areas, work areas, or similar use. Ref.—210.8(A)(2)

CODE QUIZ 11

1. Yes, 250.122(F). Ref.—Same

2. See definition of panelboard Article 100.

3. Yes, in accordance with 110.14, 300.5(E), 300.13, and 300.15. Ref.—230.46

4. Yes, for feeders or branch circuits. Ref.—590.4 (B) and (C)

5. No. However, listed prewired can be purchased in trade size ½ (16) through 1 (27) as a listed manufactured prewired assembly. Ref.—331.3(8) and 300.18

6. Yes, only one is permitted generally. Additional feeders or branch circuits are permitted in accordance with 225.30(A) through (E). Ref.—Same

7. Yes, Type AC cable with an insulated equipment grounding conductor sized to Table 250.122 is now allowed. Ref.—518.4(A)

8. Yes, they are not permitted to be installed in a face-up position in the countertops or work spaces. Ref.—406.5(E)

9. The industrial portion of a facility meeting all the conditions of 240.2 Ref.—Same

10. Listed tamper resistant; listed tamper-resistant. Ref.—517.18(C)

11. Yes, if the luminaire (lighting fixture) does not exceed 6 pounds (3 kg). Ref.—314.27(A)(1) Exception

12. At least 3 inches (75 mm) of conductor outside the opening. Ref.—300.14

13. No. Ref.—334.12(A)(1)

14. It is not required for all buildings; industrial buildings are exempted. However, where the busway vertical riser penetrates two or more floors, a 4-inch (100 mm) minimum curb must be installed to retain liquids. Ref.—368.10(C)(2)(b)

15. Two minimum where a ground rod is used and has not been tested for compliance with 250.56. Ref.—250.53(A)(3)

16. No, only medium- and low-power systems are covered by Article 830. Ref.—830.4 and Table 830.15

17. Where a general-care patient bed location is served from two transfer switches on the emergency system, they are not required to have circuits from the normal system. Ref.—517.18(A) Exception 3

18. No, a main power feeder must serve 100% of the load. Ref.—310.15(B)(7)

19. Yes. Ref.—210.8(A)(3)

20. Article 314. Ref.—314.30

PRACTICE EXAM 1

1. No. Ref.—511.12

2. No. Ref.—680.72, however, 410.4(A) and (D) limit the use of pendant-type fixtures over this area.

3. Yes. Ref.—406.4(D)(3) and no, Ref.—406.4(D) grounding-type receptacles may be used as replacements, but they must be marked "GFCI Protected"

4. No. Ref.—210.8(A)(6)—only those serving the countertops.

   Note: Only very old ranges have receptacles mounted on them.

5. No. Ref.—314.3 Exceptions 1 and 2

6. Yes. Ref.—410.36(G)

7. 15-minute finish rating. Ref.—362.10 and 12

8. No. Ref.—300.4(B)(2)

9. Where run through the framing members, ENT shall be considered supported. Ref.—362.30(B)

10. Yes. Ref.—300.4(B) and 334.17

11. Yes. Ref.—340.10(4) and 334.112

12. No. Ref.—517.18(B) and 517.19(B)(2)

13. Yes. Ref.—210.52(A)(2)

14. No. Ref.—300.13(B)

15. Yes. Ref.—210.52(C) and 210.8(A)(6)

    Note: The GFCI requirement would not apply if the receptacle was located behind the refrigerator.

16. Yes. Ref.—210.52(E)(3)

17. No. The requirement is now 6 feet 7 inches (2 m). Ref.—404.8(A)

18. When marked for the use. Ref.—410.64

19. General- and critical-care patient bed location. Ref.—517.18(B) and 517.19(B)(2)

20. No. Ref.—210.52(C)—This rule only applies to kitchens.

21. Wet 406.9(B)

22. Yes. Ref.—680.23(B)(2)(b)

23. Yes, provided it is installed and supported in accordance with Article 334. Ref.—334.6, 334.10, 334.12, and 334.30

24. No, they may be rubber or thermoplastic types. Ref.—225.4

25. Yes, if marked for use Ref.—410.64

PRACTICE EXAM 2

1. No. Ref.—Only as environmental conditions require—547.1 and 547.8(A), (B), (C)

2. No. Ref.—320.30 and 300.4

3. Extra-hard usage portable power cables listed for wet locations and sunlight resistance. Ref.—553.7(A) and (B)

4. Yes. Ref.—680.42(C)

5. Yes. Ref.—680.23(F)(1)

6. No, water heaters are not listed with an attachment cord. Ref.—400.7 and 400.8(1); 110.3(B); 422.16

7. Yes, with conditions. Ref.—312.8

8. Voltage with the highest locked rotor kVA per horsepower. Ref.—430.7(B)(3)

9. Yes. Ref.—680.25(A)

10. Yes. Ref.—210.21 and Tables 210.21(B)(2) and (B)(3)

11. Yes. Ref.—210.23(A)

12. Not enclosed, buried, or in raceway. Ref.—Table 310.15(B)(17)

13. No. Ref.—90.5(C), 352.10 Information Note is not mandatory, only explanatory

14. No, GFCI protection is still required. Ref.—590.6(A)

15. Yes. Ref.—210.8(A)(7)

16. No. Ref.—210.8(A)(7)

17. No maximum. Ref.—Table 220.14(J)

18. 6. Ref.—517.19(B)

19. Yes, if potable. Ref.—430.81(B), 430.42(C), and 430.109

20. Yes. Ref.—210.70(A) Exception 1 210.52(2)

21. Yes, one or more supplied by a 20-ampere current. Ref.—210.52(F)

22. Yes. Ref.—Article 250 and Article 410, Part V 410.40

23. Yes, flexible metal conduit must be bonded to grounded conductor. Ref.—230.43(15).

24. No, only interior metal water piping systems. Ref.—250.104(A)

25. No. Ref.—314.22

PRACTICE EXAM 3

1. No. Ref.—314.27(B)

2. No. Ref.—314.16(A) and (B)

3. No. Ref.—210.52(E)(2)

4. No, polarized or grounding type. Ref.—410.82(A)

5. Yes. Ref.—210.52(A)

   Note: Due to the arrangement of furniture, the receptacle located behind the door may be the only receptacle accessible for frequent use such as vacuuming.

6. No. Ref.—590.4(G)

7. 6, where integrated clamps, and the like, are not used. Ref.—Table 314.16(A)

8. Table 314.16(B). 314.16(A)(2) as applicable and 314.16(B)(1)

9. Yes. Ref.—501.10(A)

10. No, however, it is a raceway. Ref.—Article 358

11. Yes, up to 18 inches (450 mm). Ref.—358.30(C)

12. Yes. Ref.—410.36(B)

13. No. Ref.—Definitions, Article 100, and 410.10(A)

14. No, must be spaced 1½ inches (38 mm) from the surface. Ref.—410.136(B)

15. No. Ref.—410.115(C) Exception

16. No, only where the grounded (neutral) conductor is present. Ref.—110.15, 230.56, and 408.3(F)

17. Yes. Ref.—210.8(A)(6) and 210.8(B). All kitchen countertop receptacles are required to be GFCI protected.

18. Yes. Ref.—312.8

19. Yes. Ref.—250.112(J); Article 410, Part V

20. No. Ref.—700.12(F)

21. Yes. Ref.—210.8(A)(2)

22. Yes "all." Ref.—210.8(A)(1) and 210.8(B)

23. No; however, it cannot be counted as one of the countertop receptacles as required by 210.52(C). Ref.—210.8(A)(6)

24. No. Ref.—314.1 and 410.110

25. Within 12 inches (300 mm) of service head, goose neck, and at intervals not exceeding 30 inches (750 mm). Ref.—338.10 and 230.51(A)

PRACTICE EXAM 4

1. 24 inches (600 mm). Ref.—Table 300.5

2. Yes. Ref.—Table 210.21(B)(2)

3. No, you are not required to locate required receptacles behind furniture. However, the number required by 210.52(A) must be installed

convenient to furniture arrangement. Ref.—210.60(B) and Exceptions

4. Yes, but only if the box is listed for fan support. Ref.—314.27(D) and 422.18

5. Ref.—410.64

6. Yes, generally. Ref.—410.115(C) Exceptions 1 and 2

7. Yes. Ref.—210.6(C)(3)

8. Yes, it is required. Ref.—700.12(E)

9. 150 kVA or less. Ref.—517.41(B)

10. Yes, the GFP requirements are based on the 1200- ampere switch.—Ref. 230.95

11. Yes, 15-volt luminaires generally. Ref.—680.43 (B)(1)

12. Yes. Ref.—358.10(B)

13. No, it is not permitted. Ref.—250.58

14. No. Ref.—110.26

15. Yes. Yes. Ref.—210.8(A)

16. See 547.9 and 250.32.

17. No. Ref.—Part V, Article 230, specifically 230.70.

18. 24 inches (600 mm). Ref.—Table 300.5

19. No, unless the cord and plug are part of listed appliance. Ref.—110.3(B), 400.8(1), 422.16

20. (a) -No minimal burial depth requirements. Ref.—Article 810
    (b) -As per 810.21 and 810.15

21. Yes, provided they control the power conductors in the conduit. Ref.—300.11(B)

22. Yes. Underground water piping system must be at least 10 feet (3 m) and be supplemented by an electrode for it to qualify as an electrode. Ref.—250.52(A)(1)

23. Yes. Ref.—210.70(A) and (C)

24. Entire luminaire. Ref.—410.10(D)

25. Yes. No. Ref.—225.6, 340.10, and 340.12(11)

PRACTICE EXAM 5

1. Yes. Ref.—392.10(D)

2. Ref.—410.16(C)
   (a) 12-inch (300 mm) incandescent, 6-inch (150 mm) fluorescent
   (b) 12-inch (300 mm) incandescent, 6-inch (150 mm) fluorescent
   (c) 6-inch (150 mm) incandescent, 6-inch (150 mm) fluorescent

3. No. Ref.—314.16(B)(1) Exception

4. Yes, generally. Ref.—110.26(C)

5. 4 AWG copper. Ref.—250.66(B)

6. Yes. Ref.—700.12 (F) Exception

7. Yes, if all conditions are met. Ref.—550.32(B) (1)–(7)

8. Yes. Ref.—680.25(B)

9. Yes, when smaller than ⅝ inch (15.87 mm), must be listed. Ref.—250.52(A)(5)(b)

10. Yes, both are acceptable. Ref.—230.43

11. Exothermic welds. Ref.—250.70

12. No. Ref.—250.118

13. Yes. Ref.—518.4(A)

14. Ref.—210

15. Yes. Ref.—356.10 and 356.12

16. Yes, but bare conductors are permitted. Ref.—230.41 and Exception

17. No. Ref.— 725.136(A)

18. No, only circuits over 250 volts to ground. Ref.—250.97. See the Exception for listed boxes with a specified type of concentric knockouts that are acceptable.

19. Yes. Exceptions would permit this practice in some instances without raceway protection. Ref.—300.22(C)

20. Yes. Ref.—300.13(B)

21. 60 amperes. Ref.—430.72(B) Exception 2—Table 430.72(B), Column C

22. Yes. Ref.—250.146

23. *Code* uses nominal voltages. Ref.—220.5

24. 12 AWG THW conductor in cable or raceway. Ref.—Table 310.15(B)(16) = 25 amperes

    Four conductors in a raceway Table 310.15(B)(2)—four to six conductors in a raceway or cable shall be derated to 80% = 25 × 80% = 20 amperes the allowable ampacity.

    Ambient temperature correction factors shown at the bottom of Table 310.15(B)(16) require an additional deduction of 0.82 — 20 × 0.82 = 16.4 amperes. 210.21—outlet devices shall have an ampere rating not less than the load to be served. Table 210.21(B)(2) permits a 20-ampere branch circuit for a maximum of 16 amperes. 16.4 exceeds this requirement. The conductor will carry the maximum allowable load.

Note: 240.4(D) limits the overcurrent protection on a 12 AWG conductor to 20 amperes.

25. 36″ × 36″ (900 mm × 900 mm). Ref.—324.41

PRACTICE EXAM 6

1. Yes. Ref.—250.97

2. Six. Ref.—230.71(A), 230.40 Exception 1

3. (a) Yes. Ref.—600.10(C)(2)
   (b) No. Ref.—600.10(C)(2)—There is no exception

4. Yes. Ref.—545.2, 210.70 (a dwelling), 210.70(C) (other types of buildings)

5. Yes, if they meet all the conditions. Ref.—605.8

6. Yes. Ref.—514.9 Gasoline is a Group D, Class I hazard. Diesel is not. However, the sealing requirements are to prevent the passage of gases, vapors, and flames.

7. Yes. Ref.—210.52(A)

8. (a) 2 AWG copper or 1/0 AWG aluminum. Ref.—250.107(A) base on 300-kcmil Service-Drop Table 250.66
   (b) Bonded to the panelboard supplying each apartment size to Table 250.122 base and an overcurrent device 100 amperes 8 AWG copper or 6 AWG aluminum. Ref.—250.104
   (c) No. Ref.—250.53(D)(2)—A supplementary grounding electrode is only required where the underground water pipe is present. This installation is suggested by a nonmetallic water system.

9. Yes. Ref.—210.23(A)

10. No, generally. Yes, for interior portion of a one-family dwelling. Ref.—680.21(A)(1)

11. No, where an equipment grounding conductor and not over four fixture wires smaller than 14 AWG enter the box from the fixture. Ref.—314.16(B)(1) Exception

12. Receptacle is not required specifically. If located within 6 feet (1.8 m), a GFCI is required. Ref.—210.52(A) and 210.8(A)(7)

13. Read heading for Table 220.55. Note 3 permits add nameplates of the two ranges. Use Column C. Ref.—Table 220.55 and Note 3

14. 5 kW. Ref.—220.14(B). 220.54 is used for feeder load calculations.

15. 35 pounds (16 kg). Ref.—422.18 and 314.27(C)

16. Table 348.22 for trade size ⅜ (12). For ½ (16) through 4 (53) use Table 1, Chapter 9. Ref.—348.22

17. Yes. Ref.—314.23(D)(2).

    Note: 300.11(A) the general method requirements give limited permission for branch-circuits. However, junction boxes are covered by Article 314.

18. No, explosionproof not required or acceptable. Must be approved for Class II locations. Ref.—502.5

19. No. THHN is not acceptable in wet locations. Ref.—Table 310.104(A) and 310.10(C)

20. Authority having jurisdiction. Ref.—90.4, also see definition in Article 100, "Approved"

21. Yes. Ref.—210.52(F)

22. Yes. Ref.—210.63

23. Column A is the basic rule; the overcurrent protection shall not exceed Column A. Ref—430.72(B)(1)

24. 83. Ref.—Chapter 3 of this book

25. Qualified yes "Metal Multioutlet Assembly." Ref.—Article 380, 380.3

PRACTICE EXAM 7

1. Lighting circuit serving the area. If there are three or more lighting circuits in the area, they may be on a separate circuit. Ref.—700.12(E)

2. Yes, if they are identified for purpose and meet all conditions of 392.10. Ref.— 392.60

3. No. Ref.—388.10(1)

4. If the required workspace is unobstructed or doubles. Ref.—110.26(C)(2)(a) and (b)

5. No, the conductors in MC cable are wrapped in a nylon wrapping. AC cable conductors are individually wrapped in kraft paper. Ref.—320.40, Armored Cable; 330.40, Type MC Cable

6. Generally, no, if exposed to physical damage. Ref.—362.10 and 362.12.

    Note: If these conditions do not exist, then ENT could be installed.

7. 24 inches (450 mm). Ref.—Table 300.5

8. Large capacity multibuilding industrial installations under single management. Ref.—225.32 Exception 1

9. No for Class 2 Div. 1, Yes for Class 2 Div. 2. Ref.—502.10(A) and (B)

10. No. Ref.—410.64

11. Grounded (neutral) conductor must be insulated generally. Ref.—310.110, 250.140, and 250.142. The rules for an insulated neutral or grounded conductor for a feeder, although not clearly written, the service neutral is often bare and 250.140 permits dryers and ranges originating in the service panel to be Type SE, which has a bare grounded conductor.

12. No. Classifying structures are the responsibility of the building inspector or the fire marshal generally. They rely on the adopted building code or the NFPA 101 Life Safety Code. Ref.—Article 518

13. No. Ref.—220.61, also Table 310.15(B)(5)

14. None. Ref.—Table 310.15(B)(16)

15. Table 314.16(B)

Example: An outlet box contains a 120-volt, 20-ampere single receptacle and a 120-volt, 30-ampere receptacle. The 20-ampere receptacle is fed with a 12.2 W/GRD NM cable that extends from the outlet on to additional outlets located elsewhere. The 30-ampere receptacle is fed by an individual branch circuit with 10.2 W/GRD NM cable. What size box is needed?

Answer: 314.16(A) and Table 314.16(B)

12.2 W/GRD Feed	2–12 AWG	@ 2.25 cu. in.	4.5 cu. in.
12.2 extending on	2–12 AWG	@ 2.25 cu. in.	4.5 cu. in.
10.2 W/GRD Feed	2–10 AWG	@ 2.5 cu. in.	5.0 cu. in.
30 A Receptacle	2–10 AWG	@ 2.5 cu. in.	5.0 cu. in.
20 A Receptacle	2–2 AWG	@ 2.5 cu. in.	4.5 cu. in.
GRD	1–10 AWG–10		2.5
Clamp	1–10 AWG		2.5
			28.5

A box with 28.5 cubic inches is required.

16. 10 feet (3 m). Ref.—230.24(B)

17. No, it is considered by the *NEC*® as "Other Space Used for Environmental Air." Ref.—300.22(C) and Exceptions, definition of "Plenum" in Article 100

18. Yes, this area must be unobstructed. Ref.—410.2

19. Equipment grounding conductor required. Ref.—250.134, 250.136, and 410.40

20. Yes. Ref.—410.10(A), Article 100 Location, damp

21. By interpolation. Ref.—430.6(A)(1)

22. 53%. Ref.—Chapter 9, Table 1 and Note 6

23. Yes. Ref.—Table 310.15(B)(7)

24. No, motors are computed in accordance with Article 430. Ref.—220.14 and 680.3

25. Yes, if the garage is supplied with electricity. Ref.—210.8 (A)(2) and 210.52(G)

PRACTICE EXAM 8

1. No. Ref.—410.130(F)(2)

2. As per 547.9 in compliance with Article 250. Ref.—250.32 and 547.9

3. Yes, and they occupy a dedicated space with the individual branch circuit. Ref.—Table 220.55 "Heading." If they are plugging into one of the "small appliance" outlets, then they would not have to be added to the load calculations.

4. Yes, for controlled starting. Ref.—430.82(B)

5. No. Ref.—553.4

6. No. Ref.—Article 600 and 210.6(A)

7. Yes. Ref.—210.23(D)

8. Yes, limited application. Ref.—517.12, 517.13, and 517.30

9. No, generally. Ref.—240.4

10. Yes. Must be marked SWD. Ref.—240.83(D) and 404.11

11. There shall be two deductions for the largest conductor connected to each device, and one deduction of the largest conductor entering the box for each of the following: the clamps, the hickeys, all of equipment grounding conductors, etc. Ref.—314.16(B)

12. No. Ref.—250.92(B)

13. No, the cable can be fished; however, Article 314 does not permit the fished NM cable without a box clamp. Ref.—334.30(B)(1) and 314.17(C) Exception

14. 150 amperes will meet all conditions required for a general-use switch. Ref.—440.12 and 430.110

15. No. Ref.—348.60 and 348.20(A)(2)

16. I P/E $I = 10,000/240$ $I = 41.667$

$R = E/I$ $R = 240/41.667$ $R = 5.76$

$P = IE$ $P = 208 \times 36.1$ $P = 7.51$ kW

$I = E/R$ $I = 208/5.76$ $I = 36.1$ amperes

36.1 amperes—Chapter 3, Ohm's law

I = P/E   I = 10,000/240   I = 41.667

R = E/I   R = 240/41.667   R = 5.76

I = E/R   I = 208/5.76   I = 36.1 amperes

17. Yes. Ref.—680.5, 680.22(A)(4), and 680.23(A)(3)

18. Chapter 9, Table 1 unless durably and legibly marked by manufacturer in cubic inches. Ref.—314.16(C)

19. By the markings on the luminaire (fixture) (IC) by the listing agency. Ref.—410.116(A)(2)

20. Conductors only. Ref.—240.21

21. Table 110.26(A)(1), Condition 3. Ref.—110.26(A)(1)

22. Yes, if cord-and-plug portable dishwasher or compactor is connected but not required or advisable. Generally, these appliances are wired to the requirements in Article 422. Ref.—210.52(B)(1) and (2)

23. 10 feet (3 m). Ref.—680.8(A)

24. Yes, no depth requirement but is required to be protected from severe physical damage. Ref.—250.64

25. Water meter and all insulating joints. Bonding around valves and sweated joints is not generally required but the water meter, filters, and insulating joints would be required to be bonded around. Ref.—250.52(A)(1) and 250.53(D)(1)

PRACTICE EXAM 9

1. Yes, Ref.—680.5(B), 680.22(A)(4), 680.23(A)(3), and 680.22

2. (a) Yes. Ref.—250.24(C)
   (b) In each of the parallel raceways. Ref.—250.24(C)
   (c) 2 AWG copper in each conduit. Ref.—250.24, Table 250.66

3. 21 inches × 24 inches × 6 inches. Ref.—314.28(A)(2)

4. Yes, Ref.—312.8

5. No. Ref.—358.30

6. Either system installation is okay. Grounding electrode is not required where a three-pole transfer switch is used. Ref.—250.24(D) and 250.30

7. No. Ref.—300.5(I) Exception 2

8. This installation is a violation; you cannot mix these conductors, unless the control is a Class 1 or power circuit. Ref.— 725.136

9. Permitted. Ref.—110.26(F)(1)(C)

10. Yes. Ref.—250.50. Steel can be bonded to the water pipe within 5 feet (1.5 m) of where the water pipe enters the building. Ref.—250.50 and 250.52

11. 20-ampere breaker, 16 ampere load. Ref.—240.4(D) and 210.14

12. Yes. Also, the receptacles required by 210.52(E) still must be installed at direct grade access and be GFCI protected. Ref.—210.8(A)(3) and 210.52(E)

13. Yes, if listed for 2. Ref.—110.3(B)

14. No. Ref.—230.42(A), (B), and (C)

15. Yes. Ref.—430.109(F)

16. Ref.—Table 310.15(B)(7)

17. No, neutral is not used at the switch. Ref.—200.7(C)(2) and 300.3(B)

18. Yes; but where they enter the building, they must be enclosed in a raceway. Ref.—340.10(2)

19. Yes, the area may be unclassified by AHJ where there are four air changes per hour. Ref.—511.3(A) and 511.3(B)(1) Exception

20. No. Ref.—200.4

21. Yes. Ref.—348.20(A)

22. All areas with Exceptions 406.12

23. Yes, but it must be connected to the same circuit as supplies the lighting for the area to be served. Ref.—700.12(F)

24. Yes. 210.8(B)

25. Yes; Ref 422.30 No. Ref.—422.12

PRACTICE EXAM 10

1. Yes, but storage of combustibles should be considered. Ref.—Part II, Article 410

2. Hot tubs and spas are defined in Article 680. Ref.—680.2

3. No. Ref.—511.4, 501.15(B)(2) Exception 1; 511.9

4. 150; No. P = EI = 480 × 104 = 49,920 single phase

   49,920/(480 × 1.73) = 60.12 amperes

   60.12 multi; 250% = 150

   Ref.—Table 450.3(B)

5. Each section 5 feet (1.5 m) or fraction thereof shall be calculated at 180 VA in other than dwellings. Ref.—220.14(H)

6. No. Ref.— 430.102(B) Exception

7. Yes, possibly. Ref.—430.72 and Exceptions, Table 430.72(B)

8. Yes, except in an industrial facility. Ref.—368.17(B)

9. No. Ref.—340.12(1) and 340.12(10)

10. Yes. Ref.—250.20(B)(2) and 250.24(C)

11. No. Ref.—250.30 covers the requirements for a separately derived system.

12. A tap is defined in 240.2. No, you cannot tap a tap. Ref.—240.21

13. Yes. Ref.—240.15(B)

14. 2 AWG. Ref.—Chapter 9, Table 8, Table 250.66

15. Yes. However, consideration must be made for corrosive elements. Ref.—300.6, 501.10(A), Article 344—Non-Ferrous Rigid Metal Conduit. 344.10(3)

16. No. Ref.—700.12(F)

17. No. Ref.—250.140 Exception

18. Yes, but there are installation requirements for disconnecting overcurrent. Ref.—230.82 and 695.3

19. Bond services together. Ref.—250.58

20. No. Ref.—680.72 and 410.10(D)

21. No. Ref.—*NEC*®. Most requirements for GFCI protection appear in 210.8. Other requirements appear in Articles 511, 517, 527, 550, 551, 553, 555, and 680.

22. No. Ref.—410.56(D) and 410.64

23. Yes. Ref.—362.10(5). The 15-minute finish rating is a wall finish rating established by Underwriters Laboratories and can be found in the "UL Fire Resistant Directory."

24. Yes. Ref.—Article 314, 314.16(A) and (B), Table 314.6(B). $4 \times 2\frac{1}{8} = 30.3$ cubic inches. 8 AWG conductor = 3 cubic inches. Therefore, ten 8 AWG conductors would be allowed without considering devices or other appurtenances such as clamps.

25. They must be fastened. Ref.—410.36(C)

PRACTICE EXAM 11

1. Yes. Ref.—340.10(4)

2. No. Ref.—680.23(A)(3)

3. Yes. Ref.—700.12(F) Exception

4. No. Ref.—300.22(C) Exception. Perpendicular, yes; parallel, no.

5. No, they must be equipped with a factory-installed GFCI. Ref.—600.10(C)(2)

6. In each disconnect means. Ref.—250.64(D)

7. No. Ref.—250.64(D)(3)

8. As a wireway in accordance with Article 376 or 378. Ref.—Article 376, Part II and Article 378, Part II.

9. Yes. Ref.—250.148

10. No. It can be insulated or covered. Ref.—680.21(A)(1) Exception

11. No. This installation is in violation of the *Code*. Ref.—408.36

    240.4(B)—The 400-ampere breaker properly protects 500-kcmil conductors that permit the next higher overcurrent device.

    220.10—Conductors 500 kcmil are large enough to carry the 180-ampere load.

12. Yes, unless not required by NFPA 99. Ref.—517.25 (Informational Note)

13. Generally, no—Table 220.42, and 220.44. Ref.—220.12, 220.14, 220.40 and 220.60

14. Yes. No. Ref.—517.2 definition of patient care areas in hospitals. Ref.—517.13(A)

15. Yes, there are no conditions. Ref.—600.7 and 250.112(G)

16. No, generally, unless with listed equipment. Ref.—250.30(A) 250.30(A)(4) Ex no.2 Exceptions

17. Yes. Ref.—310.10(4)

18. They must be adjusted to 80%. Ref.—Table 310.15(B)(3)

19. Yes. Ref.—210.6(C)

20. Yes. Ref.— 210.8(A)(3)

21. Yes. Ref.— 210.11(C)(3)

22. Yes, if they do not exceed allowable load of circuit. They are not continuous loads. Ref.—210.19(C) and 210.23

23. No. Ref.—210.52(G) only applies to one family dwellings (see definition, Article 100)

24. Yes. Yes. Ref.—210.70(A) Exception 2

25. Both. 215.2(A)(1) and 215.3

PRACTICE EXAM 12

1. Only branch circuits. Ref.—220.51 and 424.3(B)

2. Yes. Table 220.55 Heading Column A applies to all notes as applicable, Columns B and C apply to Note 3. Ref.—Same

3. Yes. Ref.—225.33

4. Location is not specified; where bonded to different electrodes, they must then be bonded together. Ref.—250.58

5. Yes, but may be wired as unclassified in accordance with Chapters 1–4. Ref.—514.36

6. No clearance is required. Ref.—410.116(A)(2)

7. System ground. Ref.—810.21(F)

8. Yes. 339.10

9. Yes. Ref.—250.97 Exception

10. Yes. Ref.—680.21(A)(1) Exception

11. Yes, only as taps (from raceway system to fixture). Ref.—348.20(A)(2)

12. Yes. Ref.—680.23(C) and 680.22(A)(2)

13. No. Ref.—680.43(C) and 680.43(D)(3)

14. No. Ref.—210.70(A)(2)(b)

15. Both column B and Column C, Note 3—6 kW × 65% and 3 kW × 75% = 6.15 kW

16. Yes. Ref.—210.52(F). At least one but only for the laundry area.

17. Yes, all. Ref.—210.8(A)(1)

18. Yes. Ref.—725.136(I)

19. Yes. Ref.—820.93(A)

20. Bonding jumper shall be the same size as the grounding electrode conductor. Ref.—250.104(E)

21. Yes, for wet bar sink; yes, for laundry sink. Ref.—210.8(A)(7)

22. 25 foot tap = 1 AWG THW, 10 foot tap = 10 AWG THW. Ref.—240.21, Table 310.15(B)(16)

23. No. Ref.—210.70(A). At least one switch

24. 60°C or 75°C, because there are no circuit breakers rated 90°C. Ref.—110.14(C)

25. Those areas determined by governing body of facility. Ref.—517.2 Definitions, Patient Care Areas of a Hospital

FINAL EXAMINATION

1. C. Ref.—800.53

2. B. Ref.—General knowledge

3. D. Ref.—General knowledge, *IEEE Dictionary*

4. A. Ref.—Chapter 3

5. A. Ref.—Chapter 3

6. C. Ref.—Chapter 3

7. C. Ref.—Chapter 3
P = I(E × 1.73) P = 83 kW

8. A. Ref.—General knowledge

9. C. Ref.—Chapter 3
0.5/1 ohm = ⅕
2.5 = 1

10. A. Ref.—Chapter 3

11. A. Ref.—Chapter 3

12. A. Ref.—210.18(B)

13. C. Ref.—Chapter 3
(16 × 4) [64] × (10 × 4) [40] = 2560

14. A. Ref.—250.102(C)

15. B. Ref.—Chapter 9, Table 8
Ohm's law
Chapter 3
240 volts × 3% = 7.2 volts drop permitted
R = Ed/I   R = 7.2/58
R = 0.124 ohm for 500 feet
3 AWG = 0.254 ohm for 1000 feet

16. A. Ref.—Chapter 9, Tables 1, 5, and 4
2/0 AWG = 0.2265 square inches × 2 = 0.4530
1 AWG = 0.1590
0.4530 + 0.1590 = 0.6120 Table 4 = 1½ inches

17. B. Ref.—Chapter 9, Tables 1, 4, and 5
2—0.1590
1—0.1893
1—0.2265
1—0.7338 square inches Table 4, 1½ = 0.82

18. D. Ref.—220.80, Chapter 9 Example

19. B. Ref.—Table 250.66

20. C. Ref.—230.95(A)

21. B with Exception. Ref.—230.9(A)

22. B. Ref.—Duff and Herman, *Alternating Current Fundamentals, 4E,* Delmar Cengage Learning

23. B. Ref.—Table 310.15(B)(3)(a)

24. A. Ref.—310.15(A)(2)

25. A. Ref.—Table 310.15(B)(16) correction factors

26. A. Ref.—Chapter 9, Note 3

27. B. 342.22 states for conductor fill see Chapter 9, Table 1. Ref.—Chapter 9, Tables 1, 4, and 5

Solution:

Table 1: over 2 conduits 40%.

Table 5:

-4 AWG = 0.0824 sq. in. × 3   = 0.2472

1/0 AWG = 0.2223 sq. in. × 4 = <u>0.8892</u>

                Total = 1.1364 square inches

2″ IMC = 1.452   1½ IMC = 0.890

Therefore 2 (53) IMC would be required

28. B. Ref.—352.22, Chapter 9, Table 1, Note 1 and Annex C, Table C10

29. C. Authority having jurisdiction. Ref.—90.4

30. C. Ref.—230.70(A)(1)

31. C. Ref.—Appendix 1 (text)

32. D. Ref.—Appendix 1 (text)

33. D. Ref.—Basic Math, Chapter 3

34. B. Ref.—110.9 and 110.10

35. C. Ref.—430.52

36. D. Ref.—Article 430, 430.37, 430.38, and 430.39, Table 430.37

37. B. Ref.—430.226

38. A. Ref.—Article 430, Part XI

39. D. Ref.—430.89

40. D. Ref.—250.140 (existing installations only)

41. B. Ref.—250.110(1)

42. C. Ref.—250.53(G)

43. C. Ref.—250.114(2) Exception 2

44. D. Ref.— 430.7

45. D. Ref.—225.18(4)

46. C. Ref.—230.95(C)

47. A. Ref.—430.24

     Largest 5.6 × 1.25 = 7

               4.5 × 1.00 = 4.5

               4.5 × 1.00 = <u>4.5</u>

                    16.0 amperes— minimum ampacityof conductors

48. C. Ref.—690.4(A), Article 705, 230.82(6)

49. C. Ref.—250.24 and 250.28

50. D. Ref.—Electrician's Handbook

51. B. Ref.—404.8(A)

52. A. Ref.—430.36

53. D. Ref.—General knowledge; see Annex 2

54. D. Ref.—General knowledge; Annex 2

55. A. Ref.—General knowledge; Chapter 3

56. B. Ref.—Chapter 3; Ohm's law; Annex 2

57. C. Ref.—Article 310, Chapter 9

58. A. Ref.—General knowledge; Chapter 3

59. A. Ref.—Definitions, Article 100

60. D. Ref.—Ohm's law; Annex 2

61. D. Ref.—Ohm's law; General knowledge; $P = I2R$

62. B. Ref.—General knowledge

63. C. Ref.—Ohm's law; Annex 2

64. A. Ref.—Definitions, Article 100

65. B. Ref.—230.9(A) Exception

66. D. Not a *Code* requirement

67. A. Ref.—Chapter 3

68. B. Ref.—General knowledge

69. A. Ref.—*Direct Current Fundamentals*, Delmar Cengage Learning

70. B. Ref.—*Direct Current Fundamentals*, Delmar Cengage Learning

71. A. Ref.—Basic Math, Power Formula, Ohm's law

     Clue: Resistance is constant

     I = P/E   2400/240 = 10

     R = E/I   240/10 = 24

     I = E/R   120/24 = 5

     P = EI   5 × 120 = 600

72. B. Ref.—Basic Math, Ohm's law; Annex 2

73. C. Ref.—Basic Math; Annex 2

74. C. Ref.—Tables 430.250 and 430.252; 150% × 60 amperes = 90 amperes

75. D. Ref.—Ohm's law, 65% load, 220.56; Annex 2. Calculation is based on 1Ø units; if calculation based on 3Ø units, answer is "B".

Index

A

Accessible, definition, 87
Adequate clearances, 91
AFCI (arc fault circuit interrupter), 168, 175
    definition, 113
    dwellings, 168, 175
    exceptions, 175
AHJ (authority having jurisdiction), 30
Air-conditioning, 212
Alternating Current Fundamentals, 3
Americans with Disabilities Act (ADA), 3
Ampacity adjustment/correction, 95–97
Antenna systems, 242
Appliances, 208
    definition, 211
Arc fault circuit interrupter (AFCI). *See* AFCI
    (arc fault circuit interrupter)
Arc volt, 168
Arc-voltage testing, 168
Attachment plugs, 208
Authority having jurisdiction (AHJ). *See* AHJ
    (authority having jurisdiction)
Autotransformer, 214
Auxiliary, 77
Auxiliary gutters, 199

B

Bathroom, definition, 87
Bonded, definition, 87
Bonding jumper, 77
Boxes
    sizing, 198
    where required, 190
Branch circuits, 64–65
    air-conditioning, 108
    appliance, 108
    appliance, definition, 108
    definition, 108
    exceeding 120 volts, 108–109
    exceeding 600 volts, 109
    general purpose, 108–111
    heating, 108
    individual, definition, 108, 211
    motor, 108
    multiwire, 108
    not exceeding 277 volts, 109
    outside, 64–65, 113–114
    voltage limitations, 108–109
    welding receptacles, 108

Broadband. *See* Network powered broadband
Burial depths, 189

C

Cable tray, 192–196
    nonmetallic, 196
    not permitted, 196
    sunlight resistant, 196
Canadian Electrical Code (CEC), 176
Capacitor, 214
CATV systems, 242
Circuit breakers. *See* Overcurrent protection
Circuit impedance, 165–166
Class 2 and Class 3 circuits, 240–241
Commercial garages, 231
Communications systems, 242–243
Conductors
    for general wiring, 189
    general requirements, 192
    outside, 67
    overcurrent protection of, 146
    service-entrance, 131
    tap conductors, 69, 71–72
    temperature ratings, 91–95
Conduit
    cross-sectional area, 187
    expansion joints, 190
    fill, 186
    support wires, 190
    types, 186
Conduit bodies, 197
Continuity, 190
COPS (critical operations power systems), 240
Cord connectors, 208
Corner grounded delta, 163–164

D

Damp locations, definition, 88
DCOA (designated critical operations areas), 240
Definitions, *NEC*®, 87–89
Delta systems, 162
Designated critical operations areas (DCOA), 240
Direct Current Fundamentals, 3
Disconnecting means, identification of, 98
Dissimilar metals, 91
Ducts, 190

E

Electric signs and outline lighting, 231
Electronically activated fuse, definition, 89

Energized, definition, 88
Equipment for general use, 79, 208–214
Equipment grounding conductors, 165, 167
Equipment-grounding system, 165–167
European zone system, 231
Examinations
    admission letter, 3
    anxiety, 4
    bulletin, 3
    calculator, 5
    candidate bulletin, 3
    code book not marked, 208
    content outline, 2, 3–4
    education requirements, 3
    eligibility, 3
    entrance ticket, 5
    first choice, 6
    hands-on practical test, 3
    item bank, 2
    key words, 6
    multiple-choice, 2, 4
    national testing, 2
    only one answer, 2
    photo ID, 5
    posttest analysis, 2–3
    preparing, 3–6
    proctor, 6
    question analysis, 251–252
    question criteria, 2
    roster, 3
    state agencies, 2
    task analysis, 2
    verification of experience, 3
    waiting period, 4
    workshops, 2
Expansion joints, 190

F

Fault current path, 75
Fault current testing, 168
Fault currents, 146, 147, 165, 166
Feeders, 113–121, 147
    dwellings, 192
    outside, 64–65, 113–114
Fiber optic, 242
Fine print note (FPN), 60
Fire alarm systems, 242
Fittings, where required, 190
Fixed electric space heating, 211
Fixture wires, 208
Flash protection, 97
Flexible cords, 208
FPN (fine print note), 60
Fuses. *See* Overcurrent protection

G

General requirements, 63, 87–100
Generator, 214
GFCI (ground fault circuit interrupter), 64, 111–113
    for vending machines, 211
Ground detector lights, 162
Ground fault circuit interrupter (GFCI). *See* GFCI
    (ground fault circuit interrupter)
Ground fault path, 165
Grounded circuit conductors, 168
Grounded conductor, 164–165
Grounding, 74–77, 159–177
    AC systems by voltage, 162
    computer model, 167–175
    conductor, 76
    general requirements, 162
    history of, 159–161
    impedance grounded neutral systems, 162
    overview, 159
    separately derived systems, 163
    solidly grounded system, 161
Grounding Electrical Systems for Safety, 167
Grounding electrode, 77
Grounding electrode conductor, 76
Grounding-electrode system, 165

H

Healthcare facilities, 231
High leg, 97

I

IAEI Soares Book on Grounding, 167
IAS/IEEE, 168
IEEE (Institute of Electrical and Electronics
    Engineers), 160, 176
IEEE Dictionary, 87
Impedance, 165, 166, 168
Impedance grounded neutral systems, 162
Industrial control panels, 213
International Code Council, 2
Interrupting ratings, 147

K

Kitchen, definition, 88, 111

L

Lightning, 175
    30 volts or less, 208
Lightning arrestor, 164
Line surges, 175
Line-to-ground fault, 175, 176
Live parts, definition, 88
Load calculations, 65–66
Local codes, 30–31

Luminaires, 208
    30 volts or less, 208
    high discharge, 209
    incandescent, 208
    location in bathrooms, 208
    location in closets, 208

M

Mathematics
    calculator, 40
    decimals, 43
    fractions, 40–42
    Ohm's law, 45
    percentages, 43
    power factor, 48
    powers, 44
    square roots, 44–45
    voltage drop calculations, 45
Mobile homes, 231
Motor control center, definition, 88
Motors, 208, 212
    controllers, 212
    motor circuits, 212
Multifamily dwellings, 111

N

National Electrical Code® (*NEC*®). *See NEC*®
    (*National Electrical Code*®)
National Fire Protection Association (NFPA), 25, 160, 176
Neat and professional manner, 91
NEC® (*National Electrical Code*®)
    2011 changes, 26
    arrangement, 60–61
    communications systems, 242–243
    equipment for general use, 79, 208–214
    FPN (fine print note), 60
    general requirements, 63, 87–100
    grounding, 74–77, 159–176
    not covered, 60
    obelisk note, 192
    overview, 25–29, 59–79
    scope, 59–60
    "shall", 61
    special conditions, 240–243
    special equipment, occupancies and conditions, 79
    special occupancies, 222–232
    table of contents, 197
    units of measurement, 62–63
    wiring and protection, 64–70, 108–122,
        131–138, 146–150
    wiring methods and materials, 77–79, 186–200
NEMA (National Electrical Manufacturers Association), 168
Network powered broadband, 242
Neutral, 76

NFPA (National Fire Protection Association), 25, 160, 176
Nonlinear load, definition, 88
Nonmetallic-sheathed cable, 189

O

Obelisk note, 192
Ohm's law, 45
OSHA (Occupational Safety and Health
    Administration), 31, 78
Overcurrent protection, 65–74, 146–150
    600 volts or less, 65, 146
    adjustable-trip circuit breakers, 146
    air-conditioning, 146
    appliance, 146
    disconnecting requirements, 146
    feeders, 147
    fixed-trip circuit breakers, 146
    flexible cords and fixture wires, 146
    general requirements, 146
    interrupting capacity, 146
    interrupting ratings, 147
    location, 69–71, 146
    over 600 volts, 65, 146, 147
    point-to-point method, 147
    refrigeration, 146
    short-circuit calculations, 147–150
    small size, 71, 74
    supervised industrial installations, 147

P

Places of assembly, 231
Plenums, 190
Point-to-point method, 147
Power factor, 48, 214

Q

Qualified person, definition, 88, 108

R

Reactance, 166
Receptacle, 208
    damp or wet location, 210
    definition, 88
    weather-proof, 208
Receptacle outlets, 112
Refrigeration, 212
Resistance, 166
Review questions, 251–252
    code quizzes, 263–299
    general use equipment, 215–220
    grounding, 178–184
    mathematics and formulae, 49–57
    NEC® overview, 7–23, 33–38, 80–85, 101–106, 123–129,
        139–144, 151–157, 233–238, 244–249

Review questions, (*cont.*)
    practice exams, 253–262, 300–382
    wiring methods and materials, 201–206

S

Seals, 190
Separately derived systems, 163
Service conductors, definition, 89
Services, 65, 131–138
    600 volts or less, 131–137
    clearances, 131, 132, 137
    conductors, 131
    disconnecting means, 131, 132
    dwellings, 192
    multioccupancy, 131
    over 600 volts, 137–138
    overhead system, 131
    parts of, 131
    service cable, 131
    service conductors, 131
    service drop, 131, 132
    service-entrance conductors, 131, 132
    service equipment, 131
    service lateral, 131, 132
    service mast, 132
    service point, 131
    underground system, 131
    vault, 131
    warning signs, 138
Short-circuit calculations, 147–150
Some Fundamentals of Equipment Grounding, 167
Special conditions, 240–243
    antenna systems, 242
    CATV systems, 242
    Class 2 and Class 3 circuits, 240–241
    communications systems, 242–243
    COPS (critical operations power systems), 240
    DCOA (designated critical operations areas), 240
    emergency systems, 240
    fiber optic, 242
    fire alarm, 242
    legally required standby systems, 240
    network powered broadband, 242
    optional standby systems, 240
Special equipment, occupancies and systems, 79, 222–232
    5 thread minimum, 223, 229
    class and division flow charts, 224–226
    Class I, 223
    Class I Division 1 wiring methods, 223, 223, 227
    Class I Division 2 wiring methods, 227–228
    Class II, 223
    Class II Division 1 wiring methods, 229
    Class III, 223
    Class III Division 1 wiring methods, 230
    commercial garages, 231
    conductor fill, 228
    Division 1, 223
    Division 2, 223
    electric signs and outline lighting, 231
    European zone system, 231
    grounding and bonding, 230–231
    groups, 223
    healthcare facilities, 231
    intrinsic safe systems, 231
    mobile homes, 231
    places of assembly, 231
    pressure piling, 228
    sealing and draining requirements, 228–229
    seals, 223
    special equipment, 231
    specifically approved equipment, 222
    three classifications, 222
Sunlight resistant, 196
Supervised industrial installation, 69, 74, 147
Support wires, 190
Surge arresters, 74, 163
Surge-Protective Devices (SPDs), 74
Switchboard, definition, 89
Switches, 199

T

Tamper-resistant receptacles, 199, 208
Taps, 69, 71–74
Temperature limitations, 91
Temperature requirements, 91–95
Temporary wiring, 190
Terminations
    available, 95
    temperature ratings, 92–97
Testing laboratories, 31–32
    Applied Research Laboratories (APL), 31
    Canadian Standards Association (CSA), 31
    Electrical Testing Laboratories (ETL), 31
    Underwriters Laboratories (UL), 31
The American Electrician's Handbook, 3
Transfer switch, definition, 89
Transformer vaults, 131, 214
Transformers, 214
Transient overvoltages, 176

U

Ungrounded neutral system, 176, 176
Ungrounded systems, 162, 176
Unused openings, 91

V

Vending machines, 211
Voltage drop, 45–48, 121–122
    formula, 121–122

W

Wiring and protection, 64–70
Wiring methods and materials, 77–79, 183–200
    general requirements, 189
    over 600 volts, 190

Working clearances, 98–100
Working space, 200
Workmanship, 89–91